Fundamentals of Relativity

Fundamentals of Relativity

Edited by
Dallas Rivera

Larsen & Keller
www.larsen-keller.com

Fundamentals of Relativity
Edited by Dallas Rivera
ISBN: 978-1-63549-128-9 (Hardback)

▤ Larsen & Keller

Published by Larsen and Keller Education,
5 Penn Plaza,
19th Floor,
New York, NY 10001, USA

Cataloging-in-Publication Data

Fundamentals of relativity / edited by Dallas Rivera.
    p. cm.
Includes bibliographical references and index.
ISBN 978-1-63549-128-9
    1. General relativity (Physics). 2. Special relativity
(Physics) 3. Gravitation. 4. Relativity (Physics).
I. Rivera, Dallas.
QC173.6 .F86 2017
530.11--dc23

The publisher's policy is to use permanent paper from mills that operate a sustainable forestry policy. Furthermore, the publisher ensures that the text paper and cover boards used have met acceptable environmental accreditation standards.

Printed and bound in the United States of America.

For more information regarding Larsen and Keller Education and its products, please visit the publisher's website www.larsen-keller.com

Table of Contents

Permissions

Index

Preface

This book provides comprehensive insights into the field of relativity. It discusses in detail the different concepts and theories that have been developed in this area. Relativity is a complex and vast subject and in this text, effort has been made to make the principles and methods of the subject as easy and informative as possible. The topics introduced in it are of utmost significance and are very important for the basic understanding of this area. The book studies, analyses and upholds the pillars of relativity and its utmost significance in modern times. Different approaches, evaluations and methodologies have been included in it. This textbook is appropriate for those seeking detailed information in this area.

Given below is the chapter wise description of the book:

Chapter 1- The theory of gravitation proposed by Albert Einstein is known as general relativity. It generalizes special relativity and Newton's law of universal gravitation and provides an incorporated account of gravity in which gravity exists as a geometric element in the space-time continuum. This chapter will provide an integrated understanding of general relativity.

Chapter 2- The phenomenon that brings all things towards each another is known as gravity. It is caused by the uneven distribution of mass/energy in the universe. It also gives mass to bodily objects on earth and can even cause ocean tides. The following text also concentrates on Newton's law of universal gravitation, an introduction to general relativity and the history of gravitational theory.

Chapter 3- The key concepts of general relativity are spacetime, Galilean invariance, Lorentz covariance, cosmological constant and the wormhole. The mathematical model that incorporates space and time into a single intertwined continuum is known as spacetime whereas Galilean invariance considers the laws of motion to be the same in all frames. The key concepts of general relativity are considered in this chapter.

Chapter 4- General relativity uses a number of mathematical tools. Some of these tools are Einstein field equations, covariant derivative, parallel transport, geodesics in general relativity, and the Schwarzschild metric. Einstein field equations are a set of 10 equations, which help in the understanding of the fundamental interaction of gravitation and parallel transport is the conveying of geometrical data along even curves in a manifold.

Chapter 5- The universally accepted relational theory between space and time is known as special relativity. The term was coined by Albert Einstein. Some of the aspects of special relativity explained in the following chapter are Lorentz transformation, ladder paradox, time dilation, length contraction, mass-energy equivalence and classical electromagnetism and special relativity.

At the end, I would like to thank all those who dedicated their time and efforts for the successful completion of this book. I also wish to convey my gratitude towards my friends and family who supported me at every step.

Editor

Introduction to General Relativity

The theory of gravitation proposed by Albert Einstein is known as general relativity. It generalizes special relativity and Newton's law of universal gravitation and provides an incorporated account of gravity in which gravity exists as a geometric element in the space-time continuum. This chapter will provide an integrated understanding of general relativity.

General Relativity

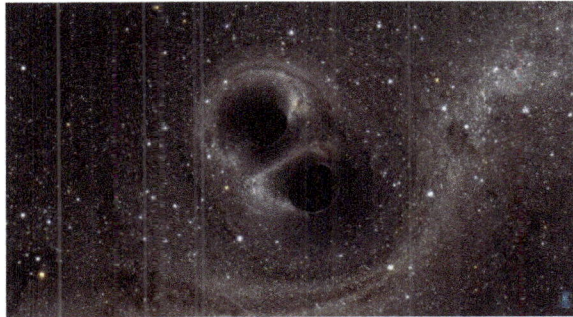

Slow motion computer simulation of the black hole binary system GW150914 as seen by a nearby observer, during 0.33 s of its final inspiral, merge, and ringdown. The star field behind the black holes is being heavily distorted and appears to rotate and move, due to extreme gravitational lensing, as space-timeitself is distorted and dragged around by the rotating black holes.

General relativity (GR, also known as the general theory of relativity or GTR) is the geometrictheory of gravitation published by Albert Einstein in 1915 and the current description of gravitation in modern physics. General relativity generalizes special relativity and Newton's law of universal gravitation, providing a unified description of gravity as a geometric property of space and time, or spacetime. In particular, the curvature of spacetime is directly related to the energy and momentum of whatever matter and radiation are present. The relation is specified by the Einstein field equations, a system of partial differential equations.

Some predictions of general relativity differ significantly from those of classical physics, especially concerning the passage of time, the geometry of space, the motion of bodies in free fall, and the propagation of light. Examples of such differences include gravitational time dilation, gravitational lensing, the gravitational redshift of light, and the gravitational time delay. The predictions of general relativity have been confirmed in all observations and experiments to date. Although general relativity is not the only relativistic theory of gravity, it is the simplest theory that is consistent with experimental data. However, unanswered questions remain, the most fundamental being how general relativity can be reconciled with the laws of quantum physics to produce a complete and self-consistent theory of quantum gravity.

Einstein's theory has important astrophysical implications. For example, it implies the existence of black holes—regions of space in which space and time are distorted in such a way that nothing, not even light, can escape—as an end-state for massive stars. There is ample evidence that the intense radiation emitted by certain kinds of astronomical objects is due to black holes; for example, microquasars and active galactic nuclei result from the presence of stellar black holes and supermassive black holes, respectively. The bending of light by gravity can lead to the phenomenon of gravitational lensing, in which multiple images of the same distant astronomical object are visible in the sky. General relativity also predicts the existence of gravitational waves, which have since been observed directly by physics collaboration LIGO. In addition, general relativity is the basis of current cosmological models of a consistently expanding universe.

History

Albert Einstein developed the theories of special and general relativity. Picture from 1921.

Soon after publishing the special theory of relativity in 1905, Einstein started thinking about how to incorporate gravity into his new relativistic framework. In 1907, beginning with a simple thought experiment involving an observer in free fall, he embarked on what would be an eight-year search for a relativistic theory of gravity. After numerous detours and false starts, his work culminated in the presentation to the Prussian Academy of Science in November 1915 of what are now known as the Einstein field equations. These equations specify how the geometry of space and time is influenced by whatever matter and radiation are present, and form the core of Einstein's general theory of relativity.

The Einstein field equations are nonlinear and very difficult to solve. Einstein used approximation methods in working out initial predictions of the theory. But as early as 1916, the astrophysicist Karl Schwarzschild found the first non-trivial exact solution to the Einstein field equations, the Schwarzschild metric. This solution laid the groundwork for the description of the final stages of gravitational collapse, and the objects known today as black holes. In the same year, the first steps towards generalizing Schwarzschild's solution to electrically charged objects were taken, which eventually resulted in the Reissner–Nordström solution, now associated with electrically charged black holes. In 1917, Einstein applied his theory to the universe as a whole, initiating the field of relativistic cosmology. In line with contemporary thinking, he assumed a static universe, adding a new parameter to his original field equations—the cosmological constant—to match that observational presumption. By 1929, however, the work of Hubble and others had shown that our universe is expanding. This is readily described by the expanding cosmological solutions found by Friedmann in 1922, which do not require a cosmological constant. Lemaître used these solutions to formulate the earliest version of the Big Bang models, in which our universe has evolved from an extremely hot and dense earlier state. Einstein later declared the cosmological constant the biggest blunder of his life.

During that period, general relativity remained something of a curiosity among physical theories. It was clearly superior to Newtonian gravity, being consistent with special relativity and accounting for several effects unexplained by the Newtonian theory. Einstein himself had shown in 1915 how his theory explained the anomalous perihelion advance of the planet Mercury without any arbitrary parameters ("fudge factors"). Similarly, a 1919 expedition led by Eddington confirmed general relativity's prediction for the deflection of starlight by the Sun during the total solar eclipse of May 29, 1919, making Einstein instantly famous. Yet the theory entered the mainstream of theoretical physics and astrophysics only with the developments between approximately 1960 and 1975, now known as the golden age of general relativity. Physicists began to understand the concept of a black hole, and to identify quasars as one of these objects' astrophysical manifestations. Ever more precise solar system tests confirmed the theory's predictive power, and relativistic cosmology, too, became amenable to direct observational tests.

From Classical Mechanics to General Relativity

General relativity can be understood by examining its similarities with and departures from classical physics. The first step is the realization that classical mechanics and Newton's law of gravity admit a geometric description. The combination of this description with the laws of special relativity results in a heuristic derivation of general relativity.

Geometry of Newtonian Gravity

According to general relativity, objects in a gravitational field behave similarly to objects within an accelerating enclosure. For example, an observer will see a ball fall the same way in a rocket (left) as it does on Earth (right), provided that the acceleration of the rocket is equal to 9.8 m/s² (the acceleration due to gravity at the surface of the Earth).

At the base of classical mechanics is the notion that a body's motion can be described as a combination of free (or inertial) motion, and deviations from this free motion. Such deviations are caused by external forces acting on a body in accordance with Newton's second law of motion, which states that the net force acting on a body is equal to that body's (inertial) mass multiplied by its acceleration. The preferred inertial motions are related to the geometry of space and time: in the standard reference frames of classical mechanics, objects in free motion move along straight lines at constant speed. In modern parlance, their paths are geodesics, straight world lines in curved spacetime.

Conversely, one might expect that inertial motions, once identified by observing the actual motions of bodies and making allowances for the external forces (such as electromagnetism or friction), can be used to define the geometry of space, as well as a time coordinate. However, there is an ambiguity once gravity comes into play. According to Newton's law of gravity, and independently verified by experiments such as that of Eötvös and its successors, there is a universality of free fall (also known as the weak equivalence principle, or the universal equality of inertial and passive-gravitational mass): the trajectory of a test body in free fall depends only on its position and initial speed, but not on any of its material properties. A simplified version of this is embodied in Einstein's elevator experiment, illustrated in the figure on the right: for an observer in a small enclosed room, it is impossible to decide, by mapping the trajectory of bodies such as a dropped ball, whether the room is at rest in a gravitational field, or in free space aboard a rocket that is accelerating at a rate equal to that of the gravitational field.

Given the universality of free fall, there is no observable distinction between inertial motion and motion under the influence of the gravitational force. This suggests the definition of a new class of inertial motion, namely that of objects in free fall under the influence of gravity. This new class of preferred motions, too, defines a geometry of space and time—in mathematical terms, it is the geodesic motion associated with a specific connection which depends on the gradient of the gravitational potential. Space, in this construction, still has the ordinary Euclidean geometry. However, space*time* as a whole is more complicated. As can be shown using simple thought experiments following the free-fall trajectories of different test particles, the result of transporting spacetime vectors that can denote a particle's velocity (time-like vectors) will vary with the particle's trajectory; mathematically speaking, the Newtonian connection is not integrable. From this, one can deduce that spacetime is curved. The resulting Newton–Cartan theory is a geometric formulation of Newtonian gravity using only covariant concepts, i.e. a description which is valid in any desired coordinate system. In this geometric description, tidal effects—the relative acceleration of bodies in free fall—are related to the derivative of the connection, showing how the modified geometry is caused by the presence of mass.

Relativistic Generalization

As intriguing as geometric Newtonian gravity may be, its basis, classical mechanics, is merely a limiting case of (special) relativistic mechanics. In the language of symmetry: where gravity can be neglected, physics is Lorentz invariant as in special relativity rather than Galilei invariant as in classical mechanics. (The defining symmetry of special relativity is the Poincaré group, which includes translations and rotations.) The differences between the two become significant when dealing with speeds approaching the speed of light, and with high-energy phenomena.

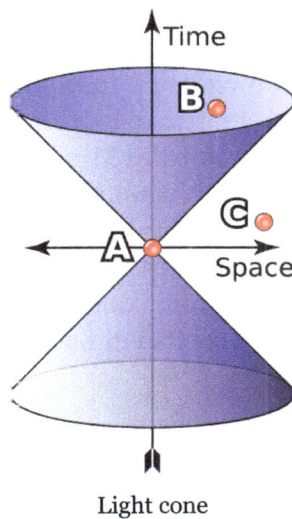

Light cone

With Lorentz symmetry, additional structures come into play. They are defined by the set of light cones. The light-cones define a causal structure: for each event A, there is a set of events that can, in principle, either influence or be influenced by A via signals or interactions that do not need to travel faster than light (such as event B in the image), and a set of events for which such an influence is impossible (such as event C in the image). These sets are observer-independent. In conjunction with the world-lines of freely falling particles, the light-cones can be used to reconstruct the space–time's semi-Riemannian metric, at least up to a positive scalar factor. In mathematical terms, this defines a Conformal structure or conformal geometry.

Special relativity is defined in the absence of gravity, so for practical applications, it is a suitable model whenever gravity can be neglected. Bringing gravity into play, and assuming the universality of free fall, an analogous reasoning as in the previous section applies: there are no global inertial frames. Instead there are approximate inertial frames moving alongside freely falling particles. Translated into the language of spacetime: the straight time-like lines that define a gravity-free inertial frame are deformed to lines that are curved relative to each other, suggesting that the inclusion of gravity necessitates a change in spacetime geometry.

A priori, it is not clear whether the new local frames in free fall coincide with the reference frames in which the laws of special relativity hold—that theory is based on the propagation of light, and thus on electromagnetism, which could have a different set of preferred frames. But using different assumptions about the special-relativistic frames (such as their being earth-fixed, or in free fall), one can derive different predictions for the gravitational redshift, that is, the way in which the frequency of light shifts as the light propagates through a gravitational field (cf. below). The actual measurements show that free-falling frames are the ones in which light propagates as it does in special relativity. The generalization of this statement, namely that the laws of special relativity hold to good approximation in freely falling (and non-rotating) reference frames, is known as the Einstein equivalence principle, a crucial guiding principle for generalizing special-relativistic physics to include gravity.

The same experimental data shows that time as measured by clocks in a gravitational field—proper time, to give the technical term—does not follow the rules of special relativity. In the language of spacetime geometry, it is not measured by the Minkowski metric. As in the Newtonian case, this is suggestive of a more general geometry. At small scales, all reference frames that are in free fall are equivalent, and approximately Minkowskian. Consequently, we are now dealing with a curved generalization of Minkowski space. The metric tensor that defines the geometry—in particular, how lengths and angles are measured—is not the Minkowski metric of special relativity, it is a generalization known as a semi- or pseudo-Riemannian metric. Furthermore, each Riemannian metric is naturally associated with one particular kind of connection, the Levi-Civita connection, and this is, in fact, the connection that satisfies the equivalence principle and makes space locally Minkowskian (that is, in suitable locally inertial coordinates, the metric is Minkowskian, and its first partial derivatives and the connection coefficients vanish).

Einstein's Equations

Having formulated the relativistic, geometric version of the effects of gravity, the question of gravity's source remains. In Newtonian gravity, the source is mass. In special relativity, mass turns out to be part of a more general quantity called the energy–momentum tensor, which includes both energy and momentumdensities as well as stress (that is, pressure and shear). Using the equivalence principle, this tensor is readily generalized to curved space-time. Drawing further upon the analogy with geometric Newtonian gravity, it is natural to assume that the field equation for gravity relates this tensor and the Ricci tensor, which describes a particular class of tidal effects: the change in volume for a small cloud of test particles that are initially at rest, and then fall freely. In special relativity, conservation of energy–momentum corresponds to the statement that the energy–momentum tensor is divergence-free. This formula, too, is readily generalized to curved spacetime by replacing partial derivatives with their curved-manifold counterparts, covariant derivatives studied in differential geometry. With this additional condition—the covariant divergence of the energy–momentum tensor, and hence of whatever is on the other side of the equation, is zero— the simplest set of equations are what are called Einstein's (field) equations:

> **Einstein's field equations**
> $$G_{\mu\nu} \equiv R_{\mu\nu} - \frac{1}{2} R g_{\mu\nu} = \frac{8\pi G}{c^4} T_{\mu\nu}$$

On the left-hand side is the Einstein tensor, a specific divergence-free combination of the Ricci tensor $R_{\mu\nu}$ and the metric. Where $G_{\mu\nu}$ is symmetric. In particular,

$$R = g^{\mu\nu} R_{\mu\nu}$$

is the curvature scalar. The Ricci tensor itself is related to the more general Riemann curvature tensor as

$$R_{\mu\nu} = R^{\alpha}{}_{\mu\alpha\nu}.$$

On the right-hand side, $T_{\mu\nu}$ is the energy–momentum tensor. All tensors are written in abstract index notation. Matching the theory's prediction to observational results for planetaryorbits (or, equivalently, assuring that the weak-gravity, low-speed limit is Newtonian mechanics), the proportionality constant can be fixed as $\kappa = 8\pi G/c^4$, with G the gravitational constant and c the speed of light. When there is no matter present, so that the energy–momentum tensor vanishes, the results are the vacuum Einstein equations,

$$R_{\mu\nu} = 0.$$

Alternatives to General Relativity

There are alternatives to general relativity built upon the same premises, which include additional rules and/or constraints, leading to different field equations. Examples are Brans–Dicke theory, teleparallelism, f(R) gravity and Einstein–Cartan theory.

Definition and Basic Applications

The derivation outlined in the previous section contains all the information needed to define general relativity, describe its key properties, and address a question of crucial importance in physics, namely how the theory can be used for model-building.

Definition and Basic Properties

General relativity is a metric theory of gravitation. At its core are Einstein's equations, which describe the relation between the geometry of a four-dimensional, pseudo-Riemannian manifold representing spacetime, and the energy–momentum contained in that spacetime. Phenomena that in classical mechanics are ascribed to the action of the force of gravity (such as free-fall, orbital motion, and spacecrafttrajectories), correspond to inertial motion within a curved geometry of spacetime in general relativity; there is no gravitational force deflecting objects from their natural, straight paths. Instead, gravity corresponds to changes in the properties of space and time, which in turn changes the straightest-possible paths that objects will naturally follow. The curvature is, in turn, caused by the energy–momentum of matter. Paraphrasing the relativist John Archibald Wheeler, spacetime tells matter how to move; matter tells spacetime how to curve.

While general relativity replaces the scalar gravitational potential of classical physics by a symmetric rank-two tensor, the latter reduces to the former in certain limiting cases. For weak gravitational fields and slow speed relative to the speed of light, the theory's predictions converge on those of Newton's law of universal gravitation.

As it is constructed using tensors, general relativity exhibits general covariance: its laws—and further laws formulated within the general relativistic framework—take on the same form in all coordinate systems. Furthermore, the theory does not contain any invariant geometric background structures, i.e. it is background independent. It thus satisfies a more stringent general principle of relativity, namely that the laws of physics are the same for all observers.Locally, as expressed in the equivalence principle, spacetime is Minkowskian, and the laws of physics exhibit local Lorentz invariance.

Model-building

The core concept of general-relativistic model-building is that of a solution of Einstein's equations. Given both Einstein's equations and suitable equations for the properties of matter, such a solution consists of a specific semi-Riemannian manifold (usually defined by giving the metric in specific coordinates), and specific matter fields defined on that manifold. Matter and geometry must satisfy Einstein's equations, so in particular, the matter's energy–momentum tensor must be divergence-free. The matter must, of course, also satisfy whatever additional equations were imposed on its properties. In short, such a solution is a model universe that satisfies the laws of general relativity, and possibly additional laws governing whatever matter might be present.

Einstein's equations are nonlinear partial differential equations and, as such, difficult to solve exactly. Nevertheless, a number of exact solutions are known, although only a few have direct physical applications. The best-known exact solutions, and also those most interesting from a physics point of view, are the Schwarzschild solution, the Reissner–Nordström solution and the Kerr metric, each corresponding to a certain type of black hole in an otherwise empty universe, and the Friedmann–Lemaître–Robertson–Walker and de Sitter universes, each describing an expanding cosmos. Exact solutions of great theoretical interest include the Gödel universe (which opens up the intriguing possibility of time travel in curved spacetimes), the Taub-NUT solution (a model universe that is homogeneous, but anisotropic), and anti-de Sitter space (which has recently come to prominence in the context of what is called the Maldacena conjecture).

Given the difficulty of finding exact solutions, Einstein's field equations are also solved frequently by numerical integration on a computer, or by considering small perturbations of exact solutions. In the field of numerical relativity, powerful computers are employed to simulate the geometry of spacetime and to solve Einstein's equations for interesting situations such as two colliding black holes. In principle, such methods may be applied to any system, given sufficient computer resources, and may address fundamental questions such as naked singularities. Approximate solutions may also be found by perturbation theories such as linearized gravity and its generalization, the post-Newtonian expansion, both of which were developed by Einstein. The latter provides a systematic approach to solving for the geometry of a spacetime that contains a distribution of matter that moves slowly compared with the speed of light. The expansion involves a series of terms; the first terms represent Newtonian gravity, whereas the later terms represent ever smaller corrections to Newton's theory due to general relativity. An extension of this expansion is the parametrized post-Newtonian (PPN) formalism, which allows quantitative comparisons between the predictions of general relativity and alternative theories.

Consequences of Einstein's Theory

General relativity has a number of physical consequences. Some follow directly from the theory's axioms, whereas others have become clear only in the course of many years of research that followed Einstein's initial publication.

Gravitational Time Dilation and Frequency Shift

Assuming that the equivalence principle holds, gravity influences the passage of time. Light sent down into a gravity well is blueshifted, whereas light sent in the opposite direction (i.e., climbing out of the gravity well) is redshifted; collectively, these two effects are known as the gravitational frequency shift. More generally, processes close to a massive body run more slowly when compared with processes taking place farther away; this effect is known as gravitational time dilation.

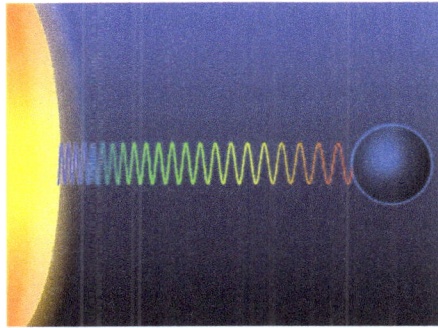

Schematic representation of the gravitational redshift of a light wave escaping from the surface of a massive body

Gravitational redshift has been measured in the laboratory and using astronomical observations. Gravitational time dilation in the Earth's gravitational field has been measured numerous times using atomic clocks, while ongoing validation is provided as a side effect of the operation of the Global Positioning System (GPS). Tests in stronger gravitational fields are provided by the observation of binary pulsars. All results are in agreement with general relativity. However, at the current level of accuracy, these observations cannot distinguish between general relativity and other theories in which the equivalence principle is valid.

Light Deflection and Gravitational Time Delay

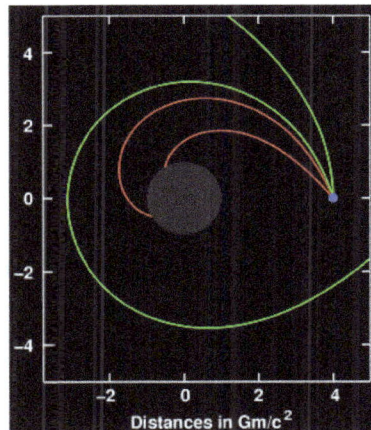

Deflection of light (sent out from the location shown in blue) near a compact body (shown in gray)

General relativity predicts that the path of light is bent in a gravitational field; light passing a massive body is deflected towards that body. This effect has been confirmed by observing the light of stars or distant quasars being deflected as it passes the Sun.

This and related predictions follow from the fact that light follows what is called a light-like or null geodesic—a generalization of the straight lines along which light travels in classical physics. Such geodesics are the generalization of the invariance of lightspeed in special relativity. As one examines suitable model spacetimes (either the exterior Schwarzschild solution or, for more than a single mass, the post-Newtonian expansion), several effects of gravity on light propagation emerge. Although the bending of light can also be derived by extending the universality of free fall to light, the angle of deflection resulting from such calculations is only half the value given by general relativity.

Closely related to light deflection is the gravitational time delay (or Shapiro delay), the phenomenon that light signals take longer to move through a gravitational field than they would in the absence of that field. There have been numerous successful tests of this prediction. In the parameterized post-Newtonian formalism (PPN), measurements of both the deflection of light and the gravitational time delay determine a parameter called γ, which encodes the influence of gravity on the geometry of space.

Gravitational Waves

Predicted in 1916 by Albert Einstein, there are gravitational waves: ripples in the metric of spacetime that propagate at the speed of light. These are one of several analogies between weak-field gravity and electromagnetism in that, they are analogous to electromagnetic waves. On February 11, 2016, the Advanced LIGO team announced that they had directly detected gravitational waves from a pair of black holes merging.

The simplest type of such a wave can be visualized by its action on a ring of freely floating particles. A sine wave propagating through such a ring towards the reader distorts the ring in a characteristic, rhythmic fashion (animated image to the right). Since Einstein's equations are non-linear, arbitrarily strong gravitational waves do not obey linear superposition, making their description difficult. However, for weak fields, a linear approximation can be made. Such linearized gravitational waves are sufficiently accurate to describe the exceedingly weak waves that are expected to arrive here on Earth from far-off cosmic events, which typically result in relative distances increasing and decreasing by 10^{-21} or less. Data analysis methods routinely make use of the fact that these linearized waves can be Fourier decomposed.

Some exact solutions describe gravitational waves without any approximation, e.g., a wave train traveling through empty space or Gowdy universes, varieties of an expanding cosmos filled with gravitational waves. But for gravitational waves produced in astrophysically relevant situations, such as the merger of two black holes, numerical methods are presently the only way to construct appropriate models.

Orbital Effects and the Relativity of Direction

General relativity differs from classical mechanics in a number of predictions concerning orbiting bodies. It predicts an overall rotation (precession) of planetary orbits, as well as orbital decay caused by the emission of gravitational waves and effects related to the relativity of direction.

Precession of Apsides

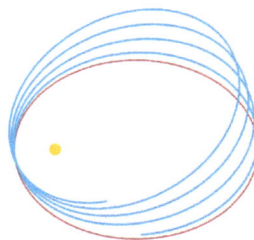

Newtonian (red) vs. Einsteinian orbit (blue) of a lone planet orbiting a star

In general relativity, the apsides of any orbit (the point of the orbiting body's closest approach to the system's center of mass) will precess—the orbit is not an ellipse, but akin to an ellipse that rotates on its focus, resulting in a rose curve-like shape. Einstein first derived this result by using an approximate metric representing the Newtonian limit and treating the orbiting body as a test particle. For him, the fact that his theory gave a straightforward explanation of the anomalous perihelion shift of the planet Mercury, discovered earlier by Urbain Le Verrier in 1859, was important evidence that he had at last identified the correct form of the gravitational field equations.

The effect can also be derived by using either the exact Schwarzschild metric (describing spacetime around a spherical mass) or the much more general post-Newtonian formalism. It is due to the influence of gravity on the geometry of space and to the contribution of self-energy to a body's gravity (encoded in the nonlinearity of Einstein's equations). Relativistic precession has been observed for all planets that allow for accurate precession measurements (Mercury, Venus, and Earth), as well as in binary pulsar systems, where it is larger by five orders of magnitude.

In general relativity the perihelion shift σ, expressed in radians per revolution, is approximately given by:

$$\sigma = \frac{24\pi^3 L^2}{T^2 c^2 (1 - e^2)},$$

where L is the semi-major axis, T is the orbital period, c is the speed of light, and e is the orbital eccentricity.

Orbital Decay

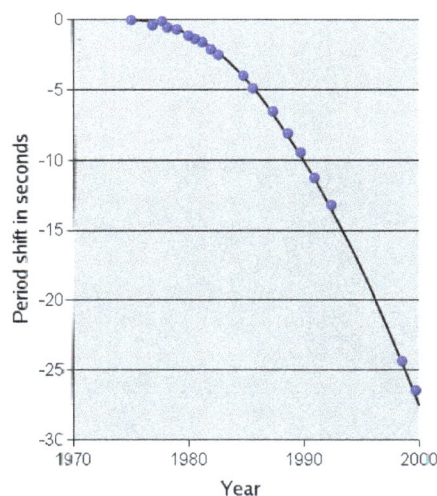

Orbital decay for PSR1913+16: time shift in seconds, tracked over three decades.

According to general relativity, a binary system will emit gravitational waves, thereby losing energy. Due to this loss, the distance between the two orbiting bodies decreases, and so does their orbital period. Within the Solar System or for ordinary double stars, the effect is too small to be observable. This is not the case for a close binary pulsar, a system of two orbiting neutron stars, one of which is a pulsar: from the pulsar, observers on Earth receive a regular series of radio pulses that can serve as a highly accurate clock, which allows precise measurements of the orbital period. Because neutron stars are immensely compact, significant amounts of energy are emitted in the form of gravitational radiation.

The first observation of a decrease in orbital period due to the emission of gravitational waves was made by Hulse and Taylor, using the binary pulsar PSR1913+16 they had discovered in 1974. This was the first detection of gravitational waves, albeit indirect, for which they were awarded the 1993 Nobel Prize in physics. Since then, several other binary pulsars have been found, in particular the double pulsar PSR J0737-3039, in which both stars are pulsars.

Geodetic Precession and Frame-dragging

Several relativistic effects are directly related to the relativity of direction. One is geodetic precession: the axis direction of a gyroscope in free fall in curved spacetime will change when compared, for instance, with the direction of light received from distant stars—even though such a gyroscope represents the way of keeping a direction as stable as possible ("parallel transport"). For the Moon−Earth system, this effect has been measured with the help of lunar laser ranging. More recently, it has been measured for test masses aboard the satellite Gravity Probe B to a precision of better than 0.3%.

Near a rotating mass, there are gravitomagnetic or frame-dragging effects. A distant observer will determine that objects close to the mass get "dragged around". This is most extreme for rotating black holes where, for any object entering a zone known as the ergosphere, rotation is inevitable. Such effects can again be tested through their influence on the orientation of gyroscopes in free fall. Somewhat controversial tests have been performed using the LAGEOS satellites, confirming the relativistic prediction. Also the Mars Global Surveyor probe around Mars has been used.

Astrophysical Applications

Gravitational Lensing

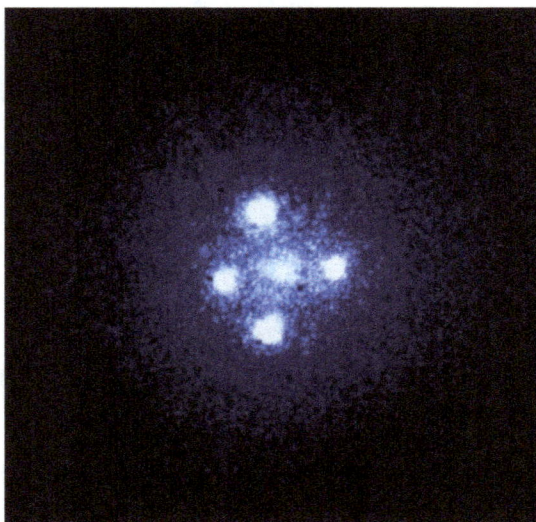

Einstein cross: four images of the same astronomical object, produced by a gravitational lens

The deflection of light by gravity is responsible for a new class of astronomical phenomena. If a massive object is situated between the astronomer and a distant target object with appropriate mass and relative distances, the astronomer will see multiple distorted images of the target. Such effects are known as gravitational lensing. Depending on the configuration, scale, and mass distribution, there can be two or more images, a bright ring known as an Einstein ring, or partial rings called arcs. The earliest example was discovered in 1979; since then, more than a hundred gravitational lenses have been observed.Even if the multiple images are too close to each other to be resolved, the effect can still be measured, e.g., as an overall brightening of the target object; a number of such "microlensing events" have been observed.

Gravitational lensing has developed into a tool of observational astronomy. It is used to detect the presence and distribution of dark matter, provide a "natural telescope" for observing distant galaxies, and to obtain an independent estimate of the Hubble constant. Statistical evaluations of lensing data provide valuable insight into the structural evolution of galaxies.

Gravitational Wave Astronomy

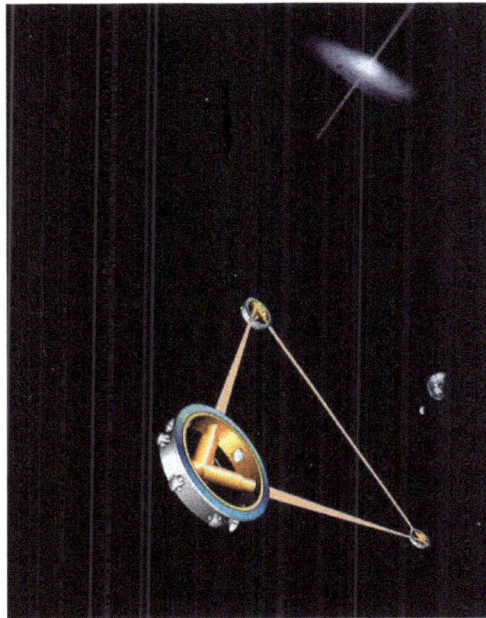

Artist's impression of the space-borne gravitational wave detector LISA

Observations of binary pulsars provide strong indirect evidence for the existence of gravitational waves. Detection of these waves is a major goal of current relativity-related research. Several land-based gravitational wave detectors are currently in operation, most notably the interferometric detectorsGEO 600, LIGO (two detectors), TAMA 300 and VIRGO. Various pulsar timing arrays are using millisecond pulsars to detect gravitational waves in the 10^{-9} to 10^{-6}Hertz frequency range, which originate from binary supermassive blackholes. A European space-based detector, eLISA / NGO, is currently under development, with a precursor mission (LISA Pathfinder) having launched in December 2015.

Observations of gravitational waves promise to complement observations in the electromagnetic spectrum. They are expected to yield information about black holes and other dense objects such as neutron stars and white dwarfs, about certain kinds of supernova implosions, and about processes in the very early universe, including the signature of certain types of hypothetical cosmic string. In February 2016, the Advanced LIGO team announced that they had detected gravitational waves from a black hole merger.

Black Holes and Other Compact Objects

Whenever the ratio of an object's mass to its radius becomes sufficiently large, general relativity predicts the formation of a black hole, a region of space from which nothing, not even light, can escape. In the currently accepted models of stellar evolution, neutron stars of around 1.4 solar masses, and stellar black holes with a few to a few dozen solar masses, are thought to be the final state for the evolution of massive stars. Usually a galaxy has one supermassive black hole with a few million to a few billion solar masses in its center, and its presence is thought to have played an important role in the formation of the galaxy and larger cosmic structures.

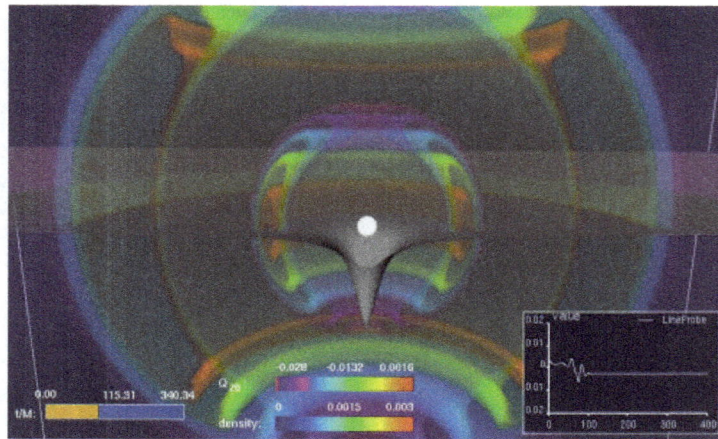

Simulation based on the equations of general relativity: a star collapsing to form a black hole while emitting gravitational waves

Astronomically, the most important property of compact objects is that they provide a supremely efficient mechanism for converting gravitational energy into electromagnetic radiation.Accretion, the falling of dust or gaseous matter onto stellar or supermassive black holes, is thought to be responsible for some spectacularly luminous astronomical objects, notably diverse kinds of active galactic nuclei on galactic scales and stellar-size objects such as microquasars. In particular, accretion can lead to relativistic jets, focused beams of highly energetic particles that are being flung into space at almost light speed. General relativity plays a central role in modelling all these phenomena, and observations provide strong evidence for the existence of black holes with the properties predicted by the theory.

Black holes are also sought-after targets in the search for gravitational waves (cf. Gravitational waves, above). Merging black hole binaries should lead to some of the strongest gravitational wave signals reaching detectors here on Earth, and the phase directly before the merger ("chirp") could be used as a "standard candle" to deduce the distance to the merger events—and hence serve as a probe of cosmic expansion at large distances.The gravitational waves produced as a stellar black hole plunges into a supermassive one should provide direct information about the supermassive black hole's geometry.

Cosmology

This blue horseshoe is a distant galaxy that has been magnified and warped into a nearly complete ring by the strong gravitational pull of the massive foreground luminous red galaxy.

The current models of cosmology are based on Einstein's field equations, which include the cosmological constant Λ since it has important influence on the large-scale dynamics of the cosmos,

$$R_{\mu\nu} - \frac{1}{2}R g_{\mu\nu} + \Lambda\, g_{\mu\nu} = \frac{8\pi G}{c^4}T_{\mu\nu}$$

where $g_{\mu\nu}$ is the spacetime metric. Isotropic and homogeneous solutions of these enhanced equations, the Friedmann–Lemaître–Robertson–Walker solutions, allow physicists to model a universe that has evolved over the past 14 billion years from a hot, early Big Bang phase. Once a small number of parameters (for example the universe's mean matterdensity) have been fixed by astronomical observation, further observational data can be used to put the models to the test. Predictions, all successful, include the initial abundance of chemical elements formed in a period of primordial nucleosynthesis, the large-scale structure of the universe, and the existence and properties of a "thermal echo" from the early cosmos, the cosmic background radiation.

Astronomical observations of the cosmological expansion rate allow the total amount of matter in the universe to be estimated, although the nature of that matter remains mysterious in part. About 90% of all matter appears to be dark matter, which has mass (or, equivalently, gravitational influence), but does not interact electromagnetically and, hence, cannot be observed directly. There is no generally accepted description of this new kind of matter, within the framework of known particle physics or otherwise. Observational evidence from redshift surveys of distant supernovae and measurements of the cosmic background radiation also show that the evolution of our universe is significantly influenced by a cosmological constant resulting in an acceleration of cosmic expansion or, equivalently, by a form of energy with an unusual equation of state, known as dark energy, the nature of which remains unclear.

An inflationary phase, an additional phase of strongly accelerated expansion at cosmic times of around 10^{-33} seconds, was hypothesized in 1980 to account for several puzzling observations that were unexplained by classical cosmological models, such as the nearly perfect homogeneity of the cosmic background radiation. Recent measurements of the cosmic background radiation have resulted in the first evidence for this scenario. However, there is a bewildering variety of possible inflationary scenarios, which cannot be restricted by current observations. An even larger question is the physics of the earliest universe, prior to the inflationary phase and close to where the classical models predict the big bang singularity. An authoritative answer would require a complete theory of quantum gravity, which has not yet been developed (cf. the section on quantum gravity, below).

Time Travel

Kurt Gödelshowed that solutions to Einstein's equations exist that contain closed timelike curves (CTCs), which allow for loops in time. The solutions require extreme physical conditions unlikely ever to occur in practice, and it remains an open question whether further laws of physics will eliminate them completely. Since then other—similarly impractical—GR solutions containing CTCs have been found, such as the Tipler cylinder and traversable wormholes.

Advanced Concepts

Causal Structure and Global Geometry

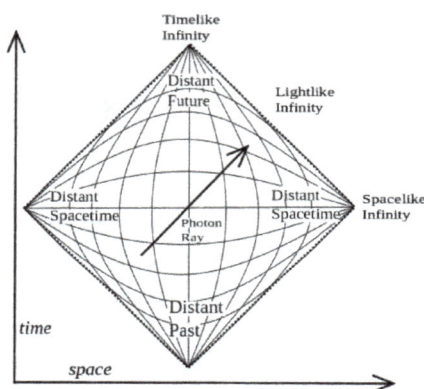

Penrose–Carter diagram of an infinite Minkowski universe

In general relativity, no material body can catch up with or overtake a light pulse. No influence from an event A can reach any other location X before light sent out at A to X. In consequence, an exploration of all light worldlines (null geodesics) yields key information about the spacetime's causal structure. This structure can be displayed using Penrose–Carter diagrams in which infinitely large regions of space and infinite time intervals are shrunk ("compactified") so as to fit onto a finite map, while light still travels along diagonals as in standard spacetime diagrams.

Aware of the importance of causal structure, Roger Penrose and others developed what is known as global geometry. In global geometry, the object of study is not one particular solution (or family of solutions) to Einstein's equations. Rather, relations that hold true for all geodesics, such as the Raychaudhuri equation, and additional non-specific assumptions about the nature of matter (usually in the form of energy conditions) are used to derive general results.

Horizons

Using global geometry, some spacetimes can be shown to contain boundaries called horizons, which demarcate one region from the rest of spacetime. The best-known examples are black holes:

if mass is compressed into a sufficiently compact region of space (as specified in the hoop conjecture, the relevant length scale is the Schwarzschild radius), no light from inside can escape to the outside. Since no object can overtake a light pulse, all interior matter is imprisoned as well. Passage from the exterior to the interior is still possible, showing that the boundary, the black hole's *horizon*, is not a physical barrier.

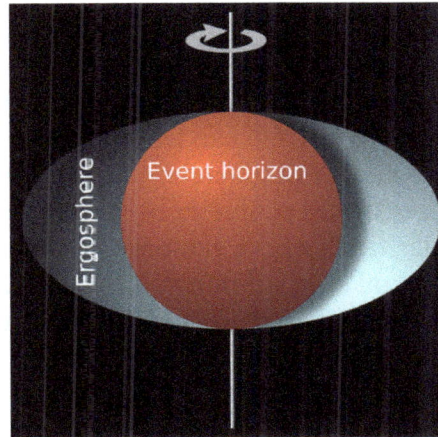

The ergosphere of a rotating black hole, which plays a key role when it comes to extracting energy from such a black hole

Early studies of black holes relied on explicit solutions of Einstein's equations, notably the spherically symmetric Schwarzschild solution (used to describe a static black hole) and the axisymmetric Kerr solution (used to describe a rotating, stationary black hole, and introducing interesting features such as the ergosphere). Using global geometry, later studies have revealed more general properties of black holes. In the long run, they are rather simple objects characterized by eleven parameters specifying energy, linear momentum, angular momentum, location at a specified time and electric charge. This is stated by the black hole uniqueness theorems: "black holes have no hair", that is, no distinguishing marks like the hairstyles of humans. Irrespective of the complexity of a gravitating object collapsing to form a black hole, the object that results (having emitted gravitational waves) is very simple.

Even more remarkably, there is a general set of laws known as black hole mechanics, which is analogous to the laws of thermodynamics. For instance, by the second law of black hole mechanics, the area of the event horizon of a general black hole will never decrease with time, analogous to the entropy of a thermodynamic system. This limits the energy that can be extracted by classical means from a rotating black hole (e.g. by the Penrose process).There is strong evidence that the laws of black hole mechanics are, in fact, a subset of the laws of thermodynamics, and that the black hole area is proportional to its entropy.This leads to a modification of the original laws of black hole mechanics: for instance, as the second law of black hole mechanics becomes part of the second law of thermodynamics, it is possible for black hole area to decrease—as long as other processes ensure that, overall, entropy increases. As thermodynamical objects with non-zero temperature, black holes should emit thermal radiation. Semi-classical calculations indicate that indeed they do, with the surface gravity playing the role of temperature in Planck's law. This radiation is known as Hawking radiation (cf. the quantum theory section, below).

There are other types of horizons. In an expanding universe, an observer may find that some regions of the past cannot be observed ("particle horizon"), and some regions of the future cannot be influenced (event horizon). Even in flat Minkowski space, when described by an accelerated observer (Rindler space), there will be horizons associated with a semi-classical radiation known as Unruh radiation.

Singularities

Another general feature of general relativity is the appearance of spacetime boundaries known as singularities. Spacetime can be explored by following up on timelike and lightlike geodesics—all possible ways that light and particles in free fall can travel. But some solutions of Einstein's equations have "ragged edges"—regions known as spacetime singularities, where the paths of light and falling particles come to an abrupt end, and geometry becomes ill-defined. In the more interesting cases, these are "curvature singularities", where geometrical quantities characterizing spacetime curvature, such as the Ricci scalar, take on infinite values. Well-known examples of spacetimes with future singularities—where worldlines end—are the Schwarzschild solution, which describes a singularity inside an eternal static black hole, or the Kerr solution with its ring-shaped singularity inside an eternal rotating black hole. The Friedmann–Lemaître–Robertson–Walker solutions and other spacetimes describing universes have past singularities on which worldlines begin, namely Big Bang singularities, and some have future singularities (Big Crunch) as well.

Given that these examples are all highly symmetric—and thus simplified—it is tempting to conclude that the occurrence of singularities is an artifact of idealization. The famous singularity theorems, proved using the methods of global geometry, say otherwise: singularities are a generic feature of general relativity, and unavoidable once the collapse of an object with realistic matter properties has proceeded beyond a certain stage and also at the beginning of a wide class of expanding universes. However, the theorems say little about the properties of singularities, and much of current research is devoted to characterizing these entities' generic structure (hypothesized e.g. by the BKL conjecture).The cosmic censorship hypothesis states that all realistic future singularities (no perfect symmetries, matter with realistic properties) are safely hidden away behind a horizon, and thus invisible to all distant observers. While no formal proof yet exists, numerical simulations offer supporting evidence of its validity.

Evolution Equations

Each solution of Einstein's equation encompasses the whole history of a universe — it is not just some snapshot of how things are, but a whole, possibly matter-filled, spacetime. It describes the state of matter and geometry everywhere and at every moment in that particular universe. Due to its general covariance, Einstein's theory is not sufficient by itself to determine the time evolution of the metric tensor. It must be combined with a coordinate condition, which is analogous to gauge fixing in other field theories.

To understand Einstein's equations as partial differential equations, it is helpful to formulate them in a way that describes the evolution of the universe over time. This is done in "3+1" formulations, where spacetime is split into three space dimensions and one time dimension. The best-known example is the ADM formalism.These decompositions show that the spacetime evolution equations of general relativity are well-behaved: solutions always exist, and are uniquely defined, once suitable initial conditions have been specified.Such formulations of Einstein's field equations are the basis of numerical relativity.

Global and Quasi-local Quantities

The notion of evolution equations is intimately tied in with another aspect of general relativistic physics. In Einstein's theory, it turns out to be impossible to find a general definition for a seemingly simple property such as a system's total mass (or energy). The main reason is that the gravitational field—like any physical field—must be ascribed a certain energy, but that it proves to be fundamentally impossible to localize that energy.

Nevertheless, there are possibilities to define a system's total mass, either using a hypothetical "infinitely distant observer" (ADM mass) or suitable symmetries (Komar mass).If one excludes from the system's total mass the energy being carried away to infinity by gravitational waves, the result is the Bondi mass at null infinity. Just as in classical physics, it can be shown that these masses are positive. Corresponding global definitions exist for momentum and angular momentum.There have also been a number of attempts to define *quasi-local* quantities, such as the mass of an isolated system formulated using only quantities defined within a finite region of space containing that system. The hope is to obtain a quantity useful for general statements about isolated systems, such as a more precise formulation of the hoop conjecture.

Relationship with Quantum Theory

If general relativity were considered to be one of the two pillars of modern physics, then quantum theory, the basis of understanding matter from elementary particles to solid state physics, would be the other. However, how to reconcile quantum theory with general relativity is still an open question.

Quantum Field Theory in Curved Spacetime

Ordinary quantum field theories, which form the basis of modern elementary particle physics, are defined in flat Minkowski space, which is an excellent approximation when it comes to describing the behavior of microscopic particles in weak gravitational fields like those found on Earth. In order to describe situations in which gravity is strong enough to influence (quantum) matter, yet not strong enough to require quantization itself, physicists have formulated quantum field theories in curved spacetime. These theories rely on general relativity to describe a curved background spacetime, and define a generalized quantum field theory to describe the behavior of quantum matter within that spacetime. Using this formalism, it can be shown that black holes emit a blackbody spectrum of particles known as Hawking radiation, leading to the possibility that they evaporate over time.As briefly mentioned above, this radiation plays an important role for the thermodynamics of black holes.

Quantum Gravity

The demand for consistency between a quantum description of matter and a geometric description of spacetime, as well as the appearance of singularities (where curvature length scales become microscopic), indicate the need for a full theory of quantum gravity: for an adequate description of the interior of black holes, and of the very early universe, a theory is required in which gravity and the associated geometry of spacetime are described in the language of quantum physics. Despite major efforts, no complete and consistent theory of quantum gravity is currently known, even though a number of promising candidates exist.

Projection of a Calabi–Yau manifold, one of the ways of compactifying the extra dimensions posited by string theory

Attempts to generalize ordinary quantum field theories, used in elementary particle physics to describe fundamental interactions, so as to include gravity have led to serious problems.Some have argued that at low energies, this approach proves successful, in that it results in an acceptable effective (quantum) field theory of gravity.At very high energies, however, the perturbative results are badly divergent and lead to models devoid of predictive power ("perturbative non-renormalizability").

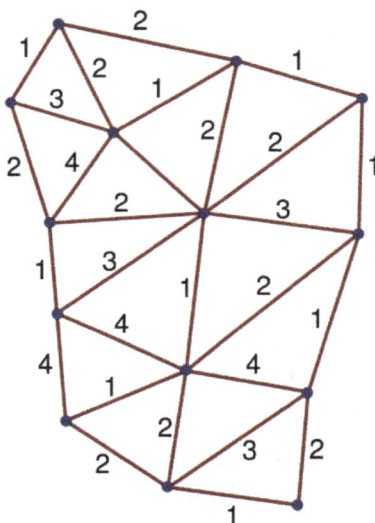

Simple spin network of the type used in loop quantum gravity

One attempt to overcome these limitations is string theory, a quantum theory not of point particles, but of minute one-dimensional extended objects. The theory promises to be a unified description of all particles and interactions, including gravity; the price to pay is unusual features such as six extra dimensions of space in addition to the usual three. In what is called the second superstring revolution, it was conjectured that both string theory and a unification of general relativity and supersymmetry known as supergravity form part of a hypothesized eleven-dimensional model known as M-theory, which would constitute a uniquely defined and consistent theory of quantum gravity.

Another approach starts with the canonical quantization procedures of quantum theory. Using the initial-value-formulation of general relativity (cf. evolution equations above), the result is the Wheeler–deWitt equation (an analogue of the Schrödinger equation) which, regrettably, turns out to be ill-defined without a proper ultraviolet (lattice) cutoff. However, with the introduction of what are now known as Ashtekar variables, this leads to a promising model known as loop quantum gravity. Space is represented by a web-like structure called a spin network, evolving over time in discrete steps.

Depending on which features of general relativity and quantum theory are accepted unchanged, and on what level changes are introduced, there are numerous other attempts to arrive at a viable theory of quantum gravity, some examples being the lattice theory of gravity based on the Feynman Path Integral approach and Regge Calculus,dynamical triangulations,causal sets,twistor models or the path-integral based models of quantum cosmology.

All candidate theories still have major formal and conceptual problems to overcome. They also face the common problem that, as yet, there is no way to put quantum gravity predictions to experimental tests (and thus to decide between the candidates where their predictions vary), although there is hope for this to change as future data from cosmological observations and particle physics experiments becomes available.

Current Status

General relativity has emerged as a highly successful model of gravitation and cosmology, which has so far passed many unambiguous observational and experimental tests. However, there are strong indications the theory is incomplete.The problem of quantum gravity and the question of the reality of spacetime singularities remain open. Observational data that is taken as evidence for dark energy and dark matter could indicate the need for new physics. Even taken as is, general relativity is rich with possibilities for further exploration. Mathematical relativists seek to understand the nature of singularities and the fundamental properties of Einstein's equations, and increasingly powerful computer simulations (such as those describing merging black holes) are run. In February 2016, it was announced that the existence of gravitational waves was directly detected by the Advanced LIGO team on September 14, 2015. A century after its publication, general relativity remains a highly active area of research.

References

- Bartusiak, Marcia (2000), Einstein's Unfinished Symphony: Listening to the Sounds of Space-Time, Berkley, ISBN 978-0-425-18620-6.

- Blair, David; McNamara, Geoff (1997), Ripples on a Cosmic Sea. The Search for Gravitational Waves, Perseus, ISBN 0-7382-0137-5.

- Blandford, R. D. (1987), "Astrophysical Black Holes", in Hawking, Stephen W.; Israel, Werner, 300 Years of Gravitation, Cambridge University Press, pp. 277–329, ISBN 0-521-37976-8.

- Bruhat, Yvonne (1962), "The Cauchy Problem", in Witten, Louis, Gravitation: An Introduction to Current Research, Wiley, p. 130, ISBN 978-1-114-29166-9.

- Carroll, Bradley W.; Ostlie, Dale A. (1996), An Introduction to Modern Astrophysics, Addison-Wesley, ISBN 0-201-54730-9.

- Chandrasekhar, Subrahmanyan (1983), The Mathematical Theory of Black Holes, Oxford University Press, ISBN 0-19-850370-9.

- Ehlers, Jürgen (1973), "Survey of general relativity theory", in Israel, Werner, Relativity, Astrophysics and Cosmology, D. Reidel, pp. 1–125, ISBN 90-277-0369-8.

- Ehlers, Jürgen; Lämmerzahl, Claus, eds. (2006), Special Relativity—Will it Survive the Next 101 Years?, Springer, ISBN 3-540-34522-1.

- Green, M. B.; Schwarz, J. H.; Witten, E. (1987), Superstring theory. Volume 1: Introduction, Cambridge University Press, ISBN 0-521-35752-7.

Gravity: A Comprehensive Study

The phenomenon that brings all things towards each another is known as gravity. It is caused by the uneven distribution of mass/energy in the universe. It also gives mass to bodily objects on earth and can even cause ocean tides. The following text also concentrates on Newton's law of universal gravitation, an introduction to general relativity and the history of gravitational theory.

Gravity

Hammer and feather drop: Apollo 15astronautDavid Scott on the Moon enacting the legend of Galileo's gravity experiment. (1.38 MB, ogg/Theora format).

Gravity, or gravitation, is a natural phenomenon by which all things with mass are brought toward (or *gravitate* toward) one another, including planets, stars and galaxies. Since energy and mass are equivalent, all forms of energy, including light, also cause gravitation and are under the influence of it. On Earth, gravity gives weight to physical objects and causes the ocean tides. The gravitational attraction of the original gaseous matter present in the Universe caused it to begin coalescing, forming stars — and the stars to group together into galaxies — so gravity is responsible for many of the large scale structures in the Universe. Gravity has an infinite range, although its effects become increasingly weaker on farther objects.

Gravity is most accurately described by the general theory of relativity (proposed by Albert Einstein in 1915) which describes gravity not as a force, but as a consequence of the curvature of spacetime caused by the uneven distribution of mass/energy. The most extreme example of this curvature of spacetime is a black hole, from which nothing can escape once past its event horizon, not even light. More gravity results in gravitational time dilation, where time lapses more slowly at a lower (stronger) gravitational potential. However, for most applications, gravity is well approximated by Newton's law of universal gravitation, which postulates that gravity causes a force where two bodies of mass are directly drawn (or 'attracted') to each other according to a mathematical relationship, where the attractive force is directly proportional to the product of their masses and inversely proportional to the square of the distance between them.

Gravity is the weakest of the four fundamental interactions of nature. The gravitational attraction is approximately 10^{38} times weaker than the strong force, 10^{36} times weaker than the electromagnetic force and 10^{29} times weaker than the weak force. As a consequence, gravity has a negligible influence on the behavior of subatomic particles, and plays no role in determining the internal properties of everyday matter. On the other hand, gravity is the dominant interaction at the macroscopic scale, and is the cause of the formation, shape and trajectory (orbit) of astronomical bodies. It is responsible for various phenomena observed on Earth and throughout the Universe; for example, it causes the Earth and the other planets to orbit the Sun, the Moon to orbit the Earth, the formation of tides, the formation and evolution of the Solar System, stars and galaxies.

The earliest instance of gravity in the Universe, possibly in the form of quantum gravity, supergravity or a gravitational singularity, along with ordinary space and time, developed during the Planck epoch (up to 10^{-43} seconds after the birth of the Universe), possibly from a primeval state, such as a false vacuum, quantum vacuum or virtual particle, in a currently unknown manner. For this reason, in part, pursuit of a theory of everything, the merging of the general theory of relativity and quantum mechanics (or quantum field theory) into quantum gravity, has become an area of research.

History of Gravitational Theory

Earlier Concepts of Gravity

While the modern European thinkers are rightly credited with development of gravitational theory, there were pre-existing ideas which had identified the force of gravity. Some of the earliest descriptions came from early mathematician-astronomers, such as Aryabhata, who had identified the force of gravity to explain why objects do not fall out when the Earth rotates.

Later, the works of Brahmagupta referred to the presence of this force.

Scientific Revolution

Modern work on gravitational theory began with the work of Galileo Galilei in the late 16th and early 17th centuries. In his famous (though possibly apocryphal) experiment dropping balls from the Tower of Pisa, and later with careful measurements of balls rolling down inclines, Galileo showed that gravitational acceleration is the same for all objects. This was a major departure from Aristotle's belief that heavier objects have a higher gravitational acceleration. Galileo postulated air resistance as the reason that objects with less mass may fall slower in an atmosphere. Galileo's work set the stage for the formulation of Newton's theory of gravity.

Newton's Theory of Gravitation

In 1687, English mathematician Sir Isaac Newton published *Principia*, which hypothesizes the inverse-square law of universal gravitation. In his own words, "I deduced that the forces which keep the planets in their orbs must [be] reciprocally as the squares of their distances from the centers about which they revolve: and thereby compared the force requisite to keep the Moon in her Orb with the force of gravity at the surface of the Earth; and found them answer pretty nearly." The equation is the following:

$$F = G \frac{m_1 m_2}{r^2}$$

Where F is the force, m_1 and m_2 are the masses of the objects interacting, r is the distance between the centers of the masses and G is the gravitational constant.

Newton's theory enjoyed its greatest success when it was used to predict the existence of Neptune based on motions of Uranus that could not be accounted for by the actions of the other planets. Calculations by both John Couch Adams and Urbain Le Verrier predicted the general position of the planet, and Le Verrier's calculations are what led Johann Gottfried Galle to the discovery of Neptune.

A discrepancy in Mercury's orbit pointed out flaws in Newton's theory. By the end of the 19th century, it was known that its orbit showed slight perturbations that could not be accounted for entirely under Newton's theory, but all searches for another perturbing body (such as a planet orbiting the Sun even closer than Mercury) had been fruitless. The issue was resolved in 1915 by Albert Einstein's new theory of general relativity, which accounted for the small discrepancy in Mercury's orbit.

Although Newton's theory has been superseded by the Einstein's general relativity, most modern non-relativistic gravitational calculations are still made using Newton's theory because it is simpler to work with and it gives sufficiently accurate results for most applications involving sufficiently small masses, speeds and energies.

Sir Isaac Newton, an English physicist who lived from 1642 to 1727

Equivalence Principle

The equivalence principle, explored by a succession of researchers including Galileo, Loránd Eötvös, and Einstein, expresses the idea that all objects fall in the same way, and that the effects of gravity are indistinguishable from certain aspects of acceleration and deceleration. The simplest way to test the weak equivalence principle is to drop two objects of different masses or compositions in a vacuum and see whether they hit the ground at the same time. Such experiments demonstrate that all objects fall at the same rate when other forces (such as air resistance and electromagnetic effects) are negligible. More sophisticated tests use a torsion balance of a type invented by Eötvös. Satellite experiments, for example STEP, are planned for more accurate experiments in space.

Formulations of the equivalence principle include:

- The weak equivalence principle: *The trajectory of a point mass in a gravitational field depends only on its initial position and velocity, and is independent of its composition.*

- The Einsteinian equivalence principle: *The outcome of any local non-gravitational experiment in a freely falling laboratory is independent of the velocity of the laboratory and its location in spacetime.*

- The strong equivalence principle requiring both of the above.

General Relativity

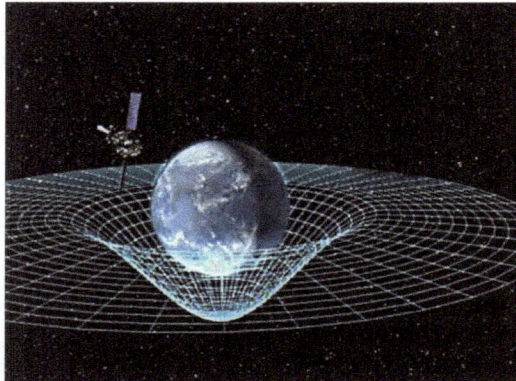

Two-dimensional analogy of spacetime distortion generated by the mass of an object. Matter changes the geometry of spacetime, this (curved) geometry being interpreted as gravity. White lines do not represent the curvature of space but instead represent the coordinate system imposed on the curved spacetime, which would be rectilinear in a flat spacetime.

In general relativity, the effects of gravitation are ascribed to spacetimecurvature instead of a force. The starting point for general relativity is the equivalence principle, which equates free fall with inertial motion and describes free-falling inertial objects as being accelerated relative to non-inertial observers on the ground. In Newtonian physics, however, no such acceleration can occur unless at least one of the objects is being operated on by a force.

Einstein proposed that spacetime is curved by matter, and that free-falling objects are moving along locally straight paths in curved spacetime. These straight paths are called geodesics. Like Newton's first law of motion, Einstein's theory states that if a force is applied on an object, it would deviate from a geodesic. For instance, we are no longer following geodesics while standing because the mechanical resistance of the Earth exerts an upward force on us, and we are non-inertial on the ground as a result. This explains why moving along the geodesics in spacetime is considered inertial.

Einstein discovered the field equations of general relativity, which relate the presence of matter and the curvature of spacetime and are named after him. The Einstein field equations are a set of 10 simultaneous, non-linear, differential equations. The solutions of the field equations are the components of the metric tensor of spacetime. A metric tensor describes a geometry of spacetime. The geodesic paths for a spacetime are calculated from the metric tensor.

Solutions

Notable solutions of the Einstein field equations include:

- The Schwarzschild solution, which describes spacetime surrounding a spherically symmetric non-rotating uncharged massive object. For compact enough objects, this solution generated a black hole with a central singularity. For radial distances from the center which are much greater than the Schwarzschild radius, the accelerations predicted by the Schwarzschild solution are practically identical to those predicted by Newton's theory of gravity.

- The Reissner-Nordström solution, in which the central object has an electrical charge. For charges with a geometrized length which are less than the geometrized length of the mass of the object, this solution produces black holes with double event horizons.

- The Kerr solution for rotating massive objects. This solution also produces black holes with multiple event horizons.

- The Kerr-Newman solution for charged, rotating massive objects. This solution also produces black holes with multiple event horizons.

- The cosmologicalFriedmann-Lemaître-Robertson-Walker solution, which predicts the expansion of the Universe.

Tests

The tests of general relativity included the following:

- General relativity accounts for the anomalous perihelion precession of Mercury.

- The prediction that time runs slower at lower potentials (gravitational time dilation) has been confirmed by the Pound–Rebka experiment (1959), the Hafele–Keating experiment, and the GPS.

- The prediction of the deflection of light was first confirmed by Arthur Stanley Eddington from his observations during the Solar eclipse of May 29, 1919. Eddington measured starlight deflections twice those predicted by Newtonian corpuscular theory, in accordance with the predictions of general relativity. However, his interpretation of the results was later disputed. More recent tests using radio interferometric measurements of quasars passing behind the Sun have more accurately and consistently confirmed the deflection of light to the degree predicted by general relativity.

- The time delay of light passing close to a massive object was first identified by Irwin I. Shapiro in 1964 in interplanetary spacecraft signals.

- Gravitational radiation has been indirectly confirmed through studies of binary pulsars. On 11 February 2016, the LIGO and Virgo collaborations announced the first observation of a gravitational wave.

- Alexander Friedmann in 1922 found that Einstein equations have non-stationary solutions (even in the presence of the cosmological constant). In 1927 Georges Lemaître showed that static solutions of the Einstein equations, which are possible in the presence of the cosmological constant, are unstable, and therefore the static Universe envisioned by Einstein could not exist. Later, in 1931, Einstein himself agreed with the results of Friedmann and Lemaître. Thus general relativity predicted that the Universe had to be non-static—it had to either expand or contract. The expansion of the Universe discovered by Edwin Hubble in 1929 confirmed this prediction.

- The theory's prediction of frame dragging was consistent with the recent Gravity Probe B results.

- General relativity predicts that light should lose its energy when traveling away from massive bodies through gravitational redshift. This was verified on earth and in the solar system around 1960.

Gravity and Quantum Mechanics

In the decades after the discovery of general relativity, it was realized that general relativity is incompatible with quantum mechanics. It is possible to describe gravity in the framework of quantum field theory like the other fundamental forces, such that the attractive force of gravity arises due to exchange of virtual gravitons, in the same way as the electromagnetic force arises from exchange of virtual photons. This reproduces general relativity in the classical limit. However, this approach fails at short distances of the order of the Planck length, where a more complete theory of quantum gravity (or a new approach to quantum mechanics) is required.

Specifics

Earth's Gravity

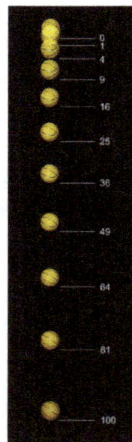

An initially-stationary object which is allowed to fall freely under gravity drops a distance which is proportional to the square of the elapsed time. This image spans half a second and was captured at 20 flashes per second.

Every planetary body (including the Earth) is surrounded by its own gravitational field, which can be conceptualized with Newtonian physics as exerting an attractive force on all objects. Assuming a spherically symmetrical planet, the strength of this field at any given point above the surface is proportional to the planetary body's mass and inversely proportional to the square of the distance from the center of the body.

The strength of the gravitational field is numerically equal to the acceleration of objects under its influence. The rate of acceleration of falling objects near the Earth's surface varies very slightly depending on latitude, surface features such as mountains and ridges, and perhaps unusually high or low sub-surface densities. For purposes of weights and measures, a standard gravity value is defined by the International Bureau of Weights and Measures, under the International System of Units (SI).

That value, denoted g, is $g = 9.80665$ m/s^2 (32.1740 ft/s^2).

The standard value of 9.80665 m/s^2 is the one originally adopted by the International Committee on Weights and Measures in 1901 for 45° latitude, even though it has been shown to be too high by about five parts in ten thousand. This value has persisted in meteorology and in some standard atmospheres as the value for 45° latitude even though it applies more precisely to latitude of 45°32'33".

Assuming the standardized value for g and ignoring air resistance, this means that an object falling freely near the Earth's surface increases its velocity by 9.80665 m/s (32.1740 ft/s or 22 mph) for each second of its descent. Thus, an object starting from rest will attain a velocity of 9.80665 m/s (32.1740 ft/s) after one second, approximately 19.62 m/s (64.4 ft/s) after two seconds, and so on, adding 9.80665 m/s (32.1740 ft/s) to each resulting velocity. Also, again ignoring air resistance, any and all objects, when dropped from the same height, will hit the ground at the same time.

According to Newton's 3rd Law, the Earth itself experiences a force equal in magnitude and opposite in direction to that which it exerts on a falling object. This means that the Earth also accelerates towards the object until they collide. Because the mass of the Earth is huge, however, the acceleration imparted to the Earth by this opposite force is negligible in comparison to the object's. If the object doesn't bounce after it has collided with the Earth, each of them then exerts a repulsive contact force on the other which effectively balances the attractive force of gravity and prevents further acceleration.

The force of gravity on Earth is the resultant (vector sum) of two forces: (a) The gravitational attraction in accordance with Newton's universal law of gravitation, and (b) the centrifugal force, which results from the choice of an earthbound, rotating frame of reference. The force of gravity is the weakest at the equator because of the centrifugal force caused by the Earth's rotation and because points on the equator are furthest from the center of the Earth. The force of gravity varies with latitude and increases from about 9.780 m/s^2 at the Equator to about 9.832 m/s^2 at the poles.

Equations for a Falling Body Near the Surface of the Earth

Under an assumption of constant gravitational attraction, Newton's law of universal gravitation simplifies to $F = mg$, where m is the mass of the body and g is a constant vector with an average magnitude of 9.81 m/s^2 on Earth. This resulting force is the object's weight. The acceleration due to gravity is equal to this g. An initially stationary object which is allowed to fall freely under gravity drops a distance which is proportional to the square of the elapsed time. The image on the right, spanning half a second, was captured with a stroboscopic flash at 20 flashes per second. During the first $\frac{1}{20}$ of a second the ball drops one unit of distance (here, a unit is about 12 mm); by $\frac{2}{20}$ it has dropped at total of 4 units; by $\frac{3}{20}$, 9 units and so on.

Under the same constant gravity assumptions, the potential energy, E_p, of a body at height h is given by $E_p = mgh$ (or $E_p = Wh$, with W meaning weight). This expression is valid only over small distances h from the surface of the Earth. Similarly, the expression $h = \frac{v^2}{2g}$ for the maximum height reached by a vertically projected body with initial velocity v is useful for small heights and small initial velocities only.

Gravity and Astronomy

Gravity acts on stars that form our Milky Way.

The application of Newton's law of gravity has enabled the acquisition of much of the detailed information we have about the planets in the Solar System, the mass of the Sun, and details of quasars; even the existence of dark matter is inferred using Newton's law of gravity. Although we have not traveled to all the planets nor to the Sun, we know their masses. These masses are obtained by applying the laws of gravity to the measured characteristics of the orbit. In space an object maintains its orbit because of the force of gravity acting upon it. Planets orbit stars, stars orbit galactic centers, galaxies orbit a center of mass in clusters, and clusters orbit in superclusters. The force of gravity exerted on one object by another is directly proportional to the product of those objects' masses and inversely proportional to the square of the distance between them.

The earliest gravity (possibly in the form of quantum gravity, supergravity or a gravitational singularity), along with ordinary space and time, developed during the Planck epoch (up to 10^{-43} seconds after the birth of the Universe), possibly from a primeval state (such as a false vacuum, quantum vacuum or virtual particle), in a currently unknown manner.

Gravitational Radiation

According to general relativity, gravitational radiation is generated in situations where the curvature of spacetime is oscillating, such as is the case with co-orbiting objects. The gravitational radiation emitted by the Solar System is far too small to measure. However, gravitational radiation has been indirectly observed as an energy loss over time in binary pulsar systems such as PSR B1913+16. It is believed that neutron star mergers and black hole formation may create detectable amounts of gravitational radiation. Gravitational radiation observatories such as the Laser Interferometer Gravitational Wave Observatory (LIGO) have been created to study the problem. In February 2016, the Advanced LIGO team announced that they had detected gravitational waves from a black hole collision. On September 14, 2015 LIGO registered gravitational waves for the first time, as a result of the collision of two black holes 1.3 billion light-years from Earth. This observation confirms the theoretical predictions of Einstein and others that such waves exist. The event confirms that binary black holes exist. It also opens the way for practical observation and understanding of the nature of gravity and events in the Universe including the Big Bang and what happened after it.

Speed of Gravity

In December 2012, a research team in China announced that it had produced measurements of the phase lag of Earth tides during full and new moons which seem to prove that the speed of gravity is equal to the speed of light. This means that if the Sun suddenly disappeared, the Earth would keep orbiting it normally for 8 minutes, which is the time light takes to travel that distance. The team's findings were released in the Chinese Science Bulletin in February 2013.

Anomalies and Discrepancies

There are some observations that are not adequately accounted for, which may point to the need for better theories of gravity or perhaps be explained in other ways.

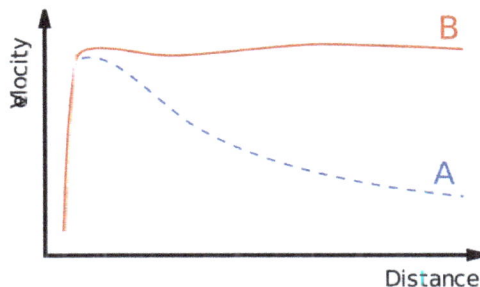

Rotation curve of a typical spiral galaxy: predicted (**A**) and observed (**B**). The discrepancy between the curves is attributed to dark matter.

- Extra-fast stars: Stars in galaxies follow a distribution of velocities where stars on the outskirts are moving faster than they should according to the observed distributions of normal matter. Galaxies within galaxy clusters show a similar pattern. Dark matter, which would interact through gravitation but not electromagnetically, would account for the discrepancy. Various modifications to Newtonian dynamics have also been proposed.

- Flyby anomaly: Various spacecraft have experienced greater acceleration than expected during gravity assist maneuvers.

- Accelerating expansion: The metric expansion of space seems to be speeding up. Dark energy has been proposed to explain this. A recent alternative explanation is that the geometry of space is not homogeneous (due to clusters of galaxies) and that when the data are reinterpreted to take this into account, the expansion is not speeding up after all, however this conclusion is disputed.

- Anomalous increase of the astronomical unit: Recent measurements indicate that planetary orbits are widening faster than if this were solely through the Sun losing mass by radiating energy.

- Extra energetic photons: Photons travelling through galaxy clusters should gain energy and then lose it again on the way out. The accelerating expansion of the Universe should stop the photons returning all the energy, but even taking this into account photons from the cosmic microwave background radiation gain twice as much energy as expected. This may indicate that gravity falls off *faster* than inverse-squared at certain distance scales.

- Extra massive hydrogen clouds: The spectral lines of the Lyman-alpha forest suggest that hydrogen clouds are more clumped together at certain scales than expected and, like dark flow, may indicate that gravity falls off *slower* than inverse-squared at certain distance scales.

- Power: Proposed extra dimensions could explain why the gravity force is so weak.

Alternative Theories

Historical Alternative Theories

- Aristotelian theory of gravity

- Le Sage's theory of gravitation (1784) also called LeSage gravity, proposed by Georges-Louis Le Sage, based on a fluid-based explanation where a light gas fills the entire Universe.

- Ritz's theory of gravitation, *Ann. Chem. Phys.* 13, 145, (1908) pp. 267–271, Weber-Gauss electrodynamics applied to gravitation. Classical advancement of perihelia.

- Nordström's theory of gravitation (1912, 1913), an early competitor of general relativity.

- Kaluza Klein theory (1921)

- Whitehead's theory of gravitation (1922), another early competitor of general relativity.

Modern Alternative Theories

- Brans–Dicke theory of gravity (1961)

- Induced gravity (1967), a proposal by Andrei Sakharov according to which general relativity might arise from quantum field theories of matter

- $f(R)$ gravity (1970)

- Horndeski theory (1974)

- Supergravity (1976)

- String theory

- In the modified Newtonian dynamics (MOND) (1981), Mordehai Milgrom proposes a modification of Newton's Second Law of motion for small accelerations

- The self-creation cosmology theory of gravity (1982) by G.A. Barber in which the Brans-Dicke theory is modified to allow mass creation

- Loop quantum gravity (1988) by Carlo Rovelli, Lee Smolin, and Abhay Ashtekar

- Nonsymmetric gravitational theory (NGT) (1994) by John Moffat

- Conformal gravity

- Tensor–vector–scalar gravity (TeVeS) (2004), a relativistic modification of MOND by Jacob Bekenstein

- Gravity as an entropic force, gravity arising as an emergent phenomenon from the thermodynamic concept of entropy.

- In the superfluid vacuum theory the gravity and curved space-time arise as a collective excitation mode of non-relativistic background superfluid.

- Chameleon theory (2004) by Justin Khoury and Amanda Weltman.

- Pressuron theory (2013) by Olivier Minazzoli and Aurélien Hees.

Newton's Law of Universal Gravitation

Newton's law of universal gravitation states that a particle attracts every other particle in the universe using a force that is directly proportional to the product of their masses and inversely proportional to the square of the distance between them. This is a general physical law derived from empiricalobservations by what Isaac Newton called induction. It is a part of classical mechanics and was formulated in Newton's work *Philosophiæ Naturalis Principia Mathematica* ("the *Principia*"), first published on 5 July 1687. (When Newton's book was presented in 1686 to the Royal Society, Robert Hooke made a claim that Newton had obtained the inverse square law from him.

In modern language, the law states: Every pointmass attracts every single other point mass by a force pointing along the line intersecting both points. The force is proportional to the product of the two masses and inversely proportional to the square of the distance between them. The first test of Newton's theory of gravitation between masses in the laboratory was the Cavendish experiment conducted by the British scientist Henry Cavendish in 1798. It took place 111 years after the publication of Newton's *Principia* and approximately 71 years after his death.

Newton's law of gravitation resembles Coulomb's law of electrical forces, which is used to calculate the magnitude of the electrical force arising between two charged bodies. Both are inverse-square laws, where force is inversely proportional to the square of the distance between the bodies. Coulomb's law has the product of two charges in place of the product of the masses, and the electrostatic constant in place of the gravitational constant.

Newton's law has since been superseded by Einstein's theory of general relativity, but it continues to be used as an excellent approximation of the effects of gravity in most applications. Relativity is required only when there is a need for extreme precision, or when dealing with very strong gravitational fields, such as those found near extremely massive and dense objects, or at very close distances (such as Mercury's orbit around the sun).

History

Early History

A recent assessment (by Ofer Gal) about the early history of the inverse square law is "by the late 1670s", the assumption of an "inverse proportion between gravity and the square of distance was rather common and had been advanced by a number of different people for different reasons". The same author does credit Hooke with a significant and even seminal contribution, but he treats Hooke's claim of priority on the inverse square point as uninteresting since several individuals besides Newton and Hooke had at least suggested it, and he points instead to the idea of "compounding the celestial motions" and the conversion of Newton's thinking away from "centrifugal" and towards "centripetal" force as Hooke's significant contributions.

Plagiarism Dispute

In 1686, when the first book of Newton's *Principia* was presented to the Royal Society, Robert Hooke accused Newton of plagiarism by claiming that he had taken from him the "notion" of "the rule of the decrease of Gravity, being reciprocally as the squares of the distances from the Center". At the same time (according to Edmond Halley's contemporary report) Hooke agreed that "the Demonstration of the Curves generated thereby" was wholly Newton's.

In this way, the question arose as to what, if anything, Newton owed to Hooke. This is a subject extensively discussed since that time and on which some points, outlined below, continue to excite controversy.

Hooke's Work and Claims

Robert Hooke published his ideas about the "System of the World" in the 1660s, when he read to the Royal Society on March 21, 1666, a paper "On gravity", "concerning the inflection of a direct motion into a curve by a supervening attractive principle", and he published them again in somewhat developed form in 1674, as an addition to "An Attempt to Prove the Motion of the Earth from Observations". Hooke announced in 1674 that he planned to "explain a System of the World differing in many particulars from any yet known", based on three "Suppositions": that "all Celestial Bodies whatsoever, have an attraction or gravitating power towards their own Centers" [and] "they do also attract all the other Celestial Bodies that are within the sphere of their activity"; that "all bodies whatsoever that are put into a direct and simple motion, will so continue to move forward in a straight line, till they are by some other effectual powers deflected and bent..."; and that "these attractive powers are so much the more powerful in operating, by how much the nearer the body wrought upon is to their own Centers". Thus Hooke clearly postulated mutual attractions between the Sun and planets, in a way that increased with nearness to the attracting body, together with a principle of linear inertia.

Hooke's statements up to 1674 made no mention, however, that an inverse square law applies or might apply to these attractions. Hooke's gravitation was also not yet universal, though it approached universality more closely than previous hypotheses. He also did not provide accompanying evidence or mathematical demonstration. On the latter two aspects, Hooke himself stated in 1674: "Now what these several degrees [of attraction] are I have not yet experimentally verified"; and as to his whole proposal: "This I only hint at present", "having my self many other things in hand which I would first compleat, and therefore cannot so well attend it" (i.e. "prosecuting this Inquiry"). It was later on, in writing on 6 January 1679|80 to Newton, that Hooke communicated his "supposition ... that the Attraction always is in a duplicate proportion to the Distance from the Center Reciprocall, and Consequently that the Velocity will be in a subduplicate proportion to the Attraction and Consequently as Kepler Supposes Reciprocall to the Distance." (The inference about the velocity was incorrect.)

Hooke's correspondence with Newton during 1679–1680 not only mentioned this inverse square supposition for the decline of attraction with increasing distance, but also, in Hooke's opening letter to Newton, of 24 November 1679, an approach of "compounding the celestial motions of the planets of a direct motion by the tangent & an attractive motion towards the central body".

Newton's Work and Claims

Newton, faced in May 1686 with Hooke's claim on the inverse square law, denied that Hooke was to be credited as author of the idea. Among the reasons, Newton recalled that the idea had been discussed with Sir Christopher Wren previous to Hooke's 1679 letter. Newton also pointed out and acknowledged prior work of others, including Bullialdus, (who suggested, but without demonstration, that there was an attractive force from the Sun in the inverse square proportion to the distance), and Borelli (who suggested, also without demonstration, that there was a centrifugal tendency in counterbalance with a gravitational attraction towards the Sun so as to make the planets move in ellipses). D T Whiteside has described the contribution to Newton's thinking that came from Borelli's book, a copy of which was in Newton's library at his death.

Newton further defended his work by saying that had he first heard of the inverse square proportion from Hooke, he would still have some rights to it in view of his demonstrations of its accuracy. Hooke, without evidence in favor of the supposition, could only guess that the inverse square law was approximately valid at great distances from the center. According to Newton, while the 'Principia' was still at pre-publication stage, there were so many a-priori reasons to doubt the accuracy of the inverse-square law (especially close to an attracting sphere) that "without my (Newton's) Demonstrations, to which Mr Hooke is yet a stranger, it cannot believed by a judicious Philosopher to be any where accurate."

This remark refers among other things to Newton's finding, supported by mathematical demonstration, that if the inverse square law applies to tiny particles, then even a large spherically symmetrical mass also attracts masses external to its surface, even close up, exactly as if all its own mass were concentrated at its center. Thus Newton gave a justification, otherwise lacking, for applying the inverse square law to large spherical planetary masses as if they were tiny particles. In addition, Newton had formulated in Propositions 43-45 of Book 1, and associated sections of Book 3, a sensitive test of the accuracy of the inverse square law, in which he showed that only where the law of force is accurately as the inverse square of the distance will the directions of orientation of the planets' orbital ellipses stay constant as they are observed to do apart from small effects attributable to inter-planetary perturbations.

In regard to evidence that still survives of the earlier history, manuscripts written by Newton in the 1660s show that Newton himself had, by 1669, arrived at proofs that in a circular case of planetary motion, "endeavour to recede" (what was later called centrifugal force) had an inverse-square relation with distance from the center. After his 1679-1680 correspondence with Hooke, Newton adopted the language of inward or centripetal force. According to Newton scholar J. Bruce Brackenridge, although much has been made of the change in language and difference of point of view, as between centrifugal or centripetal forces, the actual computations and proofs remained the same either way. They also involved the combination of tangential and radial displacements, which Newton was making in the 1660s. The lesson offered by Hooke to Newton here, although significant, was one of perspective and did not change the analysis. This background shows there was basis for Newton to deny deriving the inverse square law from Hooke.

Newton's Acknowledgment

On the other hand, Newton did accept and acknowledge, in all editions of the 'Principia', that Hooke (but not exclusively Hooke) had separately appreciated the inverse square law in the solar system. Newton acknowledged Wren, Hooke and Halley in this connection in the Scholium to Proposition 4 in Book 1. Newton also acknowledged to Halley that his correspondence with Hooke in 1679-80 had reawakened his dormant interest in astronomical matters, but that did not mean, according to Newton, that Hooke had told Newton anything new or original: "yet am I not beholden to him for any light into that business but only for the diversion he gave me from my other studies to think on these things & for his dogmaticalness in writing as if he had found the motion in the Ellipsis, which inclined me to try it ..."

Modern Priority Controversy

Since the time of Newton and Hooke, scholarly discussion has also touched on the question of whether Hooke's 1679 mention of 'compounding the motions' provided Newton with something new and valuable, even though that was not a claim actually voiced by Hooke at the time. As described above, Newton's manuscripts of the 1660s do show him actually combining tangential motion with the effects of radially directed force or endeavour, for example in his derivation of the inverse square relation for the circular case. They also show Newton clearly expressing the concept of linear inertia—for which he was indebted to Descartes' work, published in 1644 (as Hooke probably was). These matters do not appear to have been learned by Newton from Hooke.

Nevertheless, a number of authors have had more to say about what Newton gained from Hooke and some aspects remain controversial. The fact that most of Hooke's private papers had been destroyed or have disappeared does not help to establish the truth.

Newton's role in relation to the inverse square law was not as it has sometimes been represented. He did not claim to think it up as a bare idea. What Newton did was to show how the inverse-square law of attraction had many necessary mathematical connections with observable features of the motions of bodies in the solar system; and that they were related in such a way that the observational evidence and the mathematical demonstrations, taken together, gave reason to believe that the inverse square law was not just approximately true but exactly true (to the accuracy achievable in Newton's time and for about two centuries afterwards – and with some loose ends of points that could not yet be certainly examined, where the implications of the theory had not yet been adequately identified or calculated).

About thirty years after Newton's death in 1727, Alexis Clairaut, a mathematical astronomer eminent in his own right in the field of gravitational studies, wrote after reviewing what Hooke published, that "One must not think that this idea ... of Hooke diminishes Newton's glory"; and that "the example of Hooke" serves "to show what a distance there is between a truth that is glimpsed and a truth that is demonstrated".

Modern Form

In modern language, the law states the following:

Every pointmass attracts every single other point mass by a force pointing along the line intersecting both points. The force is proportional to the product of the two masses and inversely proportional to the square of the distance between them:

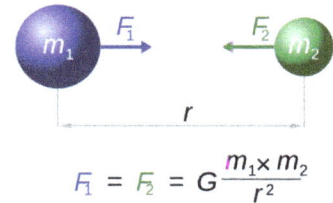

$$F = G\frac{m_1 m_2}{r^2}$$

where:

$$F_1 = F_2 = G\frac{m_1 \times m_2}{r^2}$$

- F is the force between the masses;

- G is the gravitational constant (6.674×10^{-11} N·(m/kg)²);

- m_1 is the first mass;

- m_2 is the second mass;

- r is the distance between the centers of the masses.

Assuming SI units, F is measured in newtons (N), m_1 and m_2 in kilograms (kg), r in meters (m), and the constant G is approximately equal to 6.674×10^{-11} N m² kg⁻². The value of the constant G was first accurately determined from the results of the Cavendish experiment conducted by the British scientist Henry Cavendish in 1798, although Cavendish did not himself calculate a numerical value for G. This experiment was also the first test of Newton's theory of gravitation between masses in the laboratory. It took place 111 years after the publication of Newton's *Principia* and 71 years after Newton's death, so none of Newton's calculations could use the value of G; instead he could only calculate a force relative to another force.

Bodies with Spatial Extent

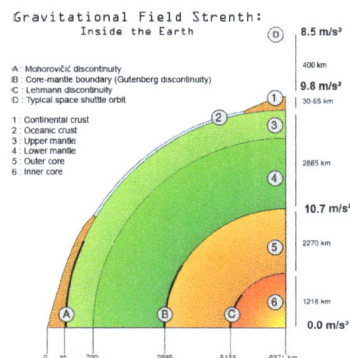

Gravitational field strength within the Earth

If the bodies in question have spatial extent (rather than being theoretical point masses), then the gravitational force between them is calculated by summing the contributions of the notional point masses which constitute the bodies. In the limit, as the component point masses become "infinitely small", this entails integrating the force over the extents of the two bodies.

In this way, it can be shown that an object with a spherically-symmetric distribution of mass exerts the same gravitational attraction on external bodies as if all the object's mass were concentrated at a point at its centre. (This is not generally true for non-spherically-symmetrical bodies.)

For points *inside* a spherically-symmetric distribution of matter, Newton's Shell theorem can be used to find the gravitational force. The theorem tells us how different parts of the mass distribution affect the gravitational force measured at a point located a distance r_0 from the center of the mass distribution:

- The portion of the mass that is located at radii $r < r_0$ causes the same force at r_0 as if all of the mass enclosed within a sphere of radius r_0 was concentrated at the center of the mass distribution (as noted above).

- The portion of the mass that is located at radii $r > r_0$ exerts *no net* gravitational force at the distance r_0 from the center. That is, the individual gravitational forces exerted by the elements of the sphere out there, on the point at r_0, cancel each other out.

As a consequence, for example, within a shell of uniform thickness and density there is *no net* gravitational acceleration anywhere within the hollow sphere.

Furthermore, inside a uniform sphere the gravity increases linearly with the distance from the center; the increase due to the additional mass is 1.5 times the decrease due to the larger distance from the center. Thus, if a spherically symmetric body has a uniform core and a uniform mantle with a density that is less than 2/3 of that of the core, then the gravity initially decreases outwardly beyond the boundary, and if the sphere is large enough, further outward the gravity increases again, and eventually it exceeds the gravity at the core/mantle boundary. The gravity of the Earth may be highest at the core/mantle boundary.

Vector Form

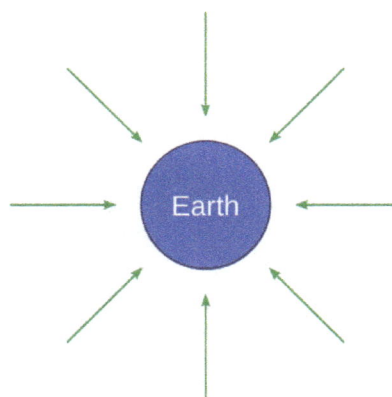

Gravity field surrounding Earth from a macroscopic perspective.

Newton's law of universal gravitation can be written as a vectorequation to account for the direction of the gravitational force as well as its magnitude. In this formula, quantities in bold represent vectors.

$$\mathbf{F}_{12} = -G\frac{m_1 m_2}{|\mathbf{r}_{12}|^2}\hat{\mathbf{r}}_{12}$$

where

> F_{12} is the force applied on object 2 due to object 1,
>
> G is the gravitational constant,
>
> m_1 and m_2 are respectively the masses of objects 1 and 2,
>
> $|r_{12}| = |r_2 - r_1|$ is the distance between objects 1 and 2, and
>
> $\hat{\mathbf{r}}_{12} \overset{\text{def}}{=} \dfrac{\mathbf{r}_2 - \mathbf{r}_1}{|\mathbf{r}_2 - \mathbf{r}_1|}$ is the unit vector from object 1 to 2.

It can be seen that the vector form of the equation is the same as the scalar form given earlier, except that F is now a vector quantity, and the right hand side is multiplied by the appropriate unit vector. Also, it can be seen that $F_{12} = -F_{21}$.

Gravitational Field

The gravitational field is a vector field that describes the gravitational force which would be applied on an object in any given point in space, per unit mass. It is actually equal to the gravitational acceleration at that point.

It is a generalisation of the vector form, which becomes particularly useful if more than 2 objects are involved (such as a rocket between the Earth and the Moon). For 2 objects (e.g. object 2 is a rocket, object 1 the Earth), we simply write r instead of r_{12} and m instead of m_2 and define the gravitational field g(r) as:

$$\mathbf{g(r)} = -G\frac{m_1}{|\mathbf{r}|^2}\hat{\mathbf{r}}$$

so that we can write:

$$\mathbf{F(r)} = m\mathbf{g(r)}.$$

This formulation is dependent on the objects causing the field. The field has units of acceleration; in SI, this is m/s².

Gravitational fields are also conservative; that is, the work done by gravity from one position to another is path-independent. This has the consequence that there exists a gravitational potential field V(r) such that

$$\mathbf{g(r)} = -\nabla V(\mathbf{r}).$$

If m_1 is a point mass or the mass of a sphere with homogeneous mass distribution, the force field g(r) outside the sphere is isotropic, i.e., depends only on the distance r from the center of the sphere. In that case

$$V(r) = -G\frac{m_1}{r}.$$

the gravitational field is on, inside and outside of symmetric masses.

As per Gauss Law, field in a symmetric body can be found by the mathematical equation:

$$\oiint_{\partial V} \mathbf{g(r)} \cdot d\mathbf{A} = -4\pi G M_{enc}$$

where ∂V is a closed surface and M_{enc} is the mass enclosed by the surface.

Hence, for a hollow sphere of radius R and total mass M,

$$|\mathbf{g(r)}| = \begin{cases} 0, & \text{if } r < R \\[2ex] \left|\dfrac{GM}{r^2}\right|, & \text{if } r \geq R \end{cases}$$

For a uniform solid sphere of radius R and total mass M,

$$|\mathbf{g(r)}| = \begin{cases} \dfrac{GMr}{R^3}, & \text{if } r < R \\[2ex] \left|\dfrac{GM}{r^2}\right|, & \text{if } r \geq R \end{cases}$$

Problematic Aspects

Newton's description of gravity is sufficiently accurate for many practical purposes and is therefore widely used. Deviations from it are small when the dimensionless quantities φ/c^2 and $(v/c)^2$ are both much less than one, where φ is the gravitational potential, v is the velocity of the objects being studied, and c is the speed of light. For example, Newtonian gravity provides an accurate description of the Earth/Sun system, since

$$\frac{\Phi}{c^2} = \frac{GM_{sun}}{r_{orbit}c^2} \sim 10^{-8}, \quad \left(\frac{v_{Earth}}{c}\right)^2 = \left(\frac{2\pi r_{orbit}}{(1\text{ yr})c}\right)^2 \sim 10^{-8}$$

whererorbit is the radius of the Earth's orbit around the Sun.

In situations where either dimensionless parameter is large, then general relativity must be used to describe the system. General relativity reduces to Newtonian gravity in the limit of small potential and low velocities, so Newton's law of gravitation is often said to be the low-gravity limit of general relativity.

Theoretical Concerns with Newton's Expression

- There is no immediate prospect of identifying the mediator of gravity. Attempts by physicists to identify the relationship between the gravitational force and other known fundamental forces are not yet resolved, although considerable headway has been made over the last 50 years. Newton himself felt that the concept of an inexplicable *action at a distance* was unsatisfactory, but that there was nothing more that he could do at the time.

- Newton's theory of gravitation requires that the gravitational force be transmitted instantaneously. Given the classical assumptions of the nature of space and time before the development of General Relativity, a significant propagation delay in gravity leads to unstable planetary and stellar orbits.

Observations Conflicting with Newton's Formula

- Newton's Theory does not fully explain the precession of the perihelion of the orbits of the planets, especially of planetMercury, which was detected long after the life of Newton. There is a 43 arcsecond per century discrepancy between the Newtonian calculation, which arises only from the gravitational attractions from the other planets, and the observed precession, made with advanced telescopes during the 19th century.

- The predicted angular deflection of light rays by gravity that is calculated by using Newton's Theory is only one-half of the deflection that is actually observed by astronomers. Calculations using General Relativity are in much closer agreement with the astronomical observations.

- In spiral galaxies, the orbiting of stars around their centers seems to strongly disobey Newton's law of universal gravitation. Astrophysicists, however, explain this spectacular phenomenon in the framework of Newton's laws, with the presence of large amounts of Dark matter.

The observed fact that the *gravitational mass* and the *inertial mass* is not the same for all objects is unexplained within Newton's Theories. The equations for Newtonian gravitation and Newtonian inertia use the same "mass", while in those of General Relativity they diverge. The Equivalence Principle is only valid in regimes where the Newtonian model is valid. Examples of this include: the experiments of Galileo Galilei that, decades before Newton, established that objects that have the same air or fluid resistance are accelerated by the force of the Earth's gravity equally, regardless of their different *inertial* masses. In these cases, the forces and energies that are required to accelerate various masses is completely dependent upon their different *inertial* masses, as can be seen from Newton's Second Law of Motion, F = ma. The modern and modified statement of Newtons Second Law gives this vector equation:

$$F = \frac{\mathrm{d}p}{\mathrm{d}t},$$

where $\vec{p}$ is the momentum of the system, and F is the net (vector sum) force

$$F = \frac{\mathrm{d}(mv)}{\mathrm{d}t}.$$

Substituting the inertial mass dilation expression from Special relativity yields the following expression.

$$F = \frac{d}{dt}\left(\frac{m}{\sqrt{1 - \frac{v^2}{c^2}}} \cdot v \right)$$

When the ratio of velocity to the speed of light is low, then the mass term is constant. When it is near to the speed of light, the denominator approaches zero and the momentum explodes because effective inertial mass explodes.

Newton's Reservations

While Newton was able to formulate his law of gravity in his monumental work, he was deeply uncomfortable with the notion of "action at a distance" that his equations implied. In 1692, in his third letter to Bentley, he wrote: *"That one body may act upon another at a distance through a vacuum without the mediation of anything else, by and through which their action and force may be conveyed from one another, is to me so great an absurdity that, I believe, no man who has in philosophic matters a competent faculty of thinking could ever fall into it."*

He never, in his words, "assigned the cause of this power". In all other cases, he used the phenomenon of motion to explain the origin of various forces acting on bodies, but in the case of gravity, he was unable to experimentally identify the motion that produces the force of gravity (although he invented two mechanical hypotheses in 1675 and 1717). Moreover, he refused to even offer a hypothesis as to the cause of this force on grounds that to do so was contrary to sound science. He lamented that "philosophers have hitherto attempted the search of nature in vain" for the source of the gravitational force, as he was convinced "by many reasons" that there were "causes hitherto unknown" that were fundamental to all the "phenomena of nature". These fundamental phenomena are still under investigation and, though hypotheses abound, the definitive answer has yet to be found. And in Newton's 1713 *General Scholium* in the second edition of *Principia*: *"I have not yet been able to discover the cause of these properties of gravity from phenomena and I feign no hypotheses…. It is enough that gravity does really exist and acts according to the laws I have explained, and that it abundantly serves to account for all the motions of celestial bodies."*

Einstein's Solution

These objections were explained by Einstein's theory of general relativity, in which gravitation is an attribute of curved spacetime instead of being due to a force propagated between bodies. In Einstein's theory, energy and momentum distort spacetime in their vicinity, and other particles move in trajectories determined by the geometry of spacetime. This allowed a description of the motions of light and mass that was consistent with all available observations. In general relativity, the gravitational force is a fictitious force due to the curvature of spacetime, because the gravitational acceleration of a body in free fall is due to its world line being a geodesic of spacetime.

Extensions

Newton was the first to consider in his Principia an extended expression of his law of gravity including an inverse-cube term of the form

$$F = G\frac{m_1 m_2}{r^2} + B\frac{m_1 m_2}{r^3} \text{ , B a constant}$$

attempting to explain the Moon's apsidal motion. Other extensions were proposed by Laplace (around 1790) and Decombes (1913):

$$F(r) = k\frac{m_1 m_2}{r^2}\exp(-\alpha r)\,(\text{Laplace})$$

$$F(r) = k\frac{m_1 m_2}{r^2}\left(1 + \frac{\alpha}{r^3}\right)(\text{Decombes})$$

In recent years, quests for non-inverse square terms in the law of gravity have been carried out by neutron interferometry.

Solutions of Newton's Law of Universal Gravitation

The n-body problem is an ancient, classical problem of predicting the individual motions of a group of celestial objects interacting with each other gravitationally. Solving this problem — from the time of the Greeks and on — has been motivated by the desire to understand the motions of the Sun, planets and the visible stars. In the 20th century, understanding the dynamics of globular cluster star systems became an important n-body problem too. The n-body problem in general relativity is considerably more difficult to solve.

The classical physical problem can be informally stated as: *given the quasi-steady orbital properties (instantaneous position, velocity and time)of a group of celestial bodies, predict their interactive forces; and consequently, predict their true orbital motions for all future times.*

The two-body problem has been completely solved, as has the *Restricted 3-Body Problem.*

Introduction to General Relativity

General relativity is a theory of gravitation that was developed by Albert Einstein between 1907 and 1915. According to general relativity, the observed gravitational effect between masses results from their warping of spacetime.

By the beginning of the 20th century, Newton's law of universal gravitation had been accepted for more than two hundred years as a valid description of the gravitational force between masses. In Newton's model, gravity is the result of an attractive force between massive objects. Although even Newton was troubled by the unknown nature of that force, the basic framework was extremely successful at describing motion.

High-precision test of general relativity by the Cassini space probe (artist's impression): radio signals sent between the Earth and the probe (green wave) are delayed by the warping of spacetime (blue lines) due to the Sun's mass.

Experiments and observations show that Einstein's description of gravitation accounts for several effects that are unexplained by Newton's law, such as minute anomalies in the orbits of Mercury and other planets. General relativity also predicts novel effects of gravity, such as gravitational waves, gravitational lensing and an effect of gravity on time known as gravitational time dilation. Many of these predictions have been confirmed by experiment or observation, most recently gravitational waves.

General relativity has developed into an essential tool in modern astrophysics. It provides the foundation for the current understanding of black holes, regions of space where the gravitational effect is so strong that even light cannot escape. Their strong gravity is thought to be responsible for the intense radiation emitted by certain types of astronomical objects (such as active galactic nuclei or microquasars). General relativity is also part of the framework of the standard Big Bang model of cosmology.

Although general relativity is not the only relativistic theory of gravity, it is the simplest such theory that is consistent with the experimental data. Nevertheless, a number of open questions remain, the most fundamental of which is how general relativity can be reconciled with the laws of quantum physics to produce a complete and self-consistent theory of quantum gravity.

From Special to General Relativity

In September 1905, Albert Einstein published his theory of special relativity, which reconciles Newton's laws of motion with electrodynamics (the interaction between objects with electric charge). Special relativity introduced a new framework for all of physics by proposing new concepts of space and time. Some then-accepted physical theories were inconsistent with that framework; a key example was Newton's theory of gravity, which describes the mutual attraction experienced by bodies due to their mass.

Albert Einstein, pictured here in 1921, developed the theories of special and general relativity.

Several physicists, including Einstein, searched for a theory that would reconcile Newton's law of gravity and special relativity. Only Einstein's theory proved to be consistent with experiments and observations. To understand the theory's basic ideas, it is instructive to follow Einstein's thinking between 1907 and 1915, from his simple thought experiment involving an observer in free fall to his fully geometric theory of gravity.

Equivalence Principle

A person in a free-falling elevator experiences weightlessness; objects either float motionless or drift at constant speed. Since everything in the elevator is falling together, no gravitational effect can be observed. In this way, the experiences of an observer in free fall are indistinguishable from those of an observer in deep space, far from any significant source of gravity. Such observers are the privileged ("inertial") observers Einstein described in his theory of special relativity: observers for whom light travels along straight lines at constant speed.

Einstein hypothesized that the similar experiences of weightless observers and inertial observers in special relativity represented a fundamental property of gravity, and he made this the corner-stone of his theory of general relativity, formalized in his equivalence principle. Roughly speaking, the principle states that a person in a free-falling elevator cannot tell that they are in free fall. Every experiment in such a free-falling environment has the same results as it would for an observer at rest or moving uniformly in deep space, far from all sources of gravity.

Gravity and Acceleration

Ball falling to the floor in an accelerating rocket (left) and on Earth (right).

Most effects of gravity vanish in free fall, but effects that seem the same as those of gravity can be *produced* by an accelerated frame of reference. An observer in a closed room cannot tell which of the following is true:

- Objects are falling to the floor because the room is resting on the surface of the Earth and the objects are being pulled down by gravity.

- Objects are falling to the floor because the room is aboard a rocket in space, which is accelerating at 9.81 m/s² and is far from any source of gravity. The objects are being pulled towards the floor by the same "inertial force" that presses the driver of an accelerating car into the back of his seat.

Conversely, any effect observed in an accelerated reference frame should also be observed in a gravitational field of corresponding strength. This principle allowed Einstein to predict several novel effects of gravity in 1907, as explained in the next section.

An observer in an accelerated reference frame must introduce what physicists call fictitious forces to account for the acceleration experienced by himself and objects around him. One example, the force pressing the driver of an accelerating car into his or her seat, has already been mentioned; another is the force you can feel pulling your arms up and out if you attempt to spin around like a top. Einstein's master insight was that the constant, familiar pull of the Earth's gravitational field is fundamentally the same as these fictitious forces. The apparent magnitude of the fictitious forces always appears to be proportional to the mass of any object on which they act - for instance, the driver's seat exerts just enough force to accelerate the driver at the same rate as the car. By analogy, Einstein proposed that an object in a gravitational field should feel a gravitational force proportional to its mass, as embodied in Newton's law of gravitation.

Physical Consequences

In 1907, Einstein was still eight years away from completing the general theory of relativity. Nonetheless, he was able to make a number of novel, testable predictions that were based on his starting point for developing his new theory: the equivalence principle.

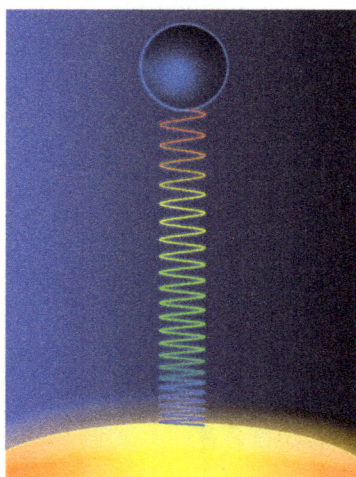

The gravitational redshift of a light wave as it moves upwards against a gravitational field (caused by the yellow star below).

The first new effect is the gravitational frequency shift of light. Consider two observers aboard an accelerating rocket-ship. Aboard such a ship, there is a natural concept of "up" and "down": the direction in which the ship accelerates is "up", and unattached objects accelerate in the opposite direction, falling "downward". Assume that one of the observers is "higher up" than the other. When the lower observer sends a light signal to the higher observer, the acceleration causes the light to be red-shifted, as may be calculated from special relativity; the second observer will measure a lower frequency for the light than the first. Conversely, light sent from the higher observer to the lower is blue-shifted, that is, shifted towards higher frequencies. Einstein argued that such frequency shifts must also be observed in a gravitational field. This is illustrated in the figure at left, which shows a light wave that is gradually red-shifted as it works its way upwards against the gravitational acceleration. This effect has been confirmed experimentally, as described below.

This gravitational frequency shift corresponds to a gravitational time dilation: Since the "higher" observer measures the same light wave to have a lower frequency than the "lower" observer, time must be passing faster for the higher observer. Thus, time runs more slowly for observers who are lower in a gravitational field.

It is important to stress that, for each observer, there are no observable changes of the flow of time for events or processes that are at rest in his or her reference frame. Five-minute-eggs as timed by each observer's clock have the same consistency; as one year passes on each clock, each observer ages by that amount; each clock, in short, is in perfect agreement with all processes happening in its immediate vicinity. It is only when the clocks are compared between separate observers that one can notice that time runs more slowly for the lower observer than for the higher. This effect is minute, but it too has been confirmed experimentally in multiple experiments, as described below.

In a similar way, Einstein predicted the gravitational deflection of light: in a gravitational field, light is deflected downward. Quantitatively, his results were off by a factor of two; the correct derivation requires a more complete formulation of the theory of general relativity, not just the equivalence principle.

Tidal Effects

Two bodies falling towards the center of the Earth accelerate towards each other as they fall.

The equivalence between gravitational and inertial effects does not constitute a complete theory of gravity. When it comes to explaining gravity near our own location on the Earth's surface, noting that our reference frame is not in free fall, so that fictitious forces are to be expected, provides a suitable explanation. But a freely falling reference frame on one side of the Earth cannot explain why the people on the opposite side of the Earth experience a gravitational pull in the opposite direction.

A more basic manifestation of the same effect involves two bodies that are falling side by side towards the Earth. In a reference frame that is in free fall alongside these bodies, they appear to hover weightlessly – but not exactly so. These bodies are not falling in precisely the same direction, but towards a single point in space: namely, the Earth's center of gravity. Consequently, there is a component of each body's motion towards the other. In a small environment such as a freely falling lift, this relative acceleration is minuscule, while for skydivers on opposite sides of the Earth, the effect is large. Such differences in force are also responsible for the tides in the Earth's oceans, so the term "tidal effect" is used for this phenomenon.

The equivalence between inertia and gravity cannot explain tidal effects – it cannot explain variations in the gravitational field. For that, a theory is needed which describes the way that matter (such as the large mass of the Earth) affects the inertial environment around it.

From Acceleration to Geometry

In exploring the equivalence of gravity and acceleration as well as the role of tidal forces, Einstein discovered several analogies with the geometry of surfaces. An example is the transition from an inertial reference frame (in which free particles coast along straight paths at constant speeds) to a rotating reference frame (in which extra terms corresponding to fictitious forces have to be introduced in order to explain particle motion): this is analogous to the transition from a Cartesian coordinate system (in which the coordinate lines are straight lines) to a curved coordinate system (where coordinate lines need not be straight).

A deeper analogy relates tidal forces with a property of surfaces called *curvature*. For gravitational fields, the absence or presence of tidal forces determines whether or not the influence of gravity can be eliminated by choosing a freely falling reference frame. Similarly, the absence or presence of curvature determines whether or not a surface is equivalent to a plane. In the summer of 1912, inspired by these analogies, Einstein searched for a geometric formulation of gravity.

The elementary objects of geometry – points, lines, triangles – are traditionally defined in three-dimensional space or on two-dimensional surfaces. In 1907, Hermann Minkowski, Einstein's former mathematics professor at the Swiss Federal Polytechnic, introduced a geometric formulation of Einstein's special theory of relativity where the geometry included not only space but also time. The basic entity of this new geometry is four-dimensional spacetime. The orbits of moving bodies are curves in spacetime; the orbits of bodies moving at constant speed without changing direction correspond to straight lines.

For surfaces, the generalization from the geometry of a plane – a flat surface – to that of a general curved surface had been described in the early 19th century by Carl Friedrich Gauss. This description had in turn been generalized to higher-dimensional spaces in a mathematical formalism introduced by Bernhard Riemann in the 1850s. With the help of Riemannian geometry, Einstein formulated a geometric description of gravity in which Minkowski's spacetime is replaced by distorted, curved spacetime, just as curved surfaces are a generalization of ordinary plane surfaces. Embedding Diagrams are used to illustrate curved spacetime in educational contexts.

After he had realized the validity of this geometric analogy, it took Einstein a further three years to find the missing cornerstone of his theory: the equations describing how matter influences spacetime's curvature. Having formulated what are now known as Einstein's equations (or, more precisely, his field equations of gravity), he presented his new theory of gravity at several sessions of the Prussian Academy of Sciences in late 1915, culminating in his final presentation on November 25, 1915.

Geometry and Gravitation

Paraphrasing John Wheeler, Einstein's geometric theory of gravity can be summarized thus: *spacetime tells matter how to move; matter tells spacetime how to curve*. What this means is addressed in the following three sections, which explore the motion of so-called test particles, examine which properties of matter serve as a source for gravity, and, finally, introduce Einstein's equations, which relate these matter properties to the curvature of spacetime.

Probing the Gravitational Field

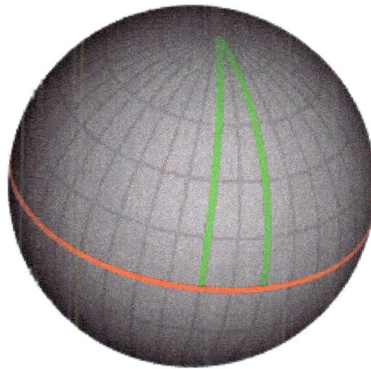

Converging geodesics: two lines of longitude (green) that start out in parallel at the equator (red) but converge to meet at the pole.

In order to map a body's gravitational influence, it is useful to think about what physicists call probe or test particles: particles that are influenced by gravity, but are so small and light that we can neglect their own gravitational effect. In the absence of gravity and other external forces, a test particle moves along a straight line at a constant speed. In the language of spacetime, this is equivalent to saying that such test particles move along straight world lines in spacetime. In the presence of gravity, spacetime is non-Euclidean, or curved, and in curved spacetime straight world lines may not exist. Instead, test particles move along lines called geodesics, which are "as straight as possible", that is, they follow the shortest path between starting and ending points, taking the curvature into consideration.

A simple analogy is the following: In geodesy, the science of measuring Earth's size and shape, a geodesic (from Greek "geo", Earth, and "daiein", to divide) is the shortest route between two points on the Earth's surface. Approximately, such a route is a segment of a great circle, such as a line of longitude or the equator. These paths are certainly not straight, simply because they must follow the curvature of the Earth's surface. But they are as straight as is possible subject to this constraint.

The properties of geodesics differ from those of straight lines. For example, on a plane, parallel lines never meet, but this is not so for geodesics on the surface of the Earth: for example, lines of longitude are parallel at the equator, but intersect at the poles. Analogously, the world lines of test particles in free fall are spacetime geodesics, the straightest possible lines in spacetime. But still there are crucial differences between them and the truly straight lines that can be traced out in the gravity-free spacetime of special relativity. In special relativity, parallel geodesics remain parallel. In a gravitational field with tidal effects, this will not, in general, be the case. If, for example, two bodies are initially at rest relative to each other, but are then dropped in the Earth's gravitational field, they will move towards each other as they fall towards the Earth's center.

Compared with planets and other astronomical bodies, the objects of everyday life (people, cars, houses, even mountains) have little mass. Where such objects are concerned, the laws governing the behavior of test particles are sufficient to describe what happens. Notably, in order to deflect a test particle from its geodesic path, an external force must be applied. A chair someone is sitting on applies an external upwards force preventing the person from falling freely towards the center of the Earth and thus following a geodesic, which they would otherwise be doing without matter in between them and the center of the Earth. In this way, general relativity explains the daily experience of gravity on the surface of the Earth *not* as the downwards pull of a gravitational force, but as the upwards push of external forces. These forces deflect all bodies resting on the Earth's surface from the geodesics they would otherwise follow. For matter objects whose own gravitational influence cannot be neglected, the laws of motion are somewhat more complicated than for test particles, although it remains true that spacetime tells matter how to move.

Sources of Gravity

In Newton's description of gravity, the gravitational force is caused by matter. More precisely, it is caused by a specific property of material objects: their mass. In Einstein's theory and related theories of gravitation, curvature at every point in spacetime is also caused by whatever matter is present. Here, too, mass is a key property in determining the gravitational influence of matter. But in a relativistic theory of gravity, mass cannot be the only source of gravity. Relativity links mass with energy, and energy with momentum.

The equivalence between mass and energy, as expressed by the formula $E = mc^2$, is the most famous consequence of special relativity. In relativity, mass and energy are two different ways of describing one physical quantity. If a physical system has energy, it also has the corresponding mass, and vice versa. In particular, all properties of a body that are associated with energy, such as its temperature or the binding energy of systems such as nuclei or molecules, contribute to that body's mass, and hence act as sources of gravity.

In special relativity, energy is closely connected to momentum. Just as space and time are, in that theory, different aspects of a more comprehensive entity called spacetime, energy and momentum are merely different aspects of a unified, four-dimensional quantity that physicists call four-momentum. In consequence, if energy is a source of gravity, momentum must be a source as well. The same is true for quantities that are directly related to energy and momentum, namely internal pressure and tension. Taken together, in general relativity it is mass, energy, momentum, pressure and tension that serve as sources of gravity: they are how matter tells spacetime how to curve. In the theory's mathematical formulation, all these quantities are but aspects of a more general physical quantity called the energy–momentum tensor.

Einstein's Equations

Einstein's equations are the centerpiece of general relativity. They provide a precise formulation of the relationship between spacetime geometry and the properties of matter, using the language of mathematics. More concretely, they are formulated using the concepts of Riemannian geometry, in which the geometric properties of a space (or a spacetime) are described by a quantity called a metric. The metric encodes the information needed to compute the fundamental geometric notions of distance and angle in a curved space (or spacetime).

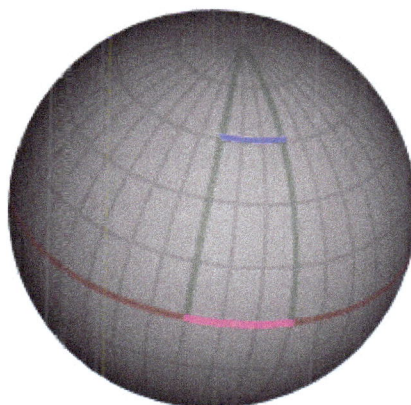

Distances, at different latitudes, corresponding to 30 degrees difference in longitude.

A spherical surface like that of the Earth provides a simple example. The location of any point on the surface can be described by two coordinates: the geographic latitude and longitude. Unlike the Cartesian coordinates of the plane, coordinate differences are not the same as distances on the surface, for someone at the equator, moving 30 degrees of longitude westward (magenta line) corresponds to a distance of roughly 3,300 kilometers (2,100 mi). On the other hand, someone at a latitude of 55 degrees, moving 30 degrees of longitude westward (blue line) covers a distance of merely 1,900 kilometers (1,200 mi). Coordinates therefore do not provide enough information to describe the geometry of a spherical surface, or indeed the geometry of any more complicated space or spacetime. That information is precisely what is encoded in the metric, which is a function defined at each point of the surface (or space, or spacetime) and relates coordinate differences to differences in distance. All other quantities that are of interest in geometry, such as the length of any given curve, or the angle at which two curves meet, can be computed from this metric function.

The metric function and its rate of change from point to point can be used to define a geometrical quantity called the Riemann curvature tensor, which describes exactly how the space or spacetime is curved at each point. In general relativity, the metric and the Riemann curvature tensor are quantities defined at each point in spacetime. As has already been mentioned, the matter content of the spacetime defines another quantity, the energy–momentum tensor T, and the principle that "spacetime tells matter how to move, and matter tells spacetime how to curve" means that these quantities must be related to each other. Einstein formulated this relation by using the Riemann curvature tensor and the metric to define another geometrical quantity G, now called the Einstein tensor, which describes some aspects of the way spacetime is curved. *Einstein's equation* then states that

$$\mathbf{G} = \frac{8\pi G}{c^4} \mathbf{T},$$

i.e., up to a constant multiple, the quantity G (which measures curvature) is equated with the quantity T (which measures matter content). Here, *G* is the gravitational constant of Newtonian gravity, and *c* is the speed of light from special relativity.

This equation is often referred to in the plural as *Einstein's equations*, since the quantities G and T are each determined by several functions of the coordinates of spacetime, and the equations equate each of these component functions. A solution of these equations describes a particular geometry of spacetime; for example, the Schwarzschild solution describes the geometry around a spherical, non-rotating mass such as a star or a black hole, whereas the Kerr solution describes a rotating black hole. Still other solutions can describe a gravitational wave or, in the case of the Friedmann–Lemaître–Robertson–Walker solution, an expanding universe. The simplest solution is the uncurved Minkowski spacetime, the spacetime described by special relativity.

Experiments

No scientific theory is apodictically true; each is a model that must be checked by experiment. Newton's law of gravity was accepted because it accounted for the motion of planets and moons in the solar system with considerable accuracy. As the precision of experimental measurements gradually improved, some discrepancies with Newton's predictions were observed, and these were accounted for in the general theory of relativity. Similarly, the predictions of general relativity must also be checked with experiment, and Einstein himself devised three tests now known as the classical tests of the theory:

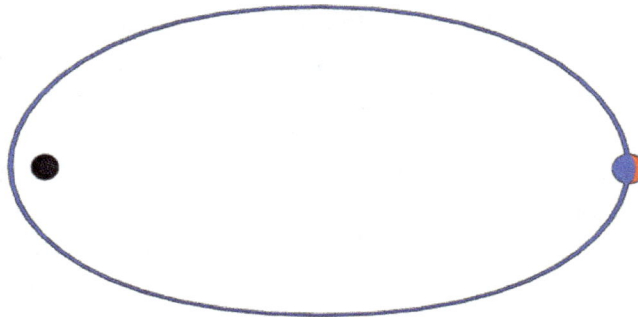

Newtonian (red) vs. Einsteinian orbit (blue) of a single planet orbiting a spherical star. (Click on the image for animation.)

- Newtonian gravity predicts that the orbit which a single planet traces around a perfectly spherical star should be an ellipse. Einstein's theory predicts a more complicated curve: the planet behaves as if it were travelling around an ellipse, but at the same time, the ellipse as a whole is rotating slowly around the star. In the diagram on the right, the ellipse predicted by Newtonian gravity is shown in red, and part of the orbit predicted by Einstein in blue. For a planet orbiting the Sun, this deviation from Newton's orbits is known as the anomalous perihelion shift. The first measurement of this effect, for the planet Mercury, dates back to 1859. The most accurate results for Mercury and for other planets to date are based on measurements which were undertaken between 1966 and 1990, using radio telescopes. General relativity predicts the correct anomalous perihelion shift for all planets where this can be measured accurately (Mercury, Venus and the Earth).

- According to general relativity, light does not travel along straight lines when it propagates in a gravitational field. Instead, it is deflected in the presence of massive bodies. In particular, starlight is deflected as it passes near the Sun, leading to apparent shifts of up 1.75 arc seconds in the stars' positions in the sky (an arc second is equal to 1/3600 of a degree). In the framework of Newtonian gravity, a heuristic argument can be made that leads to light deflection by half that amount. The different predictions can be tested by observing stars that are close to the Sun during a solar eclipse. In this way, a British expedition to West Africa in 1919, directed by Arthur Eddington, confirmed that Einstein's prediction was correct, and the Newtonian predictions wrong, via observation of the May 1919 eclipse. Eddington's results were not very accurate; subsequent observations of the deflection of the light of distant quasars by the Sun, which utilize highly accurate techniques of radio astronomy, have confirmed Eddington's results with significantly better precision (the first such measurements date from 1967, the most recent comprehensive analysis from 2004).

- Gravitational redshift was first measured in a laboratory setting in 1959 by Pound and Rebka. It is also seen in astrophysical measurements, notably for light escaping the white dwarf Sirius B. The related gravitational time dilation effect has been measured by transporting atomic clocks to altitudes of between tens and tens of thousands of kilometers (first by Hafele and Keating in 1971; most accurately to date by Gravity Probe A launched in 1976).

Of these tests, only the perihelion advance of Mercury was known prior to Einstein's final publication of general relativity in 1916. The subsequent experimental confirmation of his other predictions, especially the first measurements of the deflection of light by the sun in 1919, catapulted Einstein to international stardom. These three experiments justified adopting general relativity over Newton's theory and, incidentally, over a number of alternatives to general relativity that had been proposed.

Further tests of general relativity include precision measurements of the Shapiro effect or gravitational time delay for light, most recently in 2002 by the Cassini space probe. One set of tests focuses on effects predicted by general relativity for the behavior of gyroscopes travelling through space. One of these effects, geodetic precession, has been tested with the Lunar Laser Ranging Experiment (high-precision measurements of the orbit of the Moon). Another, which is related to rotating masses, is called frame-dragging. The geodetic and frame-dragging effects were both tested by the Gravity Probe B satellite experiment launched in 2004, with results confirming relativity to within 0.5% and 15%, respectively, as of December 2008.

Gravity Probe B with solar panels folded.

By cosmic standards, gravity throughout the solar system is weak. Since the differences between the predictions of Einstein's and Newton's theories are most pronounced when gravity is strong, physicists have long been interested in testing various relativistic effects in a setting with comparatively strong gravitational fields. This has become possible thanks to precision observations of binary pulsars. In such a star system, two highly compact neutron stars orbit each other. At least one of them is a pulsar – an astronomical object that emits a tight beam of radiowaves. These beams strike the Earth at very regular intervals, similarly to the way that the rotating beam of a lighthouse means that an observer sees the lighthouse blink, and can be observed as a highly regular series of pulses. General relativity predicts specific deviations from the regularity of these radio pulses. For instance, at times when the radio waves pass close to the other neutron star, they should be deflected by the star's gravitational field. The observed pulse patterns are impressively close to those predicted by general relativity.

One particular set of observations is related to eminently useful practical applications, namely to satellite navigation systems such as the Global Positioning System that are used both for precise positioning and timekeeping. Such systems rely on two sets of atomic clocks: clocks aboard satellites orbiting the Earth, and reference clocks stationed on the Earth's surface. General relativity predicts that these two sets of clocks should tick at slightly different rates, due to their different motions (an effect already predicted by special relativity) and their different positions within the Earth's gravitational field. In order to ensure the system's accuracy, the satellite clocks are either slowed down by a relativistic factor, or that same factor is made part of the evaluation algorithm. In turn, tests of the system's accuracy (especially the very thorough measurements that are part of the definition of universal coordinated time) are testament to the validity of the relativistic predictions.

A number of other tests have probed the validity of various versions of the equivalence principle; strictly speaking, all measurements of gravitational time dilation are tests of the weak version of that principle, not of general relativity itself. So far, general relativity has passed all observational tests.

Astrophysical Applications

Models based on general relativity play an important role in astrophysics; the success of these models is further testament to the theory's validity.

Gravitational Lensing

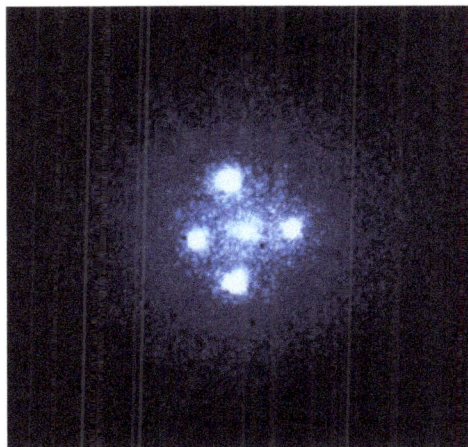

Einstein cross: four images of the same astronomical object, produced by a gravitational lens.

Since light is deflected in a gravitational field, it is possible for the light of a distant object to reach an observer along two or more paths. For instance, light of a very distant object such as a quasar can pass along one side of a massive galaxy and be deflected slightly so as to reach an observer on Earth, while light passing along the opposite side of that same galaxy is deflected as well, reaching the same observer from a slightly different direction. As a result, that particular observer will see one astronomical object in two different places in the night sky. This kind of focussing is well-known when it comes to optical lenses, and hence the corresponding gravitational effect is called gravitational lensing.

Observational astronomy uses lensing effects as an important tool to infer properties of the lensing object. Even in cases where that object is not directly visible, the shape of a lensed image provides information about the mass distribution responsible for the light deflection. In particular, gravitational lensing provides one way to measure the distribution of dark matter, which does not give off light and can be observed only by its gravitational effects. One particularly interesting application are large-scale observations, where the lensing masses are spread out over a significant fraction of the observable universe, and can be used to obtain information about the large-scale properties and evolution of our cosmos.

Gravitational Waves

Gravitational waves, a direct consequence of Einstein's theory, are distortions of geometry that propagate at the speed of light, and can be thought of as ripples in spacetime. They should not be confused with the gravity waves of fluid dynamics, which are a different concept.

In February 2016, the Advanced LIGO team announced that they had directly observed gravitational waves from a black hole merger.

Indirectly, the effect of gravitational waves had been detected in observations of specific binary stars. Such pairs of stars orbit each other and, as they do so, gradually lose energy by emitting gravitational waves. For ordinary stars like the Sun, this energy loss would be too small to be detectable, but this energy loss was observed in 1974 in a binary pulsar called PSR1913+16. In such a system, one of the orbiting stars is a pulsar. This has two consequences: a pulsar is an extremely dense object known as a neutron star, for which gravitational wave emission is much stronger than for ordinary stars. Also, a pulsar emits a narrow beam of electromagnetic radiation from its magnetic poles. As the pulsar rotates, its beam sweeps over the Earth, where it is seen as a regular series of radio pulses, just as a ship at sea observes regular flashes of light from the rotating light in a lighthouse. This regular pattern of radio pulses functions as a highly accurate "clock". It can be used to time the double star's orbital period, and it reacts sensitively to distortions of spacetime in its immediate neighborhood.

The discoverers of PSR1913+16, Russell Hulse and Joseph Taylor, were awarded the Nobel Prize in Physics in 1993. Since then, several other binary pulsars have been found. The most useful are those in which both stars are pulsars, since they provide accurate tests of general relativity.

Currently, a number of land-based gravitational wave detectors are in operation, and a mission to launch a space-based detector, LISA, is currently under development, with a precursor mission (LISA Pathfinder) which was launched in 2015. Gravitational wave observations can be used to obtain information about compact objects such as neutron stars and black holes, and also to probe the state of the early universe fractions of a second after the Big Bang.

Black Holes

Black hole-powered jet emanating from the central region of the galaxy M87.

When mass is concentrated into a sufficiently compact region of space, general relativity predicts the formation of a black hole – a region of space with a gravitational effect so strong that not even light can escape. Certain types of black holes are thought to be the final state in the evolution of massive stars. On the other hand, supermassive black holes with the mass of millions or billions of Suns are assumed to reside in the cores of most galaxies, and they play a key role in current models of how galaxies have formed over the past billions of years.

Matter falling onto a compact object is one of the most efficient mechanisms for releasing energy in the form of radiation, and matter falling onto black holes is thought to be responsible for some of the brightest astronomical phenomena imaginable. Notable examples of great interest to astronomers are quasars and other types of active galactic nuclei. Under the right conditions, falling matter accumulating around a black hole can lead to the formation of jets, in which focused beams of matter are flung away into space at speeds near that of light.

There are several properties that make black holes most promising sources of gravitational waves. One reason is that black holes are the most compact objects that can orbit each other as part of a binary system; as a result, the gravitational waves emitted by such a system are especially strong. Another reason follows from what are called black-hole uniqueness theorems: over time, black holes retain only a minimal set of distinguishing features (these theorems have become known as "no-hair" theorems, since different hairstyles are a crucial part of what gives different people their different appearances). For instance, in the long term, the collapse of a hypothetical matter cube will not result in a cube-shaped black hole. Instead, the resulting black hole will be indistinguishable from a black hole formed by the collapse of a spherical mass, but with one important difference: in its transition to a spherical shape, the black hole formed by the collapse of a cube will emit gravitational waves.

Cosmology

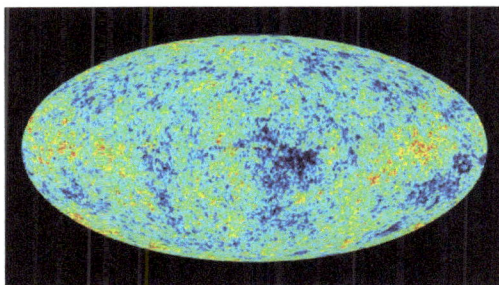

An image, created using data from the WMAP satellite telescope, of the radiation emitted no more than a few hundred thousand years after the Big Bang.

One of the most important aspects of general relativity is that it can be applied to the universe as a whole. A key point is that, on large scales, our universe appears to be constructed along very simple lines: all current observations suggest that, on average, the structure of the cosmos should be approximately the same, regardless of an observer's location or direction of observation: the universe is approximately homogeneous and isotropic. Such comparatively simple universes can be described by simple solutions of Einstein's equations. The current cosmological models of the universe are obtained by combining these simple solutions to general relativity with theories describing the properties of the universe's matter content, namely thermodynamics, nuclear- and particle physics. According to these models, our present universe emerged from an extremely dense high-temperature state – the Big Bang – roughly 14 billion years ago and has been expanding ever since.

Einstein's equations can be generalized by adding a term called the cosmological constant. When this term is present, empty space itself acts as a source of attractive (or, less commonly, repulsive) gravity. Einstein originally introduced this term in his pioneering 1917 paper on cosmology, with a very specific motivation: contemporary cosmological thought held the universe to be static, and the additional term was required for constructing static model universes within the framework of general relativity. When it became apparent that the universe is not static, but expanding, Einstein was quick to discard this additional term. Since the end of the 1990s, however, astronomical evidence indicating an accelerating expansion consistent with a cosmological constant – or, equivalently, with a particular and ubiquitous kind of dark energy – has steadily been accumulating.

Modern Research

General relativity is very successful in providing a framework for accurate models which describe an impressive array of physical phenomena. On the other hand, there are many interesting open questions, and in particular, the theory as a whole is almost certainly incomplete.

In contrast to all other modern theories of fundamental interactions, general relativity is a classical theory: it does not include the effects of quantum physics. The quest for a quantum version of general relativity addresses one of the most fundamental open questions in physics. While there are promising candidates for such a theory of quantum gravity, notably string theory and loop quantum gravity, there is at present no consistent and complete theory. It has long been hoped that a theory of quantum gravity would also eliminate another problematic feature of general relativity: the presence of spacetime singularities. These singularities are boundaries ("sharp edges") of spacetime at which geometry becomes ill-defined, with the consequence that general relativity itself loses its predictive power. Furthermore, there are so-called singularity theorems which predict that such singularities *must* exist within the universe if the laws of general relativity were to hold without any quantum modifications. The best-known examples are the singularities associated with the model universes that describe black holes and the beginning of the universe.

Other attempts to modify general relativity have been made in the context of cosmology. In the modern cosmological models, most energy in the universe is in forms that have never been detected directly, namely dark energy and dark matter. There have been several controversial proposals to remove the need for these enigmatic forms of matter and energy, by modifying the laws governing gravity and the dynamics of cosmic expansion, for example modified Newtonian dynamics.

Beyond the challenges of quantum effects and cosmology, research on general relativity is rich with possibilities for further exploration: mathematical relativists explore the nature of singularities and the fundamental properties of Einstein's equations, and ever more comprehensive computer simulations of specific spacetimes (such as those describing merging black holes) are run. More than ninety years after the theory was first published, research is more active than ever.

Gravity of Earth

The gravity of Earth, which is denoted by g, refers to the acceleration that the Earth imparts to objects on or near its surface due to gravity. In SI units this acceleration is measured in metres per second squared (in symbols, m/s^2 or $m \cdot s^{-2}$) or equivalently in newtons per kilogram (N/kg or $N \cdot kg^{-1}$). It has an approximate value of 9.8 m/s^2, which means that, ignoring the effects of air resistance, the speed of an object falling freely near the Earth's surface will increase by about 9.8 metres (32 ft) per second every second, this quantity is sometimes referred to informally as *little g* (in contrast, the gravitational constant G is referred to as *big G*).

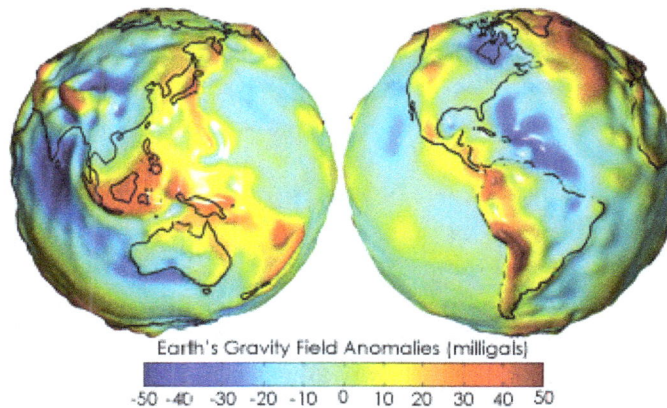

Earth's Gravity Field Anomalies (milligals)

-50 -40 -30 -20 -10 0 10 20 30 40 50

Earth's gravity measured by NASA GRACE mission, showing deviations from the theoretical gravity of an idealized smooth Earth, the so-called earth ellipsoid. Red shows the areas where gravity is stronger than the smooth, standard value, and blue reveals areas where gravity is weaker. (**Animated version.**)

There is a direct relationship between gravitational acceleration and the downwards force (weight) experienced by objects on Earth, given by the equation $F = ma$ (*force = mass × acceleration*). However, other factors such as the rotation of the Earth also contribute to the net acceleration.

The precise strength of Earth's gravity varies depending on location. The nominal "average" value at the Earth's surface, known as standard gravity is, by definition, 9.80665 m/s² (about 32.1740 ft/s²). This quantity is denoted variously as g_n, g_e (though this sometimes means the normal equatorial value on Earth, 9.78033 m/s²), g_o, gee, or simply g (which is also used for the variable local value).

Variation in Gravity and Apparent Gravity

A perfect sphere of uniform density, or whose density varies solely with distance from the centre (spherical symmetry), would produce a gravitational field of uniform magnitude at all points on its surface, always pointing directly towards the sphere's centre. The Earth is not a perfect sphere, but is slightly flatter at the poles while bulging at the Equator: an oblate spheroid. There are consequently slight deviations in both the magnitude and direction of gravity across its surface. The net force (or corresponding net acceleration) as measured by a scale and plumb bob is called "effective gravity" or "apparent gravity". Effective gravity includes other factors that affect the net force. These factors vary and include things such as centrifugal force at the surface from the Earth's rotation and the gravitational pull of the Moon and Sun.

Effective gravity on the Earth's surface varies by around 0.7%, from 9.7639 m/s² on the Nevado Huascarán mountain in Peru to 9.8337 m/s² at the surface of the Arctic Ocean. In large cities, it ranges from 9.766 in Kuala Lumpur, Mexico City, and Singapore to 9.825 in Oslo and Helsinki.

Latitude

The surface of the Earth is rotating, so it is not an inertial frame of reference. At latitudes nearer the Equator, the outward centrifugal force produced by Earth's rotation is larger than at polar latitudes. This counteracts the Earth's gravity to a small degree – up to a maximum of 0.3% at the Equator – and reduces the apparent downward acceleration of falling objects.

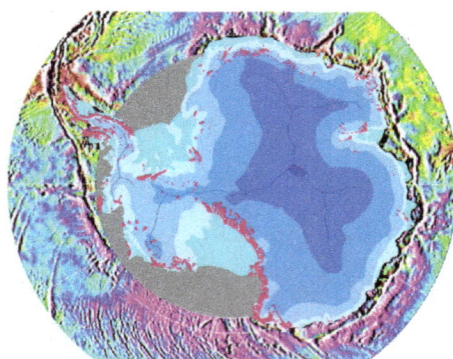

The differences of Earth's gravity around the Antarctic continent.

The second major reason for the difference in gravity at different latitudes is that the Earth's equatorial bulge (itself also caused by centrifugal force from rotation) causes objects at the Equator to be farther from the planet's centre than objects at the poles. Because the force due to gravitational attraction between two bodies (the Earth and the object being weighed) varies inversely with the square of the distance between them, an object at the Equator experiences a weaker gravitational pull than an object at the poles.

In combination, the equatorial bulge and the effects of the surface centrifugal force due to rotation mean that sea-level effective gravity increases from about 9.780 m/s² at the Equator to about 9.832 m/s² at the poles, so an object will weigh about 0.5% more at the poles than at the Equator.

The same two factors influence the direction of the effective gravity (as determined by a plumb line or as the perpendicular to the surface of water in a container). Anywhere on Earth away from the Equator or poles, effective gravity points not exactly toward the centre of the Earth, but rather perpendicular to the surface of the geoid, which, due to the flattened shape of the Earth, is somewhat toward the opposite pole. About half of the deflection is due to centrifugal force, and half because the extra mass around the Equator causes a change in the direction of the true gravitational force relative to what it would be on a spherical Earth.

Altitude

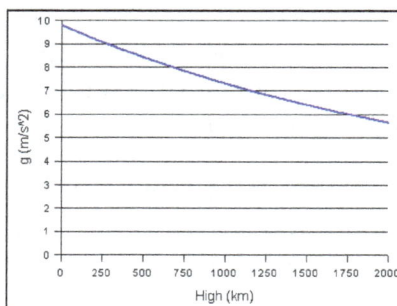

The graph shows the variation in gravity relative to the height of an object

Gravity decreases with altitude as one rises above the Earth's surface because greater altitude means greater distance from the Earth's centre. All other things being equal, an increase in altitude from sea level to 9,000 metres (30,000 ft) causes a weight decrease of about 0.29%. (An additional factor affecting apparent weight is the decrease in air density at altitude, which lessens an object's buoyancy. This would increase a person's apparent weight at an altitude of 9,000 metres by about 0.08%)

It is a common misconception that astronauts in orbit are weightless because they have flown high enough to escape the Earth's gravity. In fact, at an altitude of 400 kilometres (250 mi), equivalent to a typical orbit of the Space Shuttle, gravity is still nearly 90% as strong as at the Earth's surface. Weightlessness actually occurs because orbiting objects are in free-fall.

The effect of ground elevation depends on the density of the ground. A person flying at 30 000 ft above sea level over mountains will feel more gravity than someone at the same elevation but over the sea. However, a person standing on the earth's surface feels less gravity when the elevation is higher.

The following formula approximates the Earth's gravity variation with altitude:

$$g_h = g_0 \left(\frac{r_e}{r_e + h} \right)^2$$

Where

- g_h is the gravitational acceleration at height h above sea level.
- r_e is the Earth's mean radius.
- g_0 is the standard gravitational acceleration.

The formula treats the Earth as a perfect sphere with a radially symmetric distribution of mass; a more accurate mathematical treatment is discussed below.

Depth

An approximate value for gravity at a distance r from the centre of the Earth can be obtained by assuming that the Earth's density is spherically symmetric. The gravity depends only on the mass inside the sphere of radius r. All the contributions from outside cancel out as a consequence of the inverse-square law of gravitation. Another consequence is that the gravity is the same as if all the mass were concentrated at the centre. Thus, the gravitational acceleration at this radius is

$$g(r) = -\frac{GM(r)}{r^2}.$$

where G is the gravitational constant and $M(r)$ is the total mass enclosed within radius r. If the Earth had a constant density ρ, the mass would be $M(r) = (4/3)\pi \rho r^3$ and the dependence of gravity on depth would be

$$g(r) = \frac{4\pi}{3} G \rho r.$$

g at depth d is given by $g'=g(1-d/R)$ where g is acceleration due to gravity on surface of the earth, d is depth and R is radius of Earth. If the density decreased linearly with increasing radius from a density ρ_0 at the centre to ρ_1 at the surface, then $\rho(r) = \rho_0 - (\rho_0 - \rho_1)\, r\, /\, r_e$, and the dependence would be

$$g(r) = \frac{4\pi}{3} G\rho_0 r - \pi G (\rho_0 - \rho_1) \frac{r^2}{r_e}.$$

The actual depth dependence of density and gravity, inferred from seismic travel times.

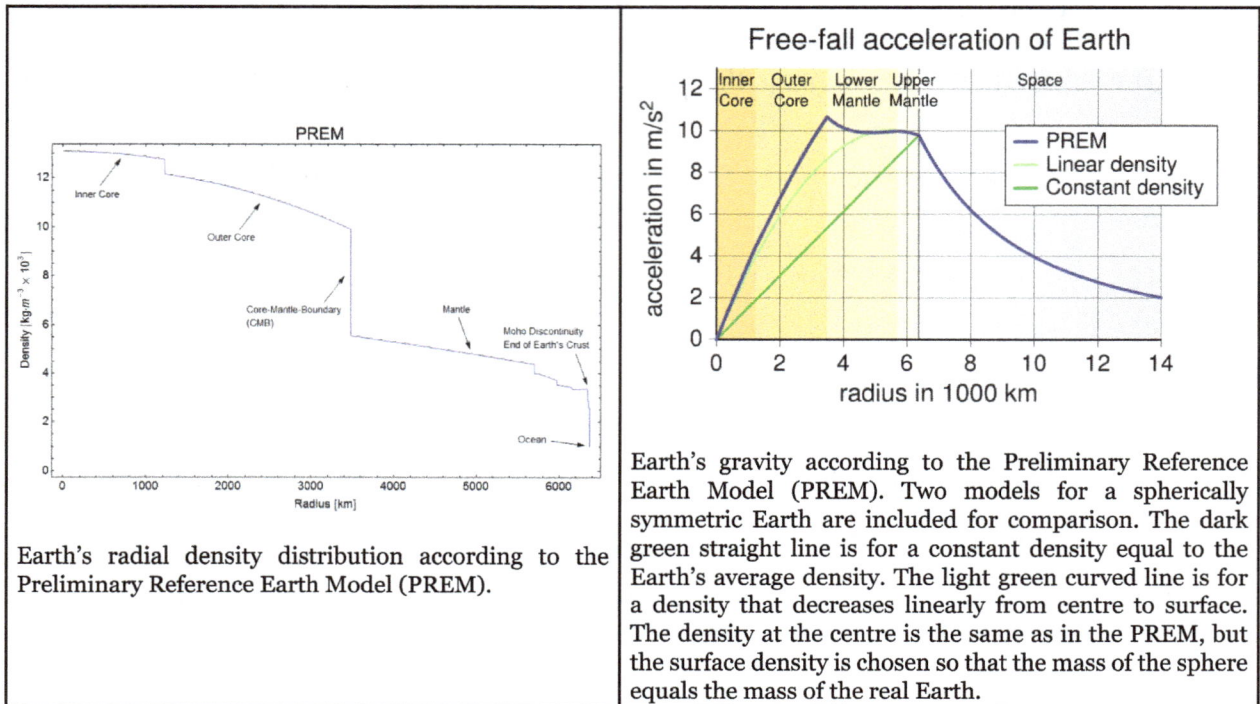

Earth's radial density distribution according to the Preliminary Reference Earth Model (PREM).

Earth's gravity according to the Preliminary Reference Earth Model (PREM). Two models for a spherically symmetric Earth are included for comparison. The dark green straight line is for a constant density equal to the Earth's average density. The light green curved line is for a density that decreases linearly from centre to surface. The density at the centre is the same as in the PREM, but the surface density is chosen so that the mass of the sphere equals the mass of the real Earth.

Local Topography and Geology

Local variations in topography (such as the presence of mountains) and geology (such as the density of rocks in the vicinity) cause fluctuations in the Earth's gravitational field, known as gravitational anomalies. Some of these anomalies can be very extensive, resulting in bulges in sea level, and throwing pendulum clocks out of synchronisation.

The study of these anomalies forms the basis of gravitational geophysics. The fluctuations are measured with highly sensitive gravimeters, the effect of topography and other known factors is subtracted, and from the resulting data conclusions are drawn. Such techniques are now used by prospectors to find oil and mineral deposits. Denser rocks (often containing mineral ores) cause higher than normal local gravitational fields on the Earth's surface. Less dense sedimentary rocks cause the opposite.

Other Factors

In air, objects experience a supporting buoyancy force which reduces the apparent strength of gravity (as measured by an object's weight). The magnitude of the effect depends on air density (and hence air pressure).

The gravitational effects of the Moon and the Sun (also the cause of the tides) have a very small effect on the apparent strength of Earth's gravity, depending on their relative positions; typical variations are 2 μm/s^2 (0.2 mGal) over the course of a day.

Comparative gravities in various cities around the world

The table below shows the gravitational acceleration in various cities around the world. The effect of latitude can be clearly seen with gravity in high-latitude cities (Anchorage, Helsinki, Oslo) being about 0.5% greater than that in cities near the equator (Kandy, Kuala Lumpur, Singapore). The effect of altitude can be seen in Mexico City (altitude 2,240 metres (7,350 ft)), and by comparing Denver (1,616 metres (5,302 ft)) with Washington, D.C. (30 metres (98 ft))(both near 39° N).

Errata, Acceleration Due to Gravity in Various Cities

Location	Quoted Value	Widget Value 01/25/2016	Measured Value 1970*
Amsterdam	9.813	9.80563	NA
Athens	9.800	9.80037	NA
Auckland	9.799	9.79936	9.79962
Bangkok	9.783	9.77958	NA
Brussels	9.811	9.8152	NA
Buenos Aires	9.797	9.79689	NA
Calcutta	9.788	9.78548	9.78816
Cape Town	9.796	9.79616	9.79659
Chicago	9.803	9.80444	NA
Copenhagen	9.815	9.82057	9.81559
Frankfurt	9.810	9.81412	NA
Greenwich	9.802	9.81597	9.81188
Havana	9.788	9.78587	NA
Helsinki	9.819	9.82519	NA
Istanbul	9.808	9.80386	NA
Jakarta	9.781	9.77669	NA
Kuwait	9.793	9.79175	NA
Lisbon	9.801	9.80131	NA
London	9.812	9.81599	9.81188 (derived from Greenwich)
Los Angeles	9.796	9.79607	NA
Madrid	9.800	9.80138	NA
Manila	9.784	9.78002	NA
Mexico City	9.779	9.77628	NA
Montréal	9.789	9.80906	9.80652
New York City	9.802	9.8036	9.80267
Nicosia	9.797	9.79709	NA
Oslo	9.819	9.825	NA

Ottawa	9.806	9.80888	9.80607
Paris	9.809	9.81289	9.80943
Rio de Janeiro	9.788	9.78575	NA
Rome	9.803	9.80491	9.80347
San Francisco	9.800	9.80035	9.79965
Singapore	9.781	9.77584	NA
Skopje	9.804	9.80444	NA
Stockholm	9.818	9.82436	NA
Sydney	9.797	9.79607	9.79683
Taipei	9.790	9.78757	NA
Tokyo	9.798	9.79805	NA
Vancouver	9.809	9.81339	NA
Washington D.C.	9.801	9.80165	9.80080 (absolute)
Wellington	9.803	9.80431	9.80292
Zurich	9.807	9.81007	9.80673

The measured values were taken from Physical and Mathematical Tables by T.M. Yarwood and F. Castle, Macmillan, revised edition 1970.

Mathematical Models

Latitude Model

If the terrain is at sea level, we can estimate $g\{\phi\}$, the acceleration at latitude ϕ:

$$g\{\phi\} = 9.780327\,\text{m}\cdot\text{s}^{-2}\left(1 + 0.0053024\sin^2\phi - 0.0000058\sin^2 2\phi\right),$$

$$= 9.780327\,\text{m}\cdot\text{s}^{-2}\left(1 + 0.0052792\sin^2\phi + 0.0000232\sin^4\phi\right),$$

$$= 9.780327\,\text{m}\cdot\text{s}^{-2}\left(1.0053024 - 0.0053256\cos^2\phi + 0.0000232\cos^4\phi\right),$$

$$= 9.780327\,\text{m}\cdot\text{s}^{-2}\left(1.0026454 - 0.0026512\cos 2\phi + 0.0000058\cos^2 2\phi\right)$$

This is the International Gravity Formula 1967, the 1967 Geodetic Reference System Formula, Helmert's equation or Clairaut's formula.

An alternate formula for g as a function of latitude is the WGS (World Geodetic System) 84 Ellipsoidal Gravity Formula:

$$g\{\phi\} = \mathbb{G}_e\left[\frac{1 + k\sin^2\phi}{\sqrt{1 - e^2\sin^2\phi}}\right],$$

where,

- a,b are the equatorial and polar semi-axes, respectively;

- $e^2 = 1 - (b/a)^2$ is the spheroid's eccentricity, squared;

- $\mathbb{G}_e, \mathbb{G}_p$ is the defined gravity at the equator and poles, respectively;

- $k = \dfrac{b\mathbb{G}_p - a\mathbb{G}_e}{a\mathbb{G}_e}$ (formula constant);

then, where $\mathbb{G}_p = 9.8321849378\,\text{m}\cdot\text{s}^{-2}$,

$$g\{\phi\} = 9.7803253359\,\text{m}\cdot\text{s}^{-2}\left[\frac{1 + 0.00193185265241\sin^2\phi}{\sqrt{1 - 0.00669437999013\sin^2\phi}}\right].$$

The difference between the WGS-84 formula and Helmert's equation is less than 0.68 μm·s⁻².

Free Air Correction

The first correction to be applied to the model is the free air correction (FAC) that accounts for heights above sea level. Near the surface of the Earth (sea level), gravity decreases with height such that linear extrapolation would give zero gravity at a height of one half of the earth's radius - (9.8 m·s⁻² per 3,200 km.)

Using the mass and radius of the Earth:

$$r_{\text{Earth}} = 6.371\cdot 10^6\,\text{m}$$

$$m_{\text{Earth}} = 5.9722\cdot 10^{24}\,\text{kg}$$

The FAC correction factor (Δg) can be derived from the definition of the acceleration due to gravity in terms of G, the Gravitational Constant (see Estimating g from the law of universal gravitation, below):

$$g_0 = Gm_{\text{Earth}} / r_{\text{Earth}}^2 = 9.8196\frac{\text{m}}{\text{s}^2}$$

where:

$$G = 6.67384\cdot 10^{-11}\frac{\text{m}^3}{\text{kg}\cdot\text{s}^2}.$$

At a height h above the nominal surface of the earth g_h is given by:

$$g_h = Gm_{\text{Earth}} / \left(r_{\text{Earth}} + h\right)^2$$

So the FAC for a height h above the nominal earth radius can be expressed:

$$\Delta g_h = \left[Gm_{\text{Earth}} / \left(r_{\text{Earth}} + h\right)^2\right] - \left[Gm_{\text{Earth}} / r_{\text{Earth}}^2\right]$$

This expression can be readily used for programming or inclusion in a spreadsheet. Collecting terms, simplifying and neglecting small terms ($h<<r_{Earth}$), however yields the good approximation:

$$\Delta g_h \approx -\frac{Gm_{Earth}}{r_{Earth}^2} \cdot \frac{2h}{r_{Earth}}$$

Using the numerical values above and for a height h in metres:

$$\Delta g_h \approx -3.086 \cdot 10^{-6} h$$

Grouping the latitude and FAC altitude factors the expression most commonly found in the literature is:

$$g\{\phi, h\} = g\{\phi\} - 3.086 \cdot 10^{-6} h$$

where $g\{\phi, h\}$ = acceleration in m·s^{-2} at latitude ϕ and altitude h in metres. Alternatively (with the same units for h) the expression can be grouped as follows:

$$g\{\phi, h\} = g\{\phi\} - 3.155 \cdot 10^{-7} h \frac{m}{s^2}$$

Slab Correction

Note: The section uses the galileo (symbol: "Gal"), which is a cgs unit for acceleration of 1 centimetre/second^2.

For flat terrain above sea level a second term is added for the gravity due to the extra mass; for this purpose the extra mass can be approximated by an infinite horizontal slab, and we get $2\pi G$ times the mass per unit area, i.e. 4.2×10^{-10} m^3·s^{-2}·kg^{-1} (0.042 µGal·kg^{-1}·m^2) (the Bouguer correction). For a mean rock density of 2.67 g·cm^{-3} this gives 1.1×10^{-6} s^{-2} (0.11 mGal·m^{-1}). Combined with the free-air correction this means a reduction of gravity at the surface of ca. 2 µm·s^{-2} (0.20 mGal) for every metre of elevation of the terrain. (The two effects would cancel at a surface rock density of 4/3 times the average density of the whole earth. The density of the whole earth is 5.515 g·cm^{-3}, so standing on a slab of something like iron whose density is over 7.35 g·cm^{-3} would increase one's weight.)

For the gravity below the surface we have to apply the free-air correction as well as a double Bouguer correction. With the infinite slab model this is because moving the point of observation below the slab changes the gravity due to it to its opposite. Alternatively, we can consider a spherically symmetrical Earth and subtract from the mass of the Earth that of the shell outside the point of observation, because that does not cause gravity inside. This gives the same result.

Estimating *g* from the Law of Universal Gravitation

From the law of universal gravitation, the force on a body acted upon by Earth's gravity is given by

$$F = G\frac{m_1 m_2}{r^2} = \left(G\frac{m_1}{r^2}\right)m_2$$

where r is the distance between the centre of the Earth and the body, and here we take m_1 to be the mass of the Earth and m_2 to be the mass of the body.

Additionally, Newton's second law, $F = ma$, where m is mass and a is acceleration, here tells us that

$$F = m_2 g$$

Comparing the two formulas it is seen that:

$$g = G\frac{m_1}{r^2}$$

So, to find the acceleration due to gravity at sea level, substitute the values of the gravitational constant, G, the Earth's mass (in kilograms), m_1, and the Earth's radius (in metres), r, to obtain the value of g:

$$g = G\frac{m_1}{r^2} = 6.67384 \cdot 10^{-11} \, \text{m}^3 \cdot \text{kg}^{-1} \cdot \text{s}^{-2} \, \frac{5.9722 \cdot 10^{24} \, \text{kg}}{(6.371 \cdot 10^6 \, \text{m})^2} = 9.8196 \, \text{m} \cdot \text{s}^{-2}$$

Note that this formula only works because of the mathematical fact that the gravity of a uniform spherical body, as measured on or above its surface, is the same as if all its mass were concentrated at a point at its centre. This is what allows us to use the Earth's radius for r.

The value obtained agrees approximately with the measured value of g. The difference may be attributed to several factors, mentioned above under "Variations":

- The Earth is not homogeneous

- The Earth is not a perfect sphere, and an average value must be used for its radius

- This calculated value of g only includes true gravity. It does not include the reduction of constraint force that we perceive as a reduction of gravity due to the rotation of Earth, and some of gravity being counteracted by centrifugal force.

There are significant uncertainties in the values of r and m_1 as used in this calculation, and the value of G is also rather difficult to measure precisely.

If G, g and r are known then a reverse calculation will give an estimate of the mass of the Earth. This method was used by Henry Cavendish.

Comparative Gravities of the Earth, Sun, Moon, and Planets

The table below shows comparative gravitational accelerations at the surface of the Sun, the Earth's moon, each of the planets in the Solar System and their major moons, Ceres, Pluto, and Eris. For gaseous bodies, the "surface" is taken to mean visible surface: the cloud tops of the gas giants (Jupiter, Saturn, Uranus and Neptune), and the Sun's photosphere. The values in the table have not been de-rated for the centrifugal force effect of planet rotation (and cloud-top wind speeds for the gas giants) and therefore, generally speaking, are similar to the actual gravity that would be experienced near the poles. For reference the time it would take an object to fall 100 metres, the height of a skyscraper, is shown, along with the maximum speed reached. Air resistance is neglected.

Body	Multiple of Earth gravity	m/s²	ft/s²	Time to fall 100 m and maximum speed reached	
Sun	27.90	274.1	899	0.85 s	843 km/h (524 mph)
Mercury	0.3770	3.703	12.15	7.4 s	98 km/h (61 mph)
Venus	0.9032	8.872	29.11	4.8 s	152 km/h (94 mph)
Earth	1	9.8067	32.174	4.5 s	159 km/h (99 mph)
Moon	0.1655	1.625	5.33	11.1 s	65 km/h (40 mph)
Mars	0.3895	3.728	12.23	7.3 s	98 km/h (61 mph)
Ceres	0.029	0.28	0.92	26.7 s	27 km/h (17 mph)
Jupiter	2.640	25.93	85.1	2.8 s	259 km/h (161 mph)
Io	0.182	1.789	5.87	10.6 s	68 km/h (42 mph)
Europa	0.134	1.314	4.31	12.3 s	58 km/h (36 mph)
Ganymede	0.145	1.426	4.68	11.8 s	61 km/h (38 mph)
Callisto	0.126	1.24	4.1	12.7 s	57 km/h (35 mph)
Saturn	1.139	11.19	36.7	4.2 s	170 km/h (110 mph)
Titan	0.138	1.3455	4.414	12.2 s	59 km/h (37 mph)
Uranus	0.917	9.01	29.6	4.7 s	153 km/h (95 mph)
Titania	0.039	0.379	1.24	23.0 s	31 km/h (19 mph)
Oberon	0.035	0.347	1.14	24.0 s	30 km/h (19 mph)
Neptune	1.148	11.28	37.0	4.2 s	171 km/h (106 mph)
Triton	0.079	0.779	2.56	16.0 s	45 km/h (28 mph)
Pluto	0.0621	0.610	2.00	18.1 s	40 km/h (25 mph)
Eris	0.0814	0.8	2.6	15.8 s	46 km/h (29 mph)

Gravitational Wave

Simulation showing gravitational waves produced during the final moments before the collision of two black holes. In the video, the waves could be seen to propagate outwards as the black holes spin past each other.

Gravitational waves are ripples in the curvature of spacetime that propagate as waves at the speed of light, generated in certain gravitational interactions that propagate outward from their source. The possibility of gravitational waves was discussed in 1893 by Oliver Heaviside using the analogy between the inverse-square law in gravitation and electricity. In 1905 Henri Poincaré first proposed gravitational waves (*ondes gravifiques*) emanating from a body and propagating at the speed of light as being required by the Lorentz transformations. Predicted in 1916 by Albert Einstein on the basis of his theory of general relativity, gravitational waves transport energy as gravitational radiation, a form of radiant energy similar to electromagnetic radiation. Gravitational waves cannot exist in the Newton's law of universal gravitation, since it is predicated on the assumption that physical interactions propagate at infinite speed.

Gravitational-wave astronomy is an emerging branch of observational astronomy which aims to use gravitational waves to collect observational data about objects such as neutron stars and black holes, events such as supernovae, and processes including those of the early universe shortly after the Big Bang.

Various gravitational-wave observatories (detectors) are under construction or in operation, such as Advanced LIGO which began observations in September 2015.

Potential sources of detectable gravitational waves include binary star systems composed of white dwarfs, neutron stars, and black holes. On February 11, 2016, the LIGO Scientific Collaboration and Virgo Collaboration teams announced that they had made the first observation of gravitational waves, originating from a pair of merging black holes using the Advanced LIGO detectors. On June 15, 2016, a second detection of gravitational waves from coalescing black holes was announced.

Introduction

In Einstein's theory of general relativity, gravity is treated as a phenomenon resulting from the curvature of spacetime. This curvature is caused by the presence of mass. Generally, the more mass that is contained within a given volume of space, the greater the curvature of spacetime will be at the boundary of its volume. As objects with mass move around in spacetime, the curvature changes to reflect the changed locations of those objects. In certain circumstances, accelerating objects generate changes in this curvature, which propagate outwards at the speed of light in a wave-like manner. These propagating phenomena are known as gravitational waves.

As a gravitational wave passes an observer, that observer will find spacetime distorted by the effects of strain. Distances between objects increase and decrease rhythmically as the wave passes, at a frequency corresponding to that of the wave. This occurs despite such free objects never being subjected to an unbalanced force. The magnitude of this effect decreases proportional to the inverse distance from the source. Inspiraling binary neutron stars are predicted to be a powerful source of gravitational waves as they coalesce, due to the very large acceleration of their masses as they orbit close to one another. However, due to the astronomical distances to these sources, the effects when measured on Earth are predicted to be very small, having strains of less than 1 part in 10^{20}. Scientists have demonstrated the existence of these waves with ever more sensitive detectors. The most sensitive detector accomplished the task possessing a sensitivity measurement of about one part in 5×10^{22} (as of 2012) provided by the LIGO and VIRGO observatories. A space based observatory, the Laser Interferometer Space Antenna, is currently under development by ESA.

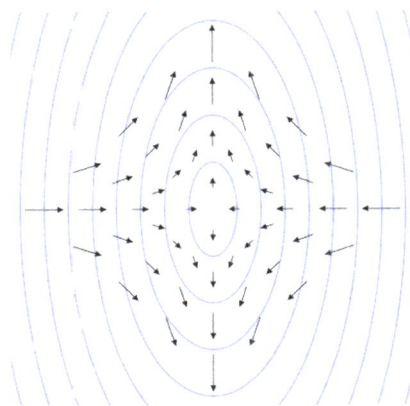

Linearly polarised gravitational wave

Gravitational waves can penetrate regions of space that electromagnetic waves cannot. They are able to allow the observation of the merger of black holes and possibly other exotic objects in the distant Universe. Such systems cannot be observed with more traditional means such as optical telescopes or radio telescopes, and so gravitational-wave astronomy gives new insights into the working of the Universe. In particular, gravitational waves could be of interest to cosmologists as they offer a possible way of observing the very early Universe. This is not possible with conventional astronomy, since before recombination the Universe was opaque to electromagnetic radiation. Precise measurements of gravitational waves will also allow scientists to test more thoroughly the general theory of relativity.

In principle, gravitational waves could exist at any frequency. However, very low frequency waves would be impossible to detect and there is no credible source for detectable waves of very high frequency. Stephen Hawking and Werner Israel list different frequency bands for gravitational waves that could plausibly be detected, ranging from 10^{-7} Hz up to 10^{11} Hz.

History

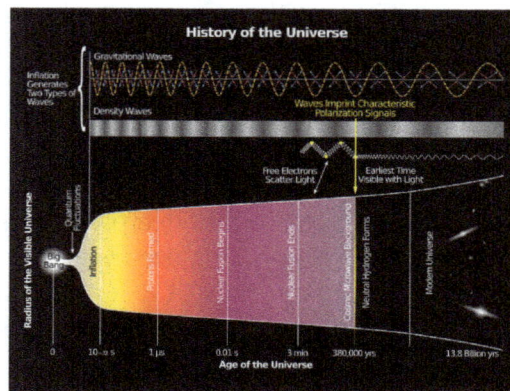

Primordial gravitational waves are hypothesized to arise from cosmic inflation, a faster-than-light expansion just after the Big Bang (2014).

In 1905, Henri Poincaré first suggested that in analogy to an accelerating electrical charge producing electromagnetic waves, accelerated masses in a relativistic field theory of gravity should produce gravitational waves. When Einstein published his theory of general relativity in 1915, he was skeptical of Poincaré's idea since the theory implied there were no "gravitational dipoles". Nonetheless, he still pursued the idea and based on various approximations came to the conclusion there must, in fact, be three types of gravitational wave (dubbed longitudinal-longitudinal, transverse-longitudinal, and transverse-transverse by Hermann Weyl).

However, the nature of Einstein's approximations led many (including Einstein himself) to doubt the result. In 1922, Arthur Eddington showed that two of Einstein's types of waves were artifacts of the coordinate system he used, and could be made to propagate at any speed by choosing appropriate coordinates, leading Eddington to jest that they "propagate at the speed of thought". This also cast doubt on the physicality of the third (transverse-transverse) type (which Eddington showed always propagate at the speed of light regardless of coordinate system). In 1936, Einstein and Nathan Rosen submitted a paper to Physical Review in which they claimed gravitational waves could not exist in the full theory of general relativity because any such solution of the field equations would have a singularity. The journal sent their manuscript to be reviewed by Howard P. Robertson, who (anonymously) reported that the singularities in question were simply the harmless coordinate singularities of the employed cylindrical coordinates. Einstein, who was unfamiliar with the concept of peer review, angrily withdrew the manuscript, never to publish in Physical Review again. Nonetheless, his assistant Leopold Infeld, who had been in contact with Robertson, convinced Einstein that the criticism was correct and the paper was rewritten with the opposite conclusion (and published elsewhere).

In 1956, Felix Pirani remedied the confusion caused by the use of various coordinate systems by rephrasing the gravitational waves in terms of the manifestly observable Riemann curvature tensor. At the time this work was mostly ignored because the community was focused on a different question: whether gravitational waves could transmit energy. This matter was settled by a thought experiment proposed by Richard Feynman during the first "GR" conference at Chapel Hill in 1957. In short, his argument (known as the "sticky bead argument") notes that if one takes a stick with beads then the effect of a passing gravitational wave would be to move the beads along the stick; friction would then produce heat, implying that the passing wave had done work. Shortly after, Hermann Bondi (a former gravitational wave skeptic) published a detailed version of the "sticky bead argument".

After the Chapel Hill conference, Joseph Weber started designing and building the first gravitational wave detectors now known as Weber bars. In 1969, Weber claimed to have detected the first gravitational waves and by 1970 he was "detecting" signals regularly; however, the frequency of detection soon raised doubts on the validity of his observations as the implied rate of energy loss of the Milky Way would drain our galaxy of energy on a timescale much shorter than its inferred age. These doubts were strengthened when, by the mid-1970s, repeat experiments from other groups building their own Weber bars across the globe failed to find any signals, and by the late 1970s general consensus was that Weber's results were spurious.

In the same period, the first indirect evidence for the existence of gravitational waves was discovered. In 1974, Russell Alan Hulse and Joseph Hooton Taylor, Jr. discovered the first binary pulsar (a discovery that earned them the 1993 Nobel Prize in Physics). In 1979, results were published detailing measurement of the gradual decay of the orbital period of the Hulse-Taylor pulsar, which fitted precisely with the loss of energy and angular momentum in gravitational radiation predicted by general relativity.

- 1962 – M. E. Gertsenshtein and V. I. Pustovoit publish the first paper describing the principles for using interferometers to detect very long wavelength gravitational waves.

- 1984 – Kip Thorne, Ronald Drever, and Rainer Weiss form a steering committee after the NSF asks MIT and Caltech to join forces to lead a LIGO project.

- 1994 – LIGO Laboratory Director Barry Barish and his team create the LIGO study, project plan, and budget, receive long-withheld NSF funding, and go-ahead for construction. Barish is appointed Principal Investigator and LIGO, with a budget of US$395 million, becomes the largest overall funded NSF project in history.

- 1997 – The LIGO Scientific Collaboration (LSC) and the Gravitational Wave International Committee (GWIC) are formed.

- 2002 – LIGO begins the search for gravitational waves

- 2004 – Advanced LIGO upgrade is approved by the National Science Board.

- 2005 - The binary black hole problem solved: three groups independently developed ground-breaking new methods to model the inspiral, merger, and ring-down of binary black holes.

- 2014 – Astronomers at the Harvard–Smithsonian Center for Astrophysics erroneously claim that they have detected and produced "the first direct image of gravitational waves" in the cosmic microwave background.

- 2015 – Advanced LIGO begins operation.

- 11 February 2016 – The LIGO Scientific Collaboration announce that they detected gravitational waves on 14 September 2015 from a 410 megaparsec (1.3 billion light years) distant merger of two black holes, 36+5
  −4 M⊙ and 29+4
  −4 M⊙, resulting in a 62+4
  −4 M⊙ black hole. The signal is named GW150914.

- 15 June 2016 – LIGO announced a second observation of gravitational waves, signal GW151226, observed on 26 December 2015, produced by the 440 megaparsec (1.4 billion light years) distant coalescence of two stellar-mass black holes, 14.2+8.3
  −3.7 M⊙ and 7.5+2.3
  −2.3 M⊙, resulting in a 20.8+6.1
  −1.7 M⊙ black hole.

Effects of Passing

The effect of a polarized gravitational wave on a ring of particles.

Gravitational waves are constantly passing Earth; however, even the strongest have a minuscule effect and their sources are generally at a great distance. For example, the waves given off by the cataclysmic final merger of GW150914 reached Earth after travelling over a billion lightyears, as a ripple in spacetime that changed the length of a 4-km LIGO arm by a ten thousandth of the width of a proton, proportionally equivalent to changing the distance to the nearest star outside the Solar System by one hair's width. This tiny effect from even extreme gravitational waves makes them undetectable on Earth by any means other than the most sophisticated detectors.

The effects of a passing gravitational wave, in an extremely exaggerated form, can be visualized by imagining a perfectly flat region of spacetime with a group of motionless test particles lying in a plane (e.g., the surface of a computer screen). As a gravitational wave passes through the particles along a line perpendicular to the plane of the particles (i.e. following the observer's line of vision into the screen), the particles will follow the distortion in spacetime, oscillating in a "cruciform" manner, as shown in the animations. The area enclosed by the test particles does not change and there is no motion along the direction of propagation.

The oscillations depicted in the animation are exaggerated for the purpose of discussion — in reality a gravitational wave has a very small amplitude (as formulated in linearized gravity). However, they help illustrate the kind of oscillations associated with gravitational waves as produced, for example, by a pair of masses in a circular orbit. In this case the amplitude of the gravitational wave is constant, but its plane of polarization changes or rotates at twice the orbital rate and so the time-varying gravitational wave size (or 'periodic spacetime strain') exhibits a variation as shown in the animation. If the orbit of the masses is elliptical then the gravitational wave's amplitude also varies with time according to Einstein's quadrupole formula.

As with other waves, there are a number of characteristics used to describe a gravitational wave:

- Amplitude: Usually denoted h, this is the size of the wave — the fraction of stretching or squeezing in the animation. The amplitude shown here is roughly $h = 0.5$ (or 50%). Gravitational waves passing through the Earth are many sextillion times weaker than this — $h \approx 10^{-20}$.

- Frequency: Usually denoted f, this is the frequency with which the wave oscillates (1 divided by the amount of time between two successive maximum stretches or squeezes)

- Wavelength: Usually denoted λ, this is the distance along the wave between points of maximum stretch or squeeze.

- Speed: This is the speed at which a point on the wave (for example, a point of maximum stretch or squeeze) travels. For gravitational waves with small amplitudes, this wave speed is equal to the speed of light (c).

The speed, wavelength, and frequency of a gravitational wave are related by the equation $c = \lambda f$, just like the equation for a light wave. For example, the animations shown here oscillate roughly once every two seconds. This would correspond to a frequency of 0.5 Hz, and a wavelength of about 600 000 km, or 47 times the diameter of the Earth.

In the above example, it is assumed that the wave is linearly polarized with a "plus" polarization, written h_+. Polarization of a gravitational wave is just like polarization of a light wave except that the polarizations of a gravitational wave are at 45 degrees, as opposed to 90 degrees. In particular, in a "cross"-polarized gravitational wave, $h_\times$, the effect on the test particles would be basically the same, but rotated by 45 degrees, as shown in the second animation. Just as with light polarization, the polarizations of gravitational waves may also be expressed in terms of circularly polarized waves. Gravitational waves are polarized because of the nature of their sources.

Sources

In general terms, gravitational waves are radiated by objects whose motion involves acceleration and its change, provided that the motion is not perfectly spherically symmetric (like an expanding or contracting sphere) or rotationally symmetric (like a spinning disk or sphere). A simple example of this principle is a spinning dumbbell. If the dumbbell spins around its axis of symmetry, it will not radiate gravitational waves; if it tumbles end over end, as in the case of two planets orbiting each other, it will radiate gravitational waves. The heavier the dumbbell, and the faster it tumbles, the greater is the gravitational radiation it will give off. In an extreme case, such as when the two weights of the dumbbell are massive stars like neutron stars or black holes, orbiting each other quickly, then significant amounts of gravitational radiation would be given off.

The Gravitational Wave Spectrum

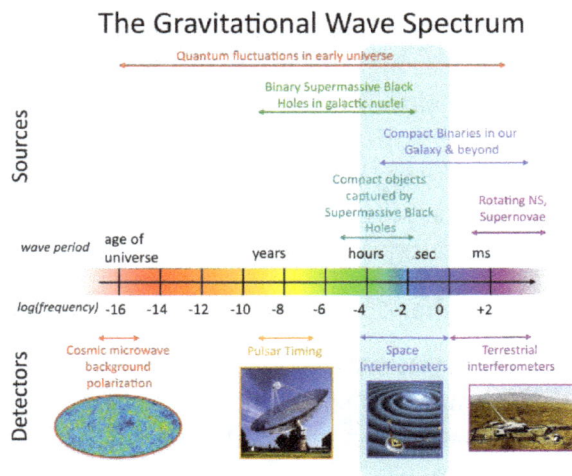

The gravitational wave spectrum with sources and detectors. *Credit: NASA Goddard Space Flight Center*

Some more detailed examples:

- Two objects orbiting each other, as a planet would orbit the Sun, *will* radiate.

- A spinning non-axisymmetric planetoid — say with a large bump or dimple on the equator — *will* radiate.

- A supernova *will* radiate except in the unlikely event that the explosion is perfectly symmetric.

- An isolated non-spinning solid object moving at a constant velocity *will not* radiate. This can be regarded as a consequence of the principle of conservation of linear momentum.

- A spinning disk *will not* radiate. This can be regarded as a consequence of the principle of conservation of angular momentum. However, it *will* show gravitomagnetic effects.

- A spherically pulsating spherical star (non-zero monopole moment or mass, but zero quadrupole moment) *will not* radiate, in agreement with Birkhoff's theorem.

More technically, the third time derivative of the quadrupole moment (or the l-th time derivative of the l-th multipole moment) of an isolated system's stress–energy tensor must be non-zero in order for it to emit gravitational radiation. This is analogous to the changing dipole moment of charge or current that is necessary for the emission of electromagnetic radiation.

Binaries

Gravitational waves carry energy away from their sources and, in the case of orbiting bodies, this is associated with an inspiral or decrease in orbit. Imagine for example a simple system of two masses — such as the Earth–Sun system — moving slowly compared to the speed of light in circular orbits. Assume that these two masses orbit each other in a circular orbit in the x–y plane. To a good approximation, the masses follow simple Keplerian orbits. However, such an orbit represents a changing quadrupole moment. That is, the system will give off gravitational waves.

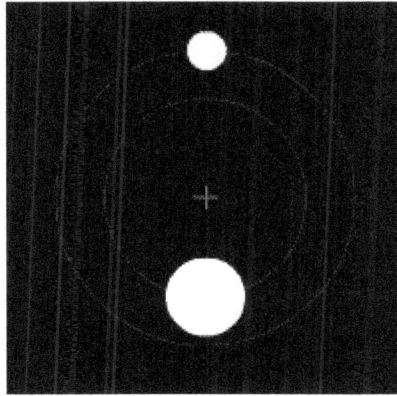

Two stars of dissimilar mass are in circular orbits. Each revolves about their common center of mass (denoted by the small red cross) in a circle with the larger mass having the smaller orbit.

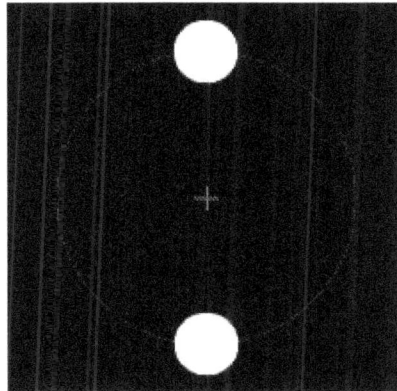

Two stars of similar mass are in circular orbits about their center of mass

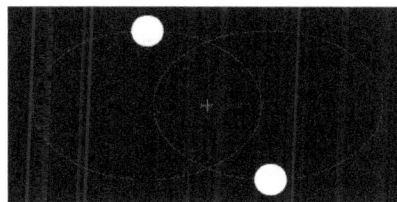

Two stars of similar mass are in highly elliptical orbits about their center of mass

In theory, the loss of energy through gravitational radiation could eventually drop the Earth into the Sun. However, the total energy of the Earth orbiting the Sun (kinetic energy + gravitational potential energy) is about 1.14×10^{36} joules of which only 200 Watts (joules per second) is lost through gravitational radiation, leading to a decay in the orbit by about 1×10^{-15} meters per day or roughly the diameter of a proton. At this rate, it would take the Earth approximately 1×10^{13} times more than the current age of the Universe to spiral onto the Sun. This estimate overlooks the decrease in r over time, but the majority of the time the bodies are far apart and only radiating slowly, so the difference is unimportant in this example.

More generally, the rate of orbital decay can be approximated by

$$\frac{dr}{dt} = -\frac{64}{5}\frac{G^3}{c^5}\frac{(m_1 m_2)(m_1 + m_2)}{r^3},$$

where r is the separation between the bodies, t time, G Newton's constant, c the speed of light, and m_1 and m_2 the masses of the bodies. This leads to an expected time to merger of

$$t = \frac{5}{256} \frac{c^5}{G^3} \frac{r^4}{(m_1 m_2)(m_1 + m_2)}.$$

For example, a pair of solar mass neutron stars in a circular orbit at a separation of 1.89×10^8 m (189,000 km) has an orbital period of 1,000 seconds, and an expected lifetime of 1.30×10^{13} seconds or about 414,000 years. Such a system could be observed by LISA if it were not too far away. A far greater number of white dwarf binaries exist with orbital periods in this range. White dwarf binaries have masses in the order of the Sun, and diameters in the order of the Earth. They cannot get much closer together than 10,000 km before they will merge and explode in a supernova which would also end the emission of gravitational waves. Until then, their gravitational radiation would be comparable to that of a neutron star binary.

When the orbit of a neutron star binary has decayed to 1.89×10^6 m (1890 km), its remaining lifetime is about 130,000 seconds or 36 hours. The orbital frequency will vary from 1 orbit per second at the start, to 918 orbits per second when the orbit has shrunk to 20 km at merger. The majority of gravitational radiation emitted will be at twice the orbital frequency. Just before merger, the in-spiral would be observed by LIGO if such a binary were close enough. LIGO has only a few minutes to observe this merger out of a total orbital lifetime that may have been billions of years. Advanced LIGO detector should be able to detect these events up to 200 megaparsec away. Within this range of the order 40 events are expected per year.

Black Holes

Black hole binaries emit gravitational waves during their in-spiral, merger, and ring-down phases. The largest amplitude of emission occurs during the merger phase, which can be modeled with the techniques of numerical relativity. The first direct detection of gravitational waves, GW150914, came from the merger of two black holes.

Supernova

A supernova is an astronomical event that occurs during the last stellar evolutionary stages of a massive star's life, whose dramatic and catastrophic destruction is marked by one final titanic explosion. This explosion can happen in one of many ways, but in all of them a significant proportion of the matter in the star is blown away into the surrounding space at extremely high velocities (up to 10% of the speed of light). Unless there is perfect spherical symmetry in these explosions (i.e., unless matter is spewed out evenly in all directions), there will be gravitational radiation from the explosion. This is because gravitational waves are generated by a changing quadrupole moment, which can happen only when there is asymmetrical movement of masses. Since the exact mechanism by which supernovae take place is not fully understood, it is not easy to model the gravitational radiation emitted by them.

Rotating Neutron Stars

As noted above, a mass distribution will emit gravitational radiation only when there is spherically asymmetric motion among the masses. A spinning neutron star will generally emit no gravitational radiation because neutron stars are highly dense objects with a strong gravitational field that keeps them almost perfectly spherical. In some cases, however, there might be slight deformities on the surface called "mountains", which are bumps extending no more than 10 centimeters (4 inches) above the surface, that make the spinning spherically asymmetric. This gives the star a quadrupole moment that changes with time, and it will emit gravitational waves until the deformities are smoothed out.

Inflation

Many models of the Universe postulate that there was an inflationary epoch in the early history of the Universe when space expanded by a large factor in a very short amount of time. If this expansion was not symmetric in all directions, it may have emitted gravitational radiation detectable today as a gravitational wave background. This background signal is too weak for any currently operational gravitational wave detector to observe, and it is thought it may be decades before such an observation can be made.

Properties and Behaviour

Energy, Momentum, and Angular Momentum

Water waves, sound waves, and electromagnetic waves are able to carry energy, momentum, and angular momentum and by doing so they carry those away from the source. Gravitational waves perform the same function. Thus, for example, a binary system loses angular momentum as the two orbiting objects spiral towards each other—the angular momentum is radiated away by gravitational waves.

The waves can also carry off linear momentum, a possibility that has some interesting implications for astrophysics. After two supermassive black holes coalesce, emission of linear momentum can produce a "kick" with amplitude as large as 4000 km/s. This is fast enough to eject the coalesced black hole completely from its host galaxy. Even if the kick is too small to eject the black hole completely, it can remove it temporarily from the nucleus of the galaxy, after which it will oscillate about the center, eventually coming to rest. A kicked black hole can also carry a star cluster with it, forming a hyper-compact stellar system. Or it may carry gas, allowing the recoiling black hole to appear temporarily as a "naked quasar". The quasar SDSS J092712.65+294344.0 is thought to contain a recoiling supermassive black hole.

Redshifting and Blueshifting

Like electromagnetic waves, gravitational waves should exhibit shifting of wavelength due to the relative velocities of the source and observer, but also due to distortions of space-time, such as cosmic expansion. This is the case even though gravity itself is a cause of distortions of space-time. Redshifting *of* gravitational waves is different from redshifting *due to* gravity.

Quantum Gravity, Wave-particle Aspects, and Graviton

At present, and unlike all other known forces in the universe, no "force carrying" particle has been identified as mediating gravitational interactions.

In the framework of quantum field theory, the graviton is the name given to a hypothetical elementary particle speculated to be the force carrier that mediates gravity. However the graviton is not yet proven to exist and no reconciliation yet exists between general relativity which describes gravity, and the Standard Model which describes all other fundamental forces.

If such a particle exists, it is expected to be massless (because the gravitational force appears to have unlimited range) and must be a spin-2 boson. It can be shown that any massless spin-2 field would give rise to a force indistinguishable from gravitation, because a massless spin-2 field must couple to (interact with) the stress–energy tensor in the same way that the gravitational field does; therefore if a massless spin-2 particle were ever discovered, it would be likely to be the graviton without further distinction from other massless spin-2 particles. Such a discovery would unite quantum theory with gravity.

Absorption, re-emission, refraction (lensing), superposition, and other wave effects

Due to the weakness of the coupling of gravity to matter, gravitational waves experience very little absorption or scattering, even as they travel over astronomical distances. In particular, gravitational waves are expected to be unaffected by the opacity of the very early universe before space became "transparent"; observations based upon light, radio waves, and other electromagnetic radiation further back into time is limited or unavailable. Therefore, gravitational waves are expected to have the potential to open a new means of observation to the very early universe.

Determining Direction of Travel

The difficulty in directly detecting gravitational waves, means it is also difficult for a single detector to identify by itself the direction of a source. Therefore, multiple detectors are used, both to distinguish signals from other "noise" by confirming the signal is not of earthly origin, and also to determine direction by means of triangulation. This technique uses the fact that the waves travel at the speed of light and will reach different detectors at different times depending on their source direction. Although the differences in arrival time may be just a few milliseconds, this is sufficient to identify the direction of the origin of the wave with considerable precision.

In the case of GW150914, only two detectors were operating at the time of the event, therefore, the direction is not so precisely defined and it could lie anywhere within an arc-shaped region of space rather than being identified as a single point.

Astrophysics Implications

During the past century, astronomy has been revolutionized by the use of new methods for observing the universe. Astronomical observations were originally made using visible light. Galileo Galilei pioneered the use of telescopes to enhance these observations. However, visible light is only a small portion of the electromagnetic spectrum, and not all objects in the distant universe shine strongly in this particular band. More useful information may be found, for example, in radio wavelengths. Using radio telescopes, astronomers have found pulsars, quasars, and other extreme objects that push the limits of our understanding of physics. Observations in the microwave band have opened our eyes to the faint imprints of the Big Bang, a discovery Stephen Hawking called the "greatest discovery of the century, if not all time". Similar advances in observations using gamma rays, x-rays, ultraviolet light, and infrared light have also brought new insights to astronomy. As each of these regions of the spectrum has opened, new discoveries have been made that could not have been made otherwise. Astronomers hope that the same holds true of gravitational waves.

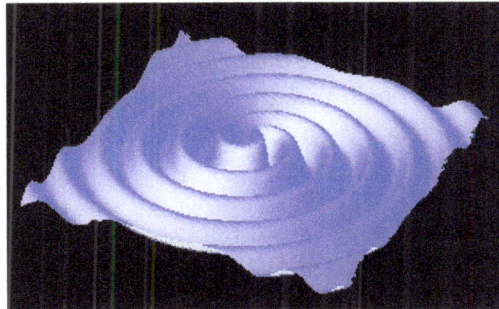

Two-dimensional representation of gravitational waves generated by two neutron stars orbiting each other.

Gravitational waves have two important and unique properties. First, there is no need for any type of matter to be present nearby in order for the waves to be generated by a binary system of uncharged black holes, which would emit no electromagnetic radiation. Second, gravitational waves can pass through any intervening matter without being scattered significantly. Whereas light from distant stars may be blocked out by interstellar dust, for example, gravitational waves will pass through essentially unimpeded. These two features allow gravitational waves to carry information about astronomical phenomena never before observed by humans.

The sources of gravitational waves described above are in the low-frequency end of the gravitational-wave spectrum (10^{-7} to 10^5 Hz). An astrophysical source at the high-frequency end of the gravitational-wave spectrum (above 10^5 Hz and probably 10^{10} Hz) generates relic gravitational waves that are theorized to be faint imprints of the Big Bang like the cosmic microwave background. At these high frequencies it is potentially possible that the sources may be "man made" that is, gravitational waves generated and detected in the laboratory.

Detection

Now disproved evidence allegedly showing gravitational waves in the infant universe was found by the BICEP2 radio telescope. The microscopic examination of the focal plane of the BICEP2 detector is shown here. In 2015, however, the BICEP2 findings were confirmed to be the result of cosmic dust.

Indirect Detection

Although the waves from the Earth–Sun system are minuscule, astronomers can point to other sources for which the radiation should be substantial. One important example is the Hulse–Taylor binary — a pair of stars, one of which is a pulsar. The characteristics of their orbit can be deduced from the Doppler shifting of radio signals given off by the pulsar. Each of the stars is about $1.4\,M_{\square}$ and the size of their orbits is about $1/75$ of the Earth–Sun orbit, just a few times larger than the diameter of our own Sun. The combination of greater masses and smaller separation means that the energy given off by the Hulse–Taylor binary will be far greater than the energy given off by the Earth–Sun system — roughly 10^{22} times as much.

The information about the orbit can be used to predict how much energy (and angular momentum) would be radiated in the form of gravitational waves. As the energy is carried off, the stars should draw closer to each other. This effect is called an inspiral, and it can be observed in the pulsar's signals. The measurements on the Hulse–Taylor system have been carried out over more than 30 years. It has been shown that the change in the orbit period, as predicted from the assumed gravitational radiation and general relativity, and the observations matched within 0.2 percent. In 1993, Russell Hulse and Joe Taylor were awarded the Nobel Prize in Physics for this work, which was the first indirect evidence for gravitational waves. The lifetime of this binary system, from the present to merger is estimated to be a few hundred million years.

Inspirals are very important sources of gravitational waves. Any time two compact objects (white dwarfs, neutron stars, or black holes) are in close orbits, they send out intense gravitational waves. As they spiral closer to each other, these waves become more intense. At some point they should become so intense that direct detection by their effect on objects on Earth or in space is possible. This direct detection is the goal of several large scale experiments.

The only difficulty is that most systems like the Hulse–Taylor binary are so far away. The amplitude of waves given off by the Hulse–Taylor binary at Earth would be roughly $h \approx 10^{-26}$. There are some sources, however, that astrophysicists expect to find that produce much greater amplitudes of $h \approx 10^{-20}$. At least eight other binary pulsars have been discovered.

Difficulties

Gravitational waves are not easily detectable. When they reach the Earth, they have a small amplitude with strain approximates 10^{-21}, meaning that an extremely sensitive detector is needed, and that other sources of noise can overwhelm the signal. Gravitational waves are expected to have frequencies 10^{-16} Hz $< f < 10^{4}$ Hz.

Ground-based Detectors

Though the Hulse–Taylor observations were very important, they give only *indirect* evidence for gravitational waves. A more conclusive observation would be a *direct* measurement of the effect of a passing gravitational wave, which could also provide more information about the system that generated it. Any such direct detection is complicated by the extraordinarily small effect the waves would produce on a detector. The amplitude of a spherical wave will fall off as the inverse of the distance from the source (the $1/R$ term in the formulas for h above). Thus, even waves from extreme systems like merging binary black holes die out to very small amplitudes by the time they reach the Earth. Astrophysicists expect that some gravitational waves passing the Earth may be as large as $h \approx 10^{-20}$, but generally no bigger.

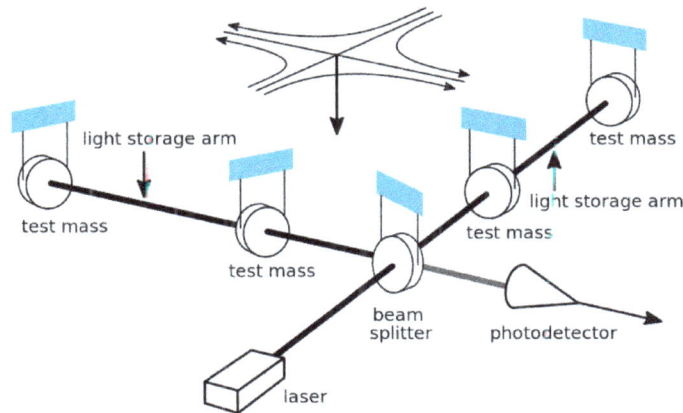

A schematic diagram of a laser interferometer

Resonant Antennae

A simple device theorised to detect the expected wave motion is called a Weber bar — a large, solid bar of metal isolated from outside vibrations. This type of instrument was the first type of gravitational wave detector. Strains in space due to an incident gravitational wave excite the bar's resonant frequency and could thus be amplified to detectable levels. Conceivably, a nearby supernova might be strong enough to be seen without resonant amplification. With this instrument, Joseph Weber claimed to have detected daily signals of gravitational waves. His results, however, were contested in 1974 by physicists Richard Garwin and David Douglass. Modern forms of the Weber bar are still operated, cryogenically cooled, with superconducting quantum interference devices to detect vibration. Weber bars are not sensitive enough to detect anything but extremely powerful gravitational waves.

MiniGRAIL is a spherical gravitational wave antenna using this principle. It is based at Leiden University, consisting of an exactingly machined 1150 kg sphere cryogenically cooled to 20 mK. The spherical configuration allows for equal sensitivity in all directions, and is somewhat experimentally simpler than larger linear devices requiring high vacuum. Events are detected by measuring deformation of the detector sphere. MiniGRAIL is highly sensitive in the 2–4 kHz range, suitable for detecting gravitational waves from rotating neutron star instabilities or small black hole mergers.

There are currently two detectors focused on the higher end of the gravitational wave spectrum (10^{-7} to 10^5 Hz): one at University of Birmingham, England, and the other at INFN Genoa, Italy. A third is under development at Chongqing University, China. The Birmingham detector measures changes in the polarization state of a microwave beam circulating in a closed loop about one meter across. Both detectors are expected to be sensitive to periodic spacetime strains of $h \sim 2 \times 10^{-13} / \sqrt{Hz}$, given as an amplitude spectral density. The INFN Genoa detector is a resonant antenna consisting of two coupled spherical superconducting harmonic oscillators a few centimeters in diameter. The oscillators are designed to have (when uncoupled) almost equal resonant frequencies. The system is currently expected to have a sensitivity to periodic spacetime strains of $h \sim 2 \times 10^{-17} / \sqrt{\text{Hz}}$, with an expectation to reach a sensitivity of $h \sim 2 \times 10^{-20} / \sqrt{\text{Hz}}$. The Chongqing University detector is planned to detect relic high-frequency gravitational waves with the predicted typical parameters $\sim 10^{11}$ Hz (100 GHz) and $h \sim 10^{-30}$ to 10^{-32}.

Interferometers

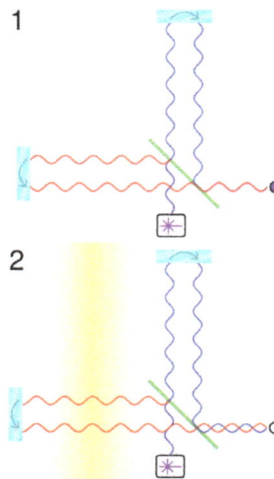

Simplified operation of a gravitational wave observatory

Figure 1: A beamsplitter (green line) splits coherent light (from the white box) into two beams which reflect off the mirrors (cyan oblongs); only one outgoing and reflected beam in each arm is shown, and separated for clarity. The reflected beams recombine and an interference pattern is detected (purple circle).

Figure 2: A gravitational wave passing over the left arm (yellow) changes its length and thus the interference pattern.

A more sensitive class of detector uses laser interferometry to measure gravitational-wave induced motion between separated 'free' masses. This allows the masses to be separated by large distances (increasing the signal size); a further advantage is that it is sensitive to a wide range of frequencies (not just those near a resonance as is the case for Weber bars). Ground-based interferometers are now operational. Currently, the most sensitive is LIGO — the Laser Interferometer Gravitational Wave Observatory. LIGO has three detectors: one in Livingston, Louisiana, one at the Hanford site in Richland, Washington and a third (formerly installed as a second detector at Hanford) that is planned to be moved to India. Each observatory has two light storage arms that are 4 kilometers in length. These are at 90 degree angles to each other, with the light passing through 1 m diameter vacuum tubes running the entire 4 kilometers. A passing gravitational wave will slightly stretch one arm as it shortens the other. This is precisely the motion to which an interferometer is most sensitive.

Even with such long arms, the strongest gravitational waves will only change the distance between the ends of the arms by at most roughly 10^{-18} meters. LIGO should be able to detect gravitational waves as small as $h \sim 5 \times 10^{-22}$. Upgrades to LIGO and other detectors such as Virgo, GEO 600, and TAMA 300 should increase the sensitivity still further; the next generation of instruments

(Advanced LIGO and Advanced Virgo) will be more than ten times more sensitive. Another highly sensitive interferometer, KAGRA, is under construction in the Kamiokande mine in Japan. A key point is that a tenfold increase in sensitivity (radius of 'reach') increases the volume of space accessible to the instrument by one thousand times. This increases the rate at which detectable signals might be seen from one per tens of years of observation, to tens per year.

Interferometric detectors are limited at high frequencies by shot noise, which occurs because the lasers produce photons randomly; one analogy is to rainfall—the rate of rainfall, like the laser intensity, is measurable, but the raindrops, like photons, fall at random times, causing fluctuations around the average value. This leads to noise at the output of the detector, much like radio static. In addition, for sufficiently high laser power, the random momentum transferred to the test masses by the laser photons shakes the mirrors, masking signals of low frequencies. Thermal noise (e.g., Brownian motion) is another limit to sensitivity. In addition to these 'stationary' (constant) noise sources, all ground-based detectors are also limited at low frequencies by seismic noise and other forms of environmental vibration, and other 'non-stationary' noise sources; creaks in mechanical structures, lightning or other large electrical disturbances, etc. may also create noise masking an event or may even imitate an event. All these must be taken into account and excluded by analysis before detection may be considered a true gravitational wave event.

Einstein@Home

The simplest gravitational waves are those with constant frequency. The waves given off by a spinning, non-axisymmetric neutron star would be approximately monochromatic: a pure tone in acoustics. Unlike signals from supernovae of binary black holes, these signals evolve little in amplitude or frequency over the period it would be observed by ground-based detectors. However, there would be some change in the measured signal, because of Doppler shifting caused by the motion of the Earth. Despite the signals being simple, detection is extremely computationally expensive, because of the long stretches of data that must be analysed.

The Einstein@Home project is a distributed computing project similar to SETI@home intended to detect this type of gravitational wave. By taking data from LIGO and GEO, and sending it out in little pieces to thousands of volunteers for parallel analysis on their home computers, Einstein@Home can sift through the data far more quickly than would be possible otherwise.

Space-based Interferometers

Space-based interferometers, such as LISA and DECIGO, are also being developed. LISA's design calls for three test masses forming an equilateral triangle, with lasers from each spacecraft to each other spacecraft forming two independent interferometers. LISA is planned to occupy a solar orbit trailing the Earth, with each arm of the triangle being five million kilometers. This puts the detector in an excellent vacuum far from Earth-based sources of noise, though it will still be susceptible to heat, shot noise, and artifacts caused by cosmic rays and solar wind.

Using Pulsar Timing Arrays

Pulsars are rapidly rotating stars. A pulsar emits beams of radio waves that, like lighthouse beams, sweep through the sky as the pulsar rotates. The signal from a pulsar can be detected by radio telescopes as a series of regularly spaced pulses, essentially like the ticks of a clock. Gravitational waves affect the time it takes the pulses to travel from the pulsar to a telescope on Earth. A pulsar timing array uses millisecond pulsars to seek out perturbations due to gravitational waves in measurements of pulse arrival times at a telescope, in other words, to look for deviations in the clock ticks. In particular, pulsar timing arrays can search for a distinct pattern of correlation and anti-correlation between the signals over an array of different pulsars (resulting in the name "pulsar timing array"). Although pulsar pulses travel through space for hundreds or thousands of years to reach us, pulsar timing arrays are sensitive to perturbations in their travel time of much less than a millionth of a second.

Globally there are three active pulsar timing array projects. The North American Nanohertz Gravitational Wave Observatory uses data collected by the Arecibo Radio Telescope and Green Bank Telescope. The Parkes Pulsar Timing Array at the Parkes radio-telescope has been collecting data since March 2005. The European Pulsar Timing Array uses data from the four largest telescopes in Europe: the Lovell Telescope, the Westerbork Synthesis Radio Telescope, the Effelsberg Telescope and the Nancay Radio Telescope. (Upon completion the Sardinia Radio Telescope will be added to the EPTA also.) These three projects have begun collaborating under the title of the International Pulsar Timing Array project.

Primordial

Primordial gravitational waves are gravitational waves observed in the cosmic microwave background. They were allegedly detected by the BICEP2 instrument, an announcement made on 17 March 2014, which was withdrawn on 30 January 2015 ("the signal can be entirely attributed to dust in the Milky Way").

LIGO Observations

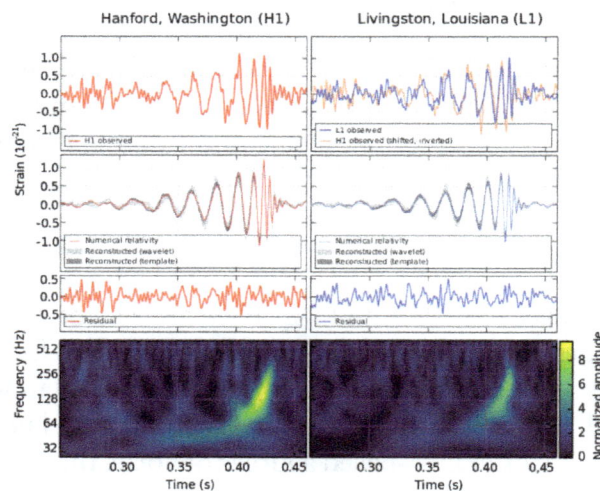

LIGO measurement of the gravitational waves at the Hanford (left) and Livingston (right) detectors, compared to the theoretical predicted values.

On 11 February 2016, the LIGO collaboration announced the detection of gravitational waves, from a signal detected at 09:50:45 GMT on 14 September 2015 of two black holes with masses of 29 and

36 solar masses merging about 1.3 billion light years away. During the final fraction of a second of the merger, it released more than 50 times the power of all the stars in the observable universe combined. The signal increased in frequency from 35 to 250 Hz over 10 cycles (5 orbits) as it rose in strength for a period of 0.2 second. The mass of the new merged black hole was 62 solar masses. Energy equivalent to three solar masses was emitted as gravitational waves. The signal was seen by both LIGO detectors in Livingston and Hanford, with a time difference of 7 milliseconds due to the angle between the two detectors and the source. The signal came from the Southern Celestial Hemisphere, in the rough direction of (but much further away than) the Magellanic Clouds. The confidence level of this being an observation of gravitational waves was 99.99994%.

On 15 June 2016, the LIGO group announced the detection of a second set of gravitational waves, which was observed at 03:38:53 GMT on 26 December 2015. The signal was seen by the Hanford LIGO detector 1.1 milliseconds after the Livingston detector. The signal rose from 35 to 450 Hz over the course of 55 cycles (27 orbits) during the period of observation of about a second. Analysis of the signal indicates that this event represented the merger of two black holes about 1.4 billion light years distant, with masses of about 14.2 and 7.5 solar masses, yielding a combined black hole of approximately of 20.8 solar masses, with one solar mass radiated away. The estimated spin parameter (ratio of angular momentum to theoretical limit) of the final black hole is 0.74, slightly higher than for the first detection (0.67); it was also found that at least one of the premerger black holes had a spin of greater than 0.2. This measurement provided additional support for general relativity.

Mathematics

Framework

Gravitational waves are presently understood to be described by Albert Einstein's theory of general relativity. In the simplest cases, and certain less-dynamic situations, the energy implications of gravitational waves can be deduced from other conservation laws such as those governing conservation of energy or conservation of momentum.

Beyond these simple cases, Einstein's equations show how the curvature of spacetime can be expressed mathematically using the metric tensor — denoted . The metric holds information regarding how distances are measured in the space under consideration. Because the propagation of gravitational waves through space and time change distances, we will need to use this to find the solution to the wave equation.

Basic Mathematics

Power Radiated by Orbiting Bodies

Suppose that the two masses are m_1 and m_2, and they are separated by a distance r. The power given off (radiated) by this system is:

$$P = \frac{\mathrm{d}E}{\mathrm{d}t} = -\frac{32}{5}\frac{G^4}{c^5}\frac{(m_1 m_2)^2 (m_1 + m_2)}{r^5},$$

where G is the gravitational constant, c is the speed of light in vacuum and where the negative sign means that power is leaving the system, rather than entering. For a system like the Sun and Earth, r is about 1.5×10^{11} m and m_1 and m_2 are about 2×10^{30} and 6×10^{24} kg respectively. In this case, the power leaving the Earth, Sun system is about 200 watts. This is truly tiny compared to the total electromagnetic radiation given off by the Sun (roughly 3.86×10^{26} watts, or almost 400 million, million, million, million watts).

Wave Amplitudes from the Earth–Sun System

We can also think in terms of the amplitude of the wave from a system in circular orbits. Let θ be the angle between the perpendicular to the plane of the orbit and the line of sight of the observer. Suppose that an observer is outside the system at a distance R from its center of mass. If R is much greater than a wavelength, the two polarizations of the wave will be

$$h_+ = -\frac{1}{R}\frac{G^2}{c^4}\frac{2m_1m_2}{r}(1+\cos^2\theta)\cos\left[2\omega(t-R/c)\right],$$

$$h_\times = -\frac{1}{R}\frac{G^2}{c^4}\frac{4m_1m_2}{r}(\cos\theta)\sin\left[2\omega(t-R/c)\right].$$

Here, we use the constant angular velocity of a circular orbit in Newtonian physics:

$$\omega = \sqrt{G(m_1+m_2)/r^3}.$$

For example, if the observer is in the x-y plane then $\theta = \pi/2$, and $\cos(\theta) = 0$, so the $h_\times$ polarization is always zero. We also see that the frequency of the wave given off is twice the rotation frequency. If we put in numbers for the Earth–Sun system, we find:

$$h_+ = -\frac{1}{R}\frac{G^2}{c^4}\frac{4m_1m_2}{r} = -\frac{1}{R}1.7\times10^{-10}\,\text{m}.$$

In this case, the minimum distance to find waves is $R \approx 1/4\pi$ light-year, so typical amplitudes will be $h \approx 10^{-25}$. That is, a ring of particles would stretch or squeeze by just one part in 10^{25}. This is well under the detectability limit of all conceivable detectors.

Advanced Mathematics

Spacetime curvature is also expressed with respect to a covariant derivative, ∇, in the form of the Einstein tensor, $G_{\mu\nu}$. This curvature is related to the stress–energy tensor, $T_{\mu\nu}$, by the key equation

$$G_{\mu\nu} = \frac{8\pi G_N}{c^4}T_{\mu\nu},$$

where G_N is Newton's gravitational constant, and c is the speed of light. We assume geometrized units, so $G_N = 1 = c.$.

With some simple assumptions, Einstein's equations can be rewritten to show explicitly that they are wave equations. To begin with, we adopt some coordinate system, like (t, r, θ, ϕ). We define the flat-space metric $\eta_{\mu\nu}$ to be the quantity that — in this coordinate system — has the components we would expect for the flat space metric. For example, in these spherical coordinates, we have

$$\eta_{\mu\nu} = \begin{bmatrix} -1 & 0 & 0 & 0 \\ 0 & 1 & 0 & 0 \\ 0 & 0 & r^2 & 0 \\ 0 & 0 & 0 & r^2 \sin^2 \theta \end{bmatrix}.$$

This flat-space metric has no physical significance; it is a purely mathematical device necessary for the analysis. Tensor indices are raised and lowered using this "flat-space metric".

Now, we can also think of the physical metric $g_{\mu\nu}$ as a matrix, and find its determinant, $\det g$. Finally, we define a quantity

$$\overline{h}^{\alpha\beta} \equiv \eta^{\alpha\beta} - g^{\alpha\beta} \sqrt{|\det g|}.$$

This is the crucial field, which will represent the radiation. It is possible (at least in an asymptotically flat spacetime) to choose the coordinates in such a way that this quantity satisfies the de Donder gauge condition (conditions on the coordinates):

$$\nabla_\beta \overline{h}^{\alpha\beta} = 0,$$

Where ∇ represents the flat-space derivative operator. These equations say that the divergence of the field is zero. The linear Einstein equations can now be written as

$$\Box \overline{h}^{\alpha\beta} = -16\pi \tau^{\alpha\beta},$$

where $\Box = -\partial_t^2 + \Delta$ represents the flat-space d'Alembertian operator, and $\tau^{\alpha\beta}$ represents the stress–energy tensor plus quadratic terms involving $\overline{h}^{\alpha\beta}$. This is just a wave equation for the field with a source, despite the fact that the source involves terms quadratic in the field itself. That is, it can be shown that solutions to this equation are waves traveling with velocity 1 in these coordinates.

Linear Approximation

The equations above are valid everywhere — near a black hole, for instance. However, because of the complicated source term, the solution is generally too difficult to find analytically. We can often assume that space is nearly flat, so the metric is nearly equal to the $\eta^{\alpha\beta}$ tensor. In this case, we can neglect terms quadratic in $\overline{h}^{\alpha\beta}$, which means that the $\tau^{\alpha\beta}$ field reduces to the usual stress–energy tensor $T^{\alpha\beta}$. That is, Einstein's equations become

$$\Box \overline{h}^{\alpha\beta} = -16\pi T^{\alpha\beta}.$$

If we are interested in the field far from a source, however, we can treat the source as a point source; everywhere else, the stress–energy tensor would be zero, so

$$\Box \overline{h}^{\alpha\beta} = 0 \ .$$

Now, this is the usual homogeneous wave equation — one for each component of $\overline{h}^{\alpha\beta}$. Solutions to this equation are well known. For a wave moving away from a point source, the radiated part (meaning the part that dies off as $1/r$ far from the source) can always be written in the form $A(t-r,\theta,\phi)/r$, where A is just some function. It can be shown that — to a linear approximation — it is always possible to make the field traceless. Now, if we further assume that the source is positioned at $r=0$, the general solution to the wave equation in spherical coordinates is

$$\overline{h}^{\alpha\beta} = \frac{1}{r}\begin{bmatrix} 0 & 0 & 0 & 0 \\ 0 & 0 & 0 & 0 \\ 0 & 0 & A_+(t-r,\theta,\phi) & A_\times(t-r,\theta,\phi) \\ 0 & 0 & A_\times(t-r,\theta,\phi) & -A_+(t-r,\theta,\phi) \end{bmatrix}$$

$$\equiv \begin{bmatrix} 0 & 0 & 0 & 0 \\ 0 & 0 & 0 & 0 \\ 0 & 0 & h_+(t-r,r,\theta,\phi) & h_\times(t-r,r,\theta,\phi) \\ 0 & 0 & h_\times(t-r,r,\theta,\phi) & -h_+(t-r,r,\theta,\phi) \end{bmatrix},$$

where we now see the origin of the two polarizations.

Relation to the Source

If we know the details of a source — for instance, the parameters of the orbit of a binary — we can relate the source's motion to the gravitational radiation observed far away. With the relation

$$\Box \overline{h}^{\alpha\beta} = -16\pi \tau^{\alpha\beta} \ ,$$

we can write the solution in terms of the tensorial Green's function for the d'Alembertian operator:

$$\overline{h}^{\alpha\beta}(t,\vec{x}) = -16\pi \int G^{\alpha\beta}_{\gamma\delta}(t,\vec{x};t',\vec{x}')\tau^{\gamma\delta}(t',\vec{x}')\mathrm{d}t'\mathrm{d}^3x' \ .$$

Though it is possible to expand the Green's function in tensor spherical harmonics, it is easier to simply use the form

$$G^{\alpha\beta}_{\gamma\delta}(t,\vec{x};t',\vec{x}') = \frac{1}{4\pi}\delta^\alpha_\gamma \delta^\beta_\delta \frac{\delta(t \pm |\vec{x}-\vec{x}'|-t')}{|\vec{x}-\vec{x}'|},$$

where the positive and negative signs correspond to ingoing and outgoing solutions, respectively. Generally, we are interested in the outgoing solutions, so

$$\overline{h}^{\alpha\beta}(t,\vec{x}) = -4\int \frac{\tau^{\alpha\beta}(t-|\vec{x}-\vec{x}'|,\vec{x}')}{|\vec{x}-\vec{x}'|}\mathrm{d}^3x' \ .$$

If the source is confined to a small region very far away, to an excellent approximation we have:

$$\overline{h}^{\alpha\beta}(t,\vec{x}) \approx -\frac{4}{r}\int \tau^{\alpha\beta}(t-r,\vec{x}')\mathrm{d}^3 x' ,$$

where $r = |\vec{x}|$.

Now, because we will eventually only be interested in the spatial components of this equation (time components can be set to zero with a coordinate transformation), and we are integrating this quantity — presumably over a region of which there is no boundary — we can put this in a different form. Ignoring divergences with the help of Stokes' theorem and an empty boundary

$$\int \tau^{ij}(t-r,\vec{x}')\mathrm{d}^3 x' = \int x'^i x'^j \nabla_k \nabla_l \tau^{kl}(t-r,\vec{x}')\mathrm{d}^3 x'.$$

Inserting this into the above equation, we arrive at

$$\overline{h}^{ij}(t,\vec{x}) \approx -\frac{4}{r}\int x'^i x'^j \nabla_k \nabla_l \tau^{kl}(t-r,\vec{x}')\mathrm{d}^3 x'.$$

Finally, because we have chosen to work in coordinates for which $\nabla_\beta \overline{h}^{\alpha\beta} = 0$, we know that $\nabla_\beta \tau^{\alpha\beta} = 0$. With a few simple manipulations, we can use this to prove that

$$\nabla_0 \nabla_0 \tau^{00} = \nabla_j \nabla_k \tau^{jk}.$$

With this relation, the expression for the radiated field is

$$\overline{h}^{ij}(t,\vec{x}) \approx -\frac{4}{r}\frac{\mathrm{d}^2}{\mathrm{d}t^2}\int x'^i x'^j \tau^{00}(t-r,\vec{x}')\mathrm{d}^3 x'.$$

In the linear case, $\tau^{00} = \rho$, the density of mass-energy.

To a very good approximation, the density of a simple binary can be described by a pair of delta-functions, which eliminates the integral. Explicitly, if the masses of the two objects are M_1 and M_2, and the positions are $\vec{x}_1$ and $\vec{x}_2$, then

$$\rho(t-r,\vec{x}') = M_1 \delta^3(\vec{x}' - \vec{x}_1(t-r)) + M_2 \delta^3(\vec{x}' - \vec{x}_2(t-r)).$$

We can use this expression to do the integral above:

$$\overline{h}^{ij}(t,\vec{x}) \approx -\frac{4}{r}\frac{\mathrm{d}^2}{\mathrm{d}t^2}\left\{M_1 x_1^i(t-r)x_1^j(t-r) + M_2 x_2^i(t-r)x_2^j(t-r)\right\}.$$

Using mass-centered coordinates, and assuming a circular binary, this is

$$\overline{h}^{ij}(t,\vec{x}) \approx -\frac{4}{r}\frac{M_1 M_2}{R}n^i(t-r)n^j(t-r),$$

where $\vec{n} = \vec{x}_1/|\vec{x}_1|$. Plugging in the known values of $\vec{x}_1(t-r)$, we obtain the expressions given above for the radiation from a simple binary.

In Fiction

An episode of the Russian science-fiction novel *Space Apprentice* by Arkady and Boris Strugatsky shows the experiment monitoring the propagation of gravitational waves at the expense of annihilating a chunk of asteroid 15 Eunomia the size of Everest. In the children's book Marie wants to be an Astronaut colliding black holes are featured.

Alternatives to General Relativity

Alternatives to general relativity are physical theories that attempt to describe the phenomena of gravitation in competition to Einstein's theory of general relativity.

There have been many different attempts at constructing an ideal theory of gravity. These attempts can be split into four broad categories:

- Straightforward alternatives to general relativity (GR), such as the Cartan, Brans–Dicke and Rosen bimetric theories.

- Those that attempt to construct a quantized gravity theory such as loop quantum gravity.

- Those that attempt to unify gravity and other forces such as Kaluza–Klein.

- Those that attempt to do several at once, such as M-theory.

Motivations

Motivations for developing new theories of gravity have changed over the years, with the first one to explain planetary orbits (Newton) and more complicated orbits (e.g. Lagrange). Then came unsuccessful attempts to combine gravity and either wave or corpuscular theories of gravity. The whole landscape of physics was changed with the discovery of Lorentz transformations, and this led to attempts to reconcile it with gravity. At the same time, experimental physicists started testing the foundations of gravity and relativity – Lorentz invariance, the gravitational deflection of light, the Eötvös experiment. These considerations led to and past the development of general relativity.

After that, motivations differ. Two major concerns were the development of quantum theory and the discovery of the strong and weak nuclear forces. Attempts to quantize and unify gravity are outside the scope of this article, and so far none has been completely successful.

After general relativity (GR), attempts were made either to improve on theories developed before GR, or to improve GR itself. Many different strategies were attempted, for example the addition of spin to GR, combining a GR-like metric with a space-time that is static with respect to the expansion of the universe, getting extra freedom by adding another parameter. At least one theory was motivated by the desire to develop an alternative to GR that is completely free from singularities.

Experimental tests improved along with the theories. Many of the different strategies that were developed soon after GR were abandoned, and there was a push to develop more general forms of the theories that survived, so that a theory would be ready the moment any test showed a disagreement with GR.

By the 1980s, the increasing accuracy of experimental tests had all led to confirmation of GR, no competitors were left except for those that included GR as a special case. Further, shortly after that, theorists switched to string theory which was starting to look promising, but has since lost popularity. In the mid-1980s a few experiments were suggesting that gravity was being modified by the addition of a fifth force (or, in one case, of a fifth, sixth and seventh force) acting on the scale of meters. Subsequent experiments eliminated these.

Motivations for the more recent alternative theories are almost all cosmological, associated with or replacing such constructs as "inflation", "dark matter" and "dark energy". Investigation of the Pioneer anomaly has caused renewed public interest in alternatives to General Relativity.

Notation in this Article

c is the speed of light, G is the gravitational constant. "Geometric variables" are not used.

Latin indexes go from 1 to 3, Greek indexes go from 0 to 3. The Einstein summation convention is used.

$\eta_{\mu\nu}$ is the Minkowski metric. $g_{\mu\nu}$ is a tensor, usually the metric tensor. These have signature $(-,+,+,+)$.

Partial differentiation is written $\partial_\mu \phi$ or $\phi_{,\mu}$. Covariant differentiation is written $\phi_{,\mu}$ or $\phi_{;\mu}$.

Classification of Theories

Theories of gravity can be classified, loosely, into several categories. Most of the theories described here have:

an 'action'

a Lagrangian density

a metric

If a theory has a Lagrangian density for gravity, say L, then the gravitational part of the action S is the integral of that.

$$S = \int L \sqrt{-g}\, \mathrm{d}^4 x$$

In this equation it is usual, though not essential, to have $g = -1$ at spatial infinity when using Cartesian coordinates. For example, the Einstein–Hilbert action uses

$$L \propto R$$

where R is the scalar curvature, a measure of the curvature of space.

Almost every theory described in this article has an action. It is the only known way to guarantee that the necessary conservation laws of energy, momentum and angular momentum are incorporated automatically; although it is easy to construct an action where those conservation laws are violated. The original 1983 version of MOND did not have an action.

A few theories have an action but not a Lagrangian density. A good example is Whitehead (1922), the action there is termed non-local.

A theory of gravity is a "metric theory" if and only if it can be given a mathematical representation in which two conditions hold:*Condition 1*: There exists a symmetric metric tensor $g_{\mu\nu}$ of signature (−, +, +, +), which governs proper-length and proper-time measurements in the usual manner of special and general relativity:

$$d\tau^2 = -g_{\mu\nu}\,dx^\mu\,dx^\nu$$

where there is a summation over indices μ and ν.*Condition 2*: Stressed matter and fields being acted upon by gravity respond in accordance with the equation:

$$0 = \nabla_\nu T^{\mu\nu} = T^{\mu\nu}{}_{,\nu} + \Gamma^\mu_{\sigma\nu}T^{\sigma\nu} + \Gamma^\nu_{\sigma\nu}T^{\mu\sigma}$$

where $T^{\mu\nu}$ is the stress–energy tensor for all matter and non-gravitational fields, and where ∇_ν is the covariant derivative with respect to the metric and $\Gamma^\alpha_{\sigma\nu}$ is the Christoffel symbol. The stress–energy tensor should also satisfy an energy condition.

Metric theories include (from simplest to most complex):

- Scalar field theories (includes Conformally flat theories & Stratified theories with conformally flat space slices)
    - Bergman
    - Coleman
    - Einstein (1912)
    - Einstein–Fokker theory
    - Lee–Lightman–Ni
    - Littlewood
    - Ni
    - Nordström's theory of gravitation (first metric theory of gravity to be developed)
    - Page–Tupper
    - Papapetrou
    - Rosen (1971)
    - Whitrow–Morduch

- Yilmaz theory of gravitation (attempted to eliminate event horizons from the theory.)
- Quasilinear theories (includes Linear fixed gauge)
  - Bollini–Giambiagi–Tiomno
  - Deser–Laurent
  - Whitehead's theory of gravity (intended to use only retarded potentials)
- Tensor theories
  - Einstein's GR
  - Fourth order gravity (allows the Lagrangian to depend on second-order contractions of the Riemann curvature tensor)
  - f(R) gravity (allows the Lagrangian to depend on higher powers of the Ricci scalar)
  - Gauss–Bonnet gravity
  - Lovelock theory of gravity (allows the Lagrangian to depend on higher-order contractions of the Riemann curvature tensor)
  - Infinite derivative theorem gravity
- Scalar-tensor theories
  - Bekenstein
  - Bergmann-Wagoner
  - Brans–Dicke theory (the most well-known alternative to GR, intended to be better at applying Mach's principle)
  - Jordan
  - Nordtvedt
  - Thiry
  - Chameleon
  - Pressuron
- Vector-tensor theories
  - Hellings–Nordtvedt
  - Will–Nordtvedt
- Bimetric theories
  - Lightman–Lee
  - Rastall

- Rosen (1975)
- Other metric theories

Non-metric theories include

- Belinfante–Swihart
- Einstein–Cartan theory (intended to handle spin-orbital angular momentum interchange)
- Kustaanheimo (1967)
- Teleparallelism
- Gauge theory gravity

A word here about Mach's principle is appropriate because a few of these theories rely on Mach's principle (e.g. Whitehead (1922)), and many mention it in passing (e.g. Einstein–Grossmann (1913), Brans–Dicke (1961)). Mach's principle can be thought of a half-way-house between Newton and Einstein. It goes this way:

- Newton: Absolute space and time.
- Mach: The reference frame comes from the distribution of matter in the universe.
- Einstein: There is no reference frame.

So far, all the experimental evidence points to Mach's principle being wrong, but it has not entirely been ruled out.

Early Theories, 1686 to 1916

Newton (1686)

In Newton's (1686) theory (rewritten using more modern mathematics) the density of mass generates a scalar field, the gravitational potential ϕ in joules per kilogram, by

$$\frac{\partial^2 \phi}{\partial x^j \partial x^j} = 4\pi G \rho.$$

Using the Nabla operator ∇ for the gradient and divergence (partial derivatives), this can be conveniently written as:

$$\nabla^2 \phi = 4\pi G \rho.$$

This scalar field governs the motion of a free-falling particle by:

$$\frac{d^2 x^j}{dt^2} = -\frac{\partial \phi}{\partial x^j}.$$

At distance, r, from an isolated mass, M, the scalar field is

$$\phi = -GM / r.$$

The theory of Newton, and Lagrange's improvement on the calculation (applying the variational principle), completely fails to take into account relativistic effects of course, and so can be rejected as a viable theory of gravity. Even so, Newton's theory is thought to be exactly correct in the limit of weak gravitational fields and low speeds and all other theories of gravity need to reproduce Newton's theory in the appropriate limits.

Mechanical explanations (1650–1900)

To explain Newton's theory, some mechanical explanations of gravitation (incl. Le Sage's theory) were created between 1650 and 1900, but they were overthrown because most of them lead to an unacceptable amount of drag, which is not observed. Other models are violating the energy conservation law and are incompatible with modern thermodynamics.

Electrostatic models (1870–1900)

At the end of the 19th century, many tried to combine Newton's force law with the established laws of electrodynamics, like those of Weber, Carl Friedrich Gauss, Bernhard Riemann and James Clerk Maxwell. Those models were used to explain the perihelion advance of Mercury. In 1890, Lévy succeeded in doing so by combining the laws of Weber and Riemann, whereby the speed of gravity is equal to the speed of light in his theory. And in another attempt, Paul Gerber (1898) even succeeded in deriving the correct formula for the Perihelion shift (which was identical to that formula later used by Einstein). However, because the basic laws of Weber and others were wrong (for example, Weber's law was superseded by Maxwell's theory), those hypothesis were rejected. In 1900, Hendrik Lorentz tried to explain gravity on the basis of his Lorentz ether theory and the Maxwell equations. He assumed, like Ottaviano Fabrizio Mossotti and Johann Karl Friedrich Zöllner, that the attraction of opposite charged particles is stronger than the repulsion of equal charged particles. The resulting net force is exactly what is known as universal gravitation, in which the speed of gravity is that of light. But Lorentz calculated that the value for the perihelion advance of Mercury was much too low.

Lorentz-invariant models (1905–1910)

Based on the principle of relativity, Henri Poincaré (1905, 1906), Hermann Minkowski (1908), and Arnold Sommerfeld (1910) tried to modify Newton's theory and to establish a Lorentz invariant gravitational law, in which the speed of gravity is that of light. However, as in Lorentz's model, the value for the perihelion advance of Mercury was much too low.

Einstein (1908, 1912)

Einstein's two part publication in 1912 (and before in 1908) is really only important for historical reasons. By then he knew of the gravitational redshift and the deflection of light. He had realized that Lorentz transformations are not generally applicable, but retained them. The theory states that the speed of light is constant in free space but varies in the presence of matter. The theory was only expected to hold when the source of the gravitational field is stationary. It includes the principle of least action:

$$\delta \int d\tau = 0$$

$$d\tau^2 = -\eta_{\mu\nu} dx^\mu dx^\nu$$

where $\eta_{\mu\nu}$ is the Minkowski metric, and there is a summation from 1 to 4 over indices μ and ν.

Einstein and Grossmann (1913) includes Riemannian geometry and tensor calculus.

$$\delta \int d\tau = 0$$

$$d\tau^2 = -g_{\mu\nu} dx^\mu dx^\nu$$

The equations of electrodynamics exactly match those of GR. The equation

$$T^{\mu\nu} = \rho \frac{dx^\mu}{d\tau} \frac{dx^\nu}{d\tau}$$

is not in GR. It expresses the stress–energy tensor as a function of the matter density.

Abraham (1912)

While this was going on, Abraham was developing an alternative model of gravity in which the speed of light depends on the gravitational field strength and so is variable almost everywhere. Abraham's 1914 review of gravitation models is said to be excellent, but his own model was poor.

Nordström (1912)

The first approach of Nordström (1912) was to retain the Minkowski metric and a constant value of c but to let mass depend on the gravitational field strength ϕ. Allowing this field strength to satisfy

$$\Box\phi = \rho$$

where ρ is rest mass energy and $\Box$ is the d'Alembertian,

$$m = m_0 \exp(\phi / c^2)$$

and

$$-\frac{\partial\phi}{\partial x^\mu} = \dot{u}_\mu + \frac{u_\mu}{c^2 \dot{\phi}}$$

where u is the four-velocity and the dot is a differential with respect to time.

The second approach of Nordström (1913) is remembered as the first logically consistent relativistic field theory of gravitation ever formulated. From (note, notation of Pais (1982) not Nordström):

$$\delta \int \psi d\tau = 0$$

$$d\tau^2 = -\eta_{\mu\nu}dx^\mu dx^\nu$$

where ψ is a scalar field,

$$-\frac{\partial T^{\mu\nu}}{\partial x^\nu} = T\frac{1}{\psi}\frac{\partial \psi}{\partial x_\mu}$$

This theory is Lorentz invariant, satisfies the conservation laws, correctly reduces to the Newtonian limit and satisfies the weak equivalence principle.

Einstein and Fokker (1914)

This theory is Einstein's first treatment of gravitation in which general covariance is strictly obeyed. Writing:

$$\delta\int ds = 0$$

$$ds^2 = g_{\mu\nu}dx^\mu dx^\nu$$

$$g_{\mu\nu} = \psi^2\eta_{\mu\nu}$$

they relate Einstein-Grossmann (1913) to Nordström (1913). They also state:

$$T \propto R.$$

That is, the trace of the stress energy tensor is proportional to the curvature of space.

Einstein (1916, 1917)

This theory is what we now call "general relativity" (included here for comparison). Discarding the Minkowski metric entirely, Einstein gets:

$$\delta\int ds = 0$$

$$ds^2 = g_{\mu\nu}dx^\mu dx^\nu$$

$$R_{\mu\nu} = \frac{8\pi G}{c^4}\left(T_{\mu\nu} - \frac{1}{2}g_{\mu\nu}T\right)$$

which can also be written

$$T^{\mu\nu} = \frac{c^4}{8\pi G}\left(R^{\mu\nu} - \frac{1}{2}g^{\mu\nu}R\right)$$

Five days before Einstein presented the last equation above, Hilbert had submitted a paper containing an almost identical equation. Hilbert was the first to correctly state the Einstein–Hilbert action for GR, which is:

$$S = \frac{c^4}{16\pi G} \int R\sqrt{-g}\ d^4x + S_m$$

where G is Newton's gravitational constant, $R = R_\mu^\mu$ is the Ricci curvature of space, $g = \det(g_{\mu\nu})$ and S_m is the action due to mass.

GR is a tensor theory, the equations all contain tensors. Nordström's theories, on the other hand, are scalar theories because the gravitational field is a scalar. Later in this article you will see scalar-tensor theories that contain a scalar field in addition to the tensors of GR, and other variants containing vector fields as well have been developed recently.

Theories from 1917 to the 1980s

This section includes alternatives to GR published after GR but before the observations of galaxy rotation that led to the hypothesis of "dark matter". Those considered here include:

Theories from 1917 to the 1980s.		
Publication year(s)	Author(s)	Theory type
1922	Whitehead	Quasilinear
1922, 1923	Cartan	Non-metric
1939	Fierz & Pauli	
1943	Birkhov	
1948	Milne	
1948	Thiry	
1954	Papapetrou	Scalar field
1953	Littlewood	Scalar field
1955	Jordan	
1956	Bergman	Scalar field
1957	Belinfante & Swihart	
1958, 1973	Yilmaz	
1961	Brans & Dicke	Scalar-tensor
1960, 1965	Whitrow & Morduch	Scalar field
1966	Kustaanheimo	
1967	Kustaanheimo & Nuotio	
1968	Deser & Laurent	Quasilinear
1968	Page & Tupper	Scalar field
1968	Bergmann	Scalar-tensor
1970	Bollini-Giambiagi-Tiomno	Quasilinear
1970	Nordtveldt	
1970	Wagoner	Scalar-tensor
1971	Rosen	Scalar field

1975	Rosen	Bimetric
1972, 1973	Wei-Tou Ni	Scalar field
1972	Will & Nordtveldt	Vector-tensor
1973	Hellings & Nordtveldt	Vector-tensor
1973	Lightman & Lee	Scalar field
1974	Lee, Lightman & Ni	
1977	Bekenstein	Scalar-tensor
1978	Barker	Scalar-tensor
1979	Rastall	Bimetric

These theories are presented here without a cosmological constant or added scalar or vector potential unless specifically noted, for the simple reason that the need for one or both of these was not recognised before the supernova observations by the Supernova Cosmology Project and High-Z Supernova Search Team. How to add a cosmological constant or quintessence to a theory is discussed under Modern Theories.

Scalar Field Theories

The scalar field theories of Nordström (1912, 1913) have already been discussed. Those of Littlewood (1953), Bergman (1956), Yilmaz (1958), Whitrow and Morduch (1960, 1965) and Page and Tupper (1968) follow the general formula give by Page and Tupper.

According to Page and Tupper (1968), who discuss all these except Nordström (1913), the general scalar field theory comes from the principle of least action:

$$\delta \int f\left(\tfrac{\phi}{c^2}\right) ds = 0$$

where the scalar field is,

$$\phi = GM / r$$

and c may or may not depend on ϕ.

In Nordström (1912),

$$f(\phi / c^2) = \exp(-\phi / c^2), \qquad c = c_\infty$$

In Littlewood (1953) and Bergmann (1956),

$$f(\phi / c^2) = \exp(-\phi / c^2 - (\phi / c^2)^2 / 2), \qquad c = c_\infty$$

In Whitrow and Morduch (1960),

$$f(\phi / c^2) = 1, \qquad c^2 = c_\infty^2 - 2\phi$$

In Whitrow and Morduch (1965),

$$f(\phi / c^2) = \exp(-\phi / c^2), \qquad c^2 = c_\infty^2 - 2\phi$$

In Page and Tupper (1968),

$$f(\phi/c^2) = \phi/c^2 + \alpha(\phi/c^2)^2, \qquad c_\infty^2/c^2 = 1 + 4(\phi/c_\infty^2) + (15 + 2\alpha)(\phi/c_\infty^2)^2$$

Page and Tupper (1968) matches Yilmaz (1958) to second order when $\alpha = -7/2$.

The gravitational deflection of light has to be zero when c is constant. Given that variable c and zero deflection of light are both in conflict with experiment, the prospect for a successful scalar theory of gravity looks very unlikely. Further, if the parameters of a scalar theory are adjusted so that the deflection of light is correct then the gravitational redshift is likely to be wrong.

Ni (1972) summarised some theories and also created two more. In the first, a pre-existing special relativity space-time and universal time coordinate acts with matter and non-gravitational fields to generate a scalar field. This scalar field acts together with all the rest to generate the metric.

The action is:

$$S = \frac{1}{16\pi G}\int d^4x\sqrt{-g}L_\phi + S_m$$

$$L_\phi = \phi R - 2g^{\mu\nu}\partial_\mu\phi\partial_\nu\phi$$

Misner et al. (1973) gives this without the ϕR term. S_m is the matter action.

$$\Box\phi = 4\pi T^{\mu\nu}\left[\eta_{\mu\nu}e^{-2\phi} + \left(e^{2\phi} + e^{-2\phi}\right)\partial_\mu t\partial_\nu t\right]$$

t is the universal time coordinate. This theory is self-consistent and complete. But the motion of the solar system through the universe leads to serious disagreement with experiment.

In the second theory of Ni (1972) there are two arbitrary functions $f(\phi)$ and $k(\phi)$ that are related vto the metric by:

$$ds^2 = e^{-2f(\phi)}dt^2 - e^{2f(\phi)}\left[dx^2 + dy^2 + dz^2\right]$$

$$\eta^{\mu\nu}\partial_\mu\partial_\nu\phi = 4\pi\rho^*k(\phi)$$

Ni (1972) quotes Rosen (1971) as having two scalar fields ϕ and ψ that are related to the metric by:

$$ds^2 = \phi^2 dt^2 - \psi^2\left[dx^2 + dy^2 + dz^2\right]$$

In Papapetrou (1954a) the gravitational part of the Lagrangian is:

$$L_\phi = e^\phi\left(\frac{1}{2}e^{-\phi}\partial_\alpha\phi\partial_\alpha\phi + \frac{3}{2}e^\phi\partial_0\phi\partial_0\phi\right)$$

In Papapetrou (1954b) there is a second scalar field χ. The gravitational part of the Lagrangian is now:

$$L_\phi = e^{\frac{1}{2}(3\phi+\chi)}\left(-\tfrac{1}{2}e^{-\phi}\partial_\alpha\phi\partial_\alpha\phi - e^{-\phi}\partial_\alpha\phi\partial_\chi\phi + \tfrac{3}{2}e^{-\chi}\partial_0\phi\partial_0\phi\right)$$

Bimetric Theories

Bimetric theories contain both the normal tensor metric and the Minkowski metric (or a metric of constant curvature), and may contain other scalar or vector fields.

Rosen (1973, 1975) Bimetric Theory The action is:

$$S = \frac{1}{64\pi G}\int d^4x\sqrt{-\eta}\,\eta^{\mu\nu}g^{\alpha\beta}g^{\gamma\delta}(g_{\alpha\gamma|\mu}g_{\alpha\delta|\nu} - \frac{1}{2}g_{\alpha\beta|\mu}g_{\gamma\delta|\nu}) + S_m$$

where the vertical line "|" denotes covariant derivative with respect to η. The field equations may be written in the form:

$$\Box_\eta g_{\mu\nu} - g^{\alpha\beta}\eta^{\gamma\delta}g_{\mu\alpha|\gamma}g_{\nu\beta|\delta} = -16\pi G\sqrt{g/\eta}(T_{\mu\nu} --g_{\mu\nu}T)$$

Lightman-Lee (1973) developed a metric theory based on the non-metric theory of Belinfante and Swihart (1957a, 1957b). The result is known as BSLL theory. Given a tensor field $B_{\mu\nu}$, $B = B_{\mu\nu}\eta^{\mu\nu}$, and two constants a and f the action is:

$$S = \frac{1}{16\pi G}\int d^4x\sqrt{-\eta}(aB^{\mu\nu|\alpha}B_{\mu\nu|\alpha} + fB_{,\alpha}B^{,\alpha}) + S_m$$

and the stress–energy tensor comes from:

$$a\Box_\eta B^{\mu\nu} + f\eta^{\mu\nu}\Box_\eta B = -4\pi G\sqrt{g/\eta}T^{\alpha\beta}(\partial g_{\alpha\beta}/\partial B_\mu\nu)$$

In Rastall (1979), the metric is an algebraic function of the Minkowski metric and a Vector field. The Action is:

$$S = \frac{1}{16\pi G}\int d^4x\sqrt{-g}F(N)K^{\mu;\nu}K_{\mu;\nu} + S_m$$

where

$$F(N) = -N/(2+N)\ \text{ and }\ N = g^{\mu\nu}K_\mu K_\nu$$

Quasilinear Theories

In Whitehead (1922), the physical metric g is constructed (by Synge) algebraically from the Minkowski metric η and matter variables, so it doesn't even have a scalar field. The construction is:

$$g_{\mu\nu}(x^\alpha) = \eta_{\mu\nu} - 2\int_{\Sigma^-} \frac{y_\mu^- y_\nu^-}{(w^-)^3}[\sqrt{-g}\,\rho u^\alpha d\Sigma_\alpha]^-$$

where the superscript (-) indicates quantities evaluated along the past η light cone of the field point x^α and

$$(y^\mu)^- = x^\mu - (x^\mu)^-, \quad (y^\mu)^-(y_\mu)^- = 0,$$

$$w^- = (y^\mu)^-(u_\mu)^-, \quad (u_\mu) = dx^\mu / d\sigma,$$

$$d\sigma^2 = \eta_{\mu\nu}dx^\mu dx^\nu$$

Nevertheless, the metric construction (from a non-metric theory) using the "length contraction" ansatz is criticised.

Deser and Laurent (1968) and Bollini-Giambiagi-Tiomno (1970) are Linear Fixed Gauge (LFG) theories. Taking an approach from quantum field theory, combine a Minkowski spacetime with the gauge invariant action of a spin-two tensor field (i.e. graviton) $h_{\mu\nu}$ to define

$$g_{\mu\nu} = \eta_{\mu\nu} + h_{\mu\nu}$$

The action is:

$$S = \frac{1}{16\pi G}\int d^4x\sqrt{-\eta}[2h_{|\nu}^{\mu\nu}h_{\mu\lambda}^{|\lambda} - 2h_{|\nu}^{\mu\nu}h_{\lambda|\mu}^\lambda + h_{\nu|\mu}^\nu h_\lambda^{\lambda|\mu} - h^{\mu\nu|\lambda}h_{\mu\nu|\lambda}] + S_m$$

The Bianchi identity associated with this partial gauge invariance is wrong. LFG theories seek to remedy this by breaking the gauge invariance of the gravitational action through the introduction of auxiliary gravitational fields that couple to $h_{\mu\nu}$.

A cosmological constant can be introduced into a quasilinear theory by the simple expedient of changing the Minkowski background to a de Sitter or anti-de Sitter spacetime, as suggested by G. Temple in 1923. Temple's suggestions on how to do this were criticized by C. B. Rayner in 1955.

Tensor Theories

Einstein's general relativity is the simplest plausible theory of gravity that can be based on just one symmetric tensor field (the metric tensor). Others include: Starobinsky (R+R^2) gravity, Gauss–Bonnet gravity, f(R) gravity, and Lovelock theory of gravity.

Starobinsky

Starobinsky gravity, proposed by Alexei Starobinsky has the Lagrangian

$$\mathcal{L} = \sqrt{-g}\left[R + \frac{R^2}{6M^2}\right]$$

and has been used to explain inflation, in the form of Starobinsky inflation.

Gauss-Bonnet

Gauss–Bonnet gravity has the action

$$\mathcal{L} = \sqrt{-g}\left[R^2 - 4R^{\mu\nu}R_{\mu\nu} + R^{\mu\nu\rho\sigma}R_{\mu\nu\rho\sigma}\right]$$

f(r)

f(R) gravity has the action

$$\mathcal{L} = \sqrt{-g}\,f(R)$$

and is a family of theories, each defined by a different function of the Ricci scalar. Starobinsky gravity is actually an $f(R)$ theory.

Lovelock

Lovelock gravity has the action

$$\mathcal{L} = \sqrt{-g}\,\left(\alpha_0 + \alpha_1 R + \alpha_2\left(R^2 + R_{\alpha\beta\mu\nu}R^{\alpha\beta\mu\nu} - 4R_{\mu\nu}R^{\mu\nu}\right) + \alpha_3\mathcal{O}(R^3)\right),$$

and can be thought of as a generalisation of GR.

Scalar-tensor Theories

These all contain at least one free parameter, as opposed to GR which has no free parameters.

Although not normally considered a Scalar-Tensor theory of gravity, the 5 by 5 metric of Kaluza–Klein reduces to a 4 by 4 metric and a single scalar. So if the 5th element is treated as a scalar gravitational field instead of an electromagnetic field then Kaluza–Klein can be considered the progenitor of Scalar-Tensor theories of gravity. This was recognised by Thiry (1948).

Scalar-Tensor theories include Thiry (1948), Jordan (1955), Brans and Dicke (1961), Bergman (1968), Nordtveldt (1970), Wagoner (1970), Bekenstein (1977) and Barker (1978).

The action S is based on the integral of the Lagrangian L_ϕ .

$$S = \frac{1}{16\pi G}\int d^4x \sqrt{-g}L_\phi + S_m$$

$$L_\phi = \phi R - \frac{\omega(\phi)}{\phi}g^{\mu\nu}\partial_\mu\phi\partial_\nu\phi + 2\phi\lambda(\phi)$$

$$S_m = \int d^4x \sqrt{g}G_N L_m$$

$$T^{\mu\nu} \overset{\text{def}}{=} \frac{2}{\sqrt{g}} \frac{\partial S_m}{\partial g_{\mu\nu}}$$

where $\omega(\phi)$ is a different dimensionless function for each different scalar-tensor theory. The function $\lambda(\phi)$ plays the same role as the cosmological constant in GR. G_N is a dimensionless normalization constant that fixes the present-day value of G. An arbitrary potential can be added for the scalar.

The full version is retained in Bergman (1968) and Wagoner (1970). Special cases are:

Nordtvedt (1970), $\lambda = 0$

Since λ was thought to be zero at the time anyway, this would not have been considered a significant difference. The role of the cosmological constant in more modern work is discussed under Cosmological constant.

Brans–Dicke (1961), ω is constant

Bekenstein (1977) Variable Mass Theory Starting with parameters r and q, found from a cosmological solution, $\phi = [1 - qf(\phi)]f(\phi)^{-r}$ determines function f then

$$\omega(\phi) = -\frac{3}{2} - \frac{1}{4} f(\phi)[(1-6q)qf(\phi) - 1][r + (1-r)qf(\phi)]^{-2}$$

Barker (1978) Constant G Theory

$$\omega(\phi) = (4 - 3\phi) / (2\phi - 2)$$

Adjustment of $\omega(\phi)$ allows Scalar Tensor Theories to tend to GR in the limit of $\omega \to \infty$ in the current epoch. However, there could be significant differences from GR in the early universe.

So long as GR is confirmed by experiment, general Scalar-Tensor theories (including Brans–Dicke) can never be ruled out entirely, but as experiments continue to confirm GR more precisely and the parameters have to be fine-tuned so that the predictions more closely match those of GR.

Vector-tensor Theories

Before we start, Will (2001) has said: "Many alternative metric theories developed during the 1970s and 1980s could be viewed as "straw-man" theories, invented to prove that such theories exist or to illustrate particular properties. Few of these could be regarded as well-motivated theories from the point of view, say, of field theory or particle physics. Examples are the vector-tensor theories studied by Will, Nordtvedt and Hellings."

Hellings and Nordtvedt (1973) and Will and Nordtvedt (1972) are both vector-tensor theories. In addition to the metric tensor there is a timelike vector field K_μ. The gravitational action is:

$$S = \frac{1}{16\pi G} \int d^4x \sqrt{-g} [R + \omega K_\mu K^\mu R + \eta K^\mu K^\nu R_{\mu\nu} - \epsilon F_{\mu\nu} F^{\mu\nu} + \tau K_{\mu;\nu} K^{\mu;\nu}] + S_m$$

where ω , η , $\dot{o}$ and τ are constants and

$$F_{\mu\nu} = K_{\nu;\mu} - K_{\mu;\nu}$$

Will and Nordtvedt (1972) is a special case where

$$\omega = \eta = \epsilon = 0 \; ; \; \tau = 1$$

Hellings and Nordtvedt (1973) is a special case where

$$\tau = 0 \; ; \; \epsilon = 1 \; ; \; \eta = -2\omega$$

These vector-tensor theories are semi-conservative, which means that they satisfy the laws of conservation of momentum and angular momentum but can have preferred frame effects. When $\omega = \eta = \epsilon = \tau = 0$ they reduce to GR so, so long as GR is confirmed by experiment, general vector-tensor theories can never be ruled out.

Other Metric Theories

Others metric theories have been proposed; that of Bekenstein (2004) is discussed under Modern Theories.

Non-metric Theories

Cartan's theory is particularly interesting both because it is a non-metric theory and because it is so old. The status of Cartan's theory is uncertain. Will (1981) claims that all non-metric theories are eliminated by Einstein's Equivalence Principle (EEP). Will (2001) tempers that by explaining experimental criteria for testing non-metric theories against EEP. Misner et al. (1973) claims that Cartan's theory is the only non-metric theory to survive all experimental tests up to that date and Turyshev (2006) lists Cartan's theory among the few that have survived all experimental tests up to that date. The following is a quick sketch of Cartan's theory as restated by Trautman (1972).

Cartan (1922, 1923) suggested a simple generalization of Einstein's theory of gravitation. He proposed a model of space time with a metric tensor and a linear "connection" compatible with the metric but not necessarily symmetric. The torsion tensor of the connection is related to the density of intrinsic angular momentum. Independently of Cartan, similar ideas were put forward by Sciama, by Kibble in the years 1958 to 1966, culminating in a 1976 review by Hehl et al.

The original description is in terms of differential forms, but for the present article that is replaced by the more familiar language of tensors (risking loss of accuracy). As in GR, the Lagrangian is made up of a massless and a mass part. The Lagrangian for the massless part is:

$$L = \frac{1}{32\pi G} \Omega_{\nu}^{\mu} g^{\nu\xi} x^{\eta} x^{\zeta} \varepsilon_{\xi\mu\eta\zeta}$$

$$\Omega_{\nu}^{\mu} = d\omega_{\nu}^{\mu} + \omega_{\xi}^{\eta}$$

$$\nabla x^{\mu} = -\omega_{v}^{\mu} x^{v}$$

The ω_{v}^{μ} is the linear connection. $\varepsilon_{\xi\mu\eta\zeta}$ is the completely antisymmetric pseudo-tensor (Levi-Civita symbol) with $\varepsilon_{0123} = \sqrt{-g}$, and $g^{v\xi}$ is the metric tensor as usual. By assuming that the linear connection is metric, it is possible to remove the unwanted freedom inherent in the non-metric theory. The stress–energy tensor is calculated from:

$$T^{\mu v} = \frac{1}{16\pi G}(g^{\mu v}\eta_{\eta}^{\xi} - g^{\xi\mu}\eta_{\eta}^{v} - g^{\xi v}\eta_{\eta}^{\mu})\Omega_{\xi}^{\eta}$$

The space curvature is not Riemannian, but on a Riemannian space-time the Lagrangian would reduce to the Lagrangian of GR.

Some equations of the non-metric theory of Belinfante and Swihart (1957a, 1957b) have already been discussed in the section on bimetric theories.

A distinctively non-metric theory is given by gauge theory gravity, which replaces the metric in its field equations with a pair of gauge fields in flat spacetime. On the one hand, the theory is quite conservative because it is substantially equivalent to Einstein–Cartan theory (or general relativity in the limit of vanishing spin), differing mostly in the nature of its global solutions. On the other hand, it is radical because it replaces differential geometry with geometric algebra.

Modern Theories 1980s to Present

This section includes alternatives to GR published after the observations of galaxy rotation that led to the hypothesis of "dark matter".

There is no known reliable list of comparison of these theories.

Those considered here include: Beckenstein (2004), Moffat (1995), Moffat (2002), Moffat (2005a, b).

These theories are presented with a cosmological constant or added scalar or vector potential.

Motivations

Motivations for the more recent alternatives to GR are almost all cosmological, associated with or replacing such constructs as "inflation", "dark matter" and "dark energy". The basic idea is that gravity agrees with GR at the present epoch but may have been quite different in the early universe.

There was a slow dawning realisation in the physics world that there were several problems inherent in the then big bang scenario, two of these were the horizon problem and the observation that at early times when quarks were first forming there was not enough space on the universe to contain even one quark. Inflation theory was developed to overcome these. Another alternative was constructing an alternative to GR in which the speed of light was larger in the early universe.

The discovery of unexpected rotation curves for galaxies took everyone by surprise. Could there be more mass in the universe than we are aware of, or is the theory of gravity itself wrong? The consensus now is that the missing mass is "cold dark matter", but that consensus was only reached after trying alternatives to general relativity and some physicists still believe that alternative models of gravity might hold the answer.

The discovery of the accelerated expansion of the universe by the supernova surveys led to the rapid reinstatement of Einstein's cosmological constant, and quintessence arrived as an alternative to the cosmological constant. At least one new alternative to GR attempted to explain the supernova surveys' results in a completely different way.

Another observation that sparked recent interest in alternatives to General Relativity is the Pioneer anomaly. It was quickly discovered that alternatives to GR could explain this anomaly. This is now believed to be accounted for by non-uniform thermal radiation.

Cosmological Constant and Quintessence

The cosmological constant Ë is a very old idea, going back to Einstein in 1917. The success of the Friedmann model of the universe in which Ë = 0 led to the general acceptance that it is zero, but the use of a non-zero value came back with a vengeance when data from supernovae indicated that the expansion of the universe is accelerating

First, let's see how it influences the equations of Newtonian gravity and General Relativity.

In Newtonian gravity, the addition of the cosmological constant changes the Newton-Poisson equation from:

$$\nabla^2 \phi = 4\pi\rho\, G;$$

to

$$\nabla^2 \phi - \Lambda\phi = 4\pi\rho\, G;$$

In GR, it changes the Einstein–Hilbert action from

$$S = \frac{1}{16\pi G} \int R\sqrt{-g}\, d^4x + S_m$$

to

$$S = \frac{1}{16\pi G} \int (R - 2\Lambda)\sqrt{-g}\, d^4x + S_m$$

which changes the field equation

$$T^{\mu\nu} = \frac{1}{8\pi G}\left(R^{\mu\nu} - \frac{1}{2}g^{\mu\nu}R\right)$$

to

$$T^{\mu\nu} = \frac{1}{8\pi G}\left(R^{\mu\nu} - \frac{1}{2}g^{\mu\nu}R + g^{\mu\nu}\Lambda\right)$$

In alternative theories of gravity, a cosmological constant can be added to the action in exactly the same way.

The cosmological constant is not the only way to get an accelerated expansion of the universe in alternatives to GR. We've already seen how the scalar potential $\lambda(\phi)$ can be added to scalar tensor theories. This can also be done in every alternative the GR that contains a scalar field ϕ by adding the term $\lambda(\phi)$ inside the Lagrangian for the gravitational part of the action, the L_ϕ part of

$$S = \frac{1}{16\pi G} \int d^4 x \sqrt{-g} L_\phi + S_m$$

Because $\lambda(\phi)$ is an arbitrary function of the scalar field, it can be set to give an acceleration that is large in the early universe and small at the present epoch. This is known as quintessence.

A similar method can be used in alternatives to GR that use vector fields, including Rastall (1979) and vector-tensor theories. A term proportional to

$$K^\mu K^\nu g_{\mu\nu}$$

is added to the Lagrangian for the gravitational part of the action.

Relativistic MOND

The original theory of MOND by Milgrom was developed in 1983 as an alternative to "dark matter". Departures from Newton's law of gravitation are governed by an acceleration scale, not a distance scale. MOND successfully explains the Tully-Fisher observation that the luminosity of a galaxy should scale as the fourth power of the rotation speed. It also explains why the rotation discrepancy in dwarf galaxies is particularly large.

There were several problems with MOND in the beginning.

1. It did not include relativistic effects

2. It violated the conservation of energy, momentum and angular momentum

3. It was inconsistent in that it gives different galactic orbits for gas and for stars

4. It did not state how to calculate gravitational lensing from galaxy clusters.

By 1984, problems 2 and 3 had been solved by introducing a Lagrangian (AQUAL). A relativistic version of this based on scalar-tensor theory was rejected because it allowed waves in the scalar field to propagate faster than light. The Lagrangian of the non-relativistic form is:

$$L = -\frac{a_0^2}{8\pi G} f\left[\frac{|\nabla\phi|^2}{a_0^2}\right] - \rho\phi$$

The relativistic version of this has:

$$L = -\frac{a_0^2}{8\pi G} \tilde{f}\left(l_0^2 g^{\mu\nu} \partial_\mu\phi\partial_\nu\phi\right)$$

with a nonstandard mass action. Here f and $\tilde{f}$ are arbitrary functions selected to give Newtonian and MOND behaviour in the correct limits, and $l_0 = c^2 / a_0$ is the MOND length scale.

By 1988, a second scalar field (PCC) fixed problems with the earlier scalar-tensor version but is in conflict with the perihelion precession of Mercury and gravitational lensing by galaxies and clusters.

By 1997, MOND had been successfully incorporated in a stratified relativistic theory [Sanders], but as this is a preferred frame theory it has problems of its own.

Bekenstein (2004) introduced a tensor-vector-scalar model (TeVeS). This has two scalar fields ϕ and σ and vector field U_α. The action is split into parts for gravity, scalars, vector and mass.

$$S = S_g + S_s + S_v + S_m$$

The gravity part is the same as in GR.

$$S_s = -\frac{1}{2}\int\left[\sigma^2 h^{\alpha\beta}\phi_{,\alpha}\phi_{,\beta} + \frac{1}{2}Gl_0^{-2}\sigma^4 F(kG\sigma^2)\right]\sqrt{-g}\,d^4x$$

$$S_v = -\frac{K}{32\pi G}\int\left[g^{\alpha\beta}g^{\mu\nu}U_{[\alpha,\mu]}U_{[\beta,\nu]} - \frac{2\lambda}{K}\left(g^{\mu\nu}U_\mu U_\nu + 1\right)\right]\sqrt{-g}\,d^4x$$

$$S_m = \int L\left(\tilde{g}_{\mu\nu}, f^\alpha, \tilde{f}^\alpha_{|\mu}, \cdots\right)\sqrt{-g}\,d^4x$$

where

$$h^{\alpha\beta} = g^{\alpha\beta} - U^\alpha U^\beta$$

$$\tilde{g}^{\alpha\beta} = e^{2\phi}g^{\alpha\beta} + 2U^\alpha U^\beta \sinh(2\phi)$$

k, K are constants, square brackets in indices $U_{[\alpha,\mu]}$ represent anti-symmetrization, λ is a Lagrange multiplier (calculated elsewhere), and L is a Lagrangian translated from flat spacetime onto the metric $\tilde{g}^{\alpha\beta}$. Note that G need not equal the observed gravitational constant G_{Newton}.

F is an arbitrary function, and

$$F(\mu) = \frac{3}{4}\frac{\mu^2(\mu-2)^2}{1-\mu}$$

is given as an example with the right asymptotic behaviour; note how it becomes undefined when $\mu = 1$

The PPN parameters of this theory are calculated in, which shows that all its parameters are equal to GR's, except for

$$\alpha_1 = \frac{4G}{K}\left((2K-1)e^{-4\phi_0} - e^{4\phi_0} + 8\right) - 8$$

$$\alpha_2 = \frac{6G}{2-K} - \frac{2G(K+4)e^{4\phi_0}}{(2-K)^2} - 1$$

both of which expressed in geometric units where $c = G_{Newtonian} = 1$; so

$$G^{-1} = \frac{2}{2-K} + \frac{k}{4\pi}.$$

The parameter ϕ_0 measures the value of the scalar field ϕ at infinity, and is given by

$$\frac{K}{2-K} = e^{-4\phi_0} - 1.$$

Milgrom proposed a "bimetric MOND" or "BIMOND" theory, with action

$$S - S_M - \hat{S}_M = -\frac{c^4}{16\pi G}\int\left[\beta g^{1/2}R + \alpha \hat{g}^{1/2}\hat{R} - 2(g\hat{g})^{1/4}f(\kappa)l_0^{-2}\mathcal{M}\left(l_0^m Y^{(m)}\right)\right]d^4x$$

with S_M and $\hat{S}_M$ the (noninteracting) matter actions attached to the two metrics, Y a tensor derived from the difference in the metrics' connections, $\kappa = (g/\hat{g})^{\bar{4}}$ the ratio between the two metric traces, and α, β are free parameters. $\mathcal{M}$ is a function which depends on some contractions of the Y tensors.

Assuming that $\mathcal{M}$ depends only on the scalar contraction of Y , Milgrom obtained as a nonrelativistic limit his bi-potential version of MOND with action

$$S - S_M = -\frac{1}{8\pi G}\int\left[\beta\left(\nabla\phi\right)^2 + \alpha\left(\nabla\hat{\phi}\right)^2 - a_0^2\mathcal{M}\left(\left(\nabla\phi - \nabla\hat{\phi}\right)^2 a_0^{-2}\right)\right]d^4x$$

$$S_M = \rho(v^2/2 - \phi)$$

Here $\mathcal{M}(z)$ should scale as $z^{-1/4}$ in the deep-MOND limit and as z in the Newtonian limit.

Moffat's Theories

J. W. Moffat (1995) developed a non-symmetric gravitation theory (NGT). This is not a metric theory. It was first claimed that it does not contain a black hole horizon, but Burko and Ori (1995) have found that NGT can contain black holes. Later, Moffat claimed that it has also been applied to explain rotation curves of galaxies without invoking "dark matter". Damour, Deser & MaCarthy (1993) have criticised NGT, saying that it has unacceptable asymptotic behaviour.

The mathematics is not difficult but is intertwined so the following is only a brief sketch. Starting with a non-symmetric tensor , the Lagrangian density is split into

$$L = L_R + L_M$$

where L_M is the same as for matter in GR.

$$L_R = \sqrt{-g}\left[R(W) - 2\lambda - \frac{1}{4}\mu^2 g^{\mu\nu} g_{[\mu\nu]}\right] - \frac{1}{6}g^{\mu\nu}W_\mu W_\nu$$

where $R(W)$ is a curvature term analogous to but not equal to the Ricci curvature in GR, λ and μ^2 are cosmological constants, $g_{[\nu\mu]}$ is the antisymmetric part of $g_{\nu\mu}$. W_μ is a connection, and is a bit difficult to explain because it's defined recursively. However, $W_\mu \approx - g_{[\mu\nu]}$

Moffat's (2002) theory is a scalar-tensor bimetric gravity theory (BGT) and is one of the many theories of gravity in which the speed of light is faster in the early universe. These theories were motivated partly be the desire to avoid the "horizon problem" without invoking inflation. It has a variable G. The theory also attempts to explain the dimming of supernovae from a perspective other than the acceleration of the universe and so runs the risk of predicting an age for the universe that is too small.

Moffat's (2005a) metric-skew-tensor-gravity (MSTG) theory is able to predict rotation curves for galaxies without either dark matter or MOND, and claims that it can also explain gravitational lensing of galaxy clusters without dark matter. It has variable G, increasing to a final constant value about a million years after the big bang.

The theory seems to contain an asymmetric tensor $A_{\mu\nu}$ field and a source current J_μ vector. The action is split into:

$$S = S_G + S_F + S_{FM} + S_M$$

Both the gravity and mass terms match those of GR with cosmological constant. The skew field action and the skew field matter coupling are:

$$S_F = \int d^4x\sqrt{-g}\left(\frac{1}{12}F_{\mu\nu\rho}F^{\mu\nu\rho} - \frac{1}{4}\mu^2 A_{\mu\nu}A^{\mu\nu}\right)$$

$$S_{FM} = \int d^4x\epsilon^{\alpha\beta\mu\nu}A_{\alpha\beta}\partial_\mu J_\nu$$

where

$$F_{\mu\nu\rho} = \partial_\mu A_{\nu\rho} + \partial_\rho A_{\mu\nu}$$

and $\partial^{\alpha\beta\mu\nu}$ is the Levi-Civita symbol. The skew field coupling is a Pauli coupling and is gauge invariant for any source current. The source current looks like a matter fermion field associated with baryon and lepton number.

Moffat (2005b) Scalar-tensor-vector gravity (SVTG) theory.

The theory contains a tensor, vector and three scalar fields. But the equations are quite straightforward. The action is split into: $S = S_G + S_K + S_S + S_M$ with terms for gravity, vector field K_μ, scalar fields G , ω & μ , and mass. S_G is the standard gravity term with the exception that G is moved inside the integral.

$$S_K = -\int d^4x\sqrt{-g}\,\omega\left(\frac{1}{4}B_{\mu\nu}B^{\mu\nu} + V(K)\right)$$

where $B_{\mu\nu} = \partial_\mu K_\nu - \partial_\nu K_\mu$

$$S_S = -\int d^4x\sqrt{-g}\,\frac{1}{G^3}\left(\frac{1}{2}g^{\mu\nu}\nabla_\mu G\nabla_\nu G - V(G)\right)$$
$$+ \frac{1}{G}\left(\frac{1}{2}g^{\mu\nu}\nabla_\mu\omega\nabla_\nu\omega - V(\omega)\right) + \frac{1}{\mu^2 G}\left(\frac{1}{2}g^{\mu\nu}\nabla_\mu\mu\nabla_\nu\mu - V(\mu)\right)$$

The potential function for the vector field is chosen to be:

$$V(K) = -\frac{1}{2}\mu^2\phi^\mu\phi_\mu - \frac{1}{4}g(\phi^\mu\phi_\mu)^2$$

where g is a coupling constant. The functions assumed for the scalar potentials are not stated.

Infinite Derivative Gravity

In order to remove ghosts in the modified propagator, as well as to obtain asymptotic freedom, Biswas, Mazumdar and Siegel (2005) considered a string-inspired infinite set of higher derivative terms

$$S = \int d^4x\sqrt{-g}\left(\frac{R}{2} + RF_1(\Box)R + R^{\mu\nu}F_2(\Box)R_{\mu\nu} + C^{\mu\nu\lambda\sigma}F_3(\Box)C_{\mu\nu\lambda\sigma}\right)$$

where each $F_i(\Box)$ is the exponential of an entire function of the D'Alembertian operator and $C^{\mu\nu\lambda\sigma}$ is the Weyl tensor. This avoids a black hole singularity near the origin, while recovering the 1/r fall of the GR potential at large distances.

Testing of Alternatives to General Relativity

Any putative alternative to general relativity would need to meet a variety of tests for it to become accepted. Most such tests can be categorized as in the following subsections.

Self-consistency

Self-consistency among non-metric theories includes eliminating theories allowing tachyons, ghost poles and higher order poles, and those that have problems with behaviour at infinity.

Among metric theories, self-consistency is best illustrated by describing several theories that fail this test. The classic example is the spin-two field theory of Fierz and Pauli (1939); the field equations imply that gravitating bodies move in straight lines, whereas the equations of motion insist that gravity deflects bodies away from straight line motion. Yilmaz (1971, 1973) contains a tensor gravitational field used to construct a metric; it is mathematically inconsistent because the functional dependence of the metric on the tensor field is not well defined.

Completeness

To be complete, a theory of gravity must be capable of analysing the outcome of every experiment of interest. It must therefore mesh with electromagnetism and all other physics. For instance, any theory that cannot predict from first principles the movement of planets or the behaviour of atomic clocks is incomplete.

Many early theories are incomplete in that it is unclear whether the density ρ used by the theory should be calculated from the stress–energy tensor T as $\rho = T_{\mu\nu}u^{\mu}u^{\nu}$ or as $\rho = T_{\mu\nu}\delta^{\mu\nu}$, where u is the four-velocity, and δ is the Kronecker delta.

The theories of Thirry (1948) and Jordan (1955) are incomplete unless Jordan's parameter η is set to -1, in which case they match the theory of Brans–Dicke (1961) and so are worthy of further consideration.

Milne (1948) is incomplete because it makes no gravitational red-shift prediction.

The theories of Whitrow and Morduch (1960, 1965), Kustaanheimo (1966) and Kustaanheimo and Nuotio (1967) are either incomplete or inconsistent. The incorporation of Maxwell's equations is incomplete unless it is assumed that they are imposed on the flat background space-time, and when that is done they are inconsistent, because they predict zero gravitational redshift when the wave version of light (Maxwell theory) is used, and nonzero redshift when the particle version (photon) is used. Another more obvious example is Newtonian gravity with Maxwell's equations; light as photons is deflected by gravitational fields (by twice that of GR) but light as waves is not.

Classical Tests

There are three "classical" tests (dating back to the 1910s or earlier) of the ability of gravity theories to handle relativistic effects; they are:

- gravitational redshift
- gravitational lensing (generally tested around the Sun)
- anomalous perihelion advance of the planets

Each theory should reproduce the observed results in these areas, which have to date always aligned with the predictions of general relativity.

In 1964, Irwin I. Shapiro found a fourth test, called the Shapiro delay. It is usually regarded as a "classical" test as well.

Agreement with Newtonian Mechanics and Special Relativity

As an example of disagreement with Newtonian experiments, Birkhoff (1943) theory predicts relativistic effects fairly reliably but demands that sound waves travel at the speed of light. This was the consequence of an assumption made to simplify handling the collision of masses.

The Einstein Equivalence Principle (EEP)

The EEP has three components.

The first is the uniqueness of free fall, also known as the Weak Equivalence Principle (WEP). This is satisfied if inertial mass is equal to gravitational mass. η is a parameter used to test the maximum allowable violation of the WEP. The first tests of the WEP were done by Eötvös before 1900 and limited η to less than 5×10^{-9}. Modern tests have reduced that to less than 5×10^{-13}.

The second is Lorentz invariance. In the absence of gravitational effects the speed of light is constant. The test parameter for this is δ. The first tests of Lorentz invariance were done by Michelson and Morley before 1890 and limited δ to less than 5×10^{-3}. Modern tests have reduced this to less than 1×10^{-21}.

The third is local position invariance, which includes spatial and temporal invariance. The outcome of any local non-gravitational experiment is independent of where and when it is performed. Spatial local position invariance is tested using gravitational redshift measurements. The test parameter for this is α. Upper limits on this found by Pound and Rebka in 1960 limited α to less than 0.1. Modern tests have reduced this to less than 1×10^{-4}.

Schiff's conjecture states that any complete, self-consistent theory of gravity that embodies the WEP necessarily embodies EEP. This is likely to be true if the theory has full energy conservation.

Metric theories satisfy the Einstein Equivalence Principle. Extremely few non-metric theories satisfy this. For example, the non-metric theory of Belinfante & Swihart (1957) is eliminated by the $TH\varepsilon\mu$ formalism for testing EEP. Gauge theory gravity is a notable exception, where the strong equivalence principle is essentially the minimal coupling of the gauge covariant derivative.

Parametric Post-Newtonian (PPN) Formalism

Work on developing a standardized rather than ad-hoc set of tests for evaluating alternative gravitation models began with Eddington in 1922 and resulted in a standard set of PPN numbers in Nordtvedt and Will (1972) and Will and Nordtvedt (1972). Each parameter measures a different aspect of how much a theory departs from Newtonian gravity. Because we are talking about deviation from Newtonian theory here, these only measure weak-field effects. The effects of strong gravitational fields are examined later.

These ten are called : γ , β , η , α_1 , α_2 , α_3 , ζ_1 , ζ_2 , ζ_3 , ζ_4

γ is a measure of space curvature, being zero for Newtonian gravity and one for GR.

β is a measure of nonlinearity in the addition of gravitational fields, one for GR.

η is a check for preferred location effects.

α_1 , α_2 , α_3 measure the extent and nature of "preferred-frame effects". Any theory of gravity with at least one α nonzero is called a preferred-frame theory.

ζ_1 , ζ_2 , ζ_3 , ζ_4 , α_3 measure the extent and nature of breakdowns in global conservation laws. A theory of gravity possesses 4 conservation laws for energy-momentum and 6 for angular momentum only if all five are zero.

Strong Gravity and Gravitational Waves

PPN is only a measure of weak field effects. Strong gravity effects can be seen in compact objects such as white dwarfs, neutron stars, and black holes. Experimental tests such as the stability of white dwarfs, spin-down rate of pulsars, orbits of binary pulsars and the existence of a black hole horizon can be used as tests of alternative to GR.

GR predicts that gravitational waves travel at the speed of light. Many alternatives to GR say that gravitational waves travel faster than light. If true, this could result in failure of causality.

Cosmological Tests

Many of these have been developed recently. For those theories that aim to replace dark matter, the galaxy rotation curve, the Tully-Fisher relation, the faster rotation rate of dwarf galaxies, and the gravitational lensing due to galactic clusters act as constraints.

For those theories that aim to replace inflation, the size of ripples in the spectrum of the cosmic microwave background radiation is the strictest test.

For those theories that incorporate or aim to replace dark energy, the supernova brightness results and the age of the universe can be used as tests.

Another test is the flatness of the universe. With GR, the combination of baryonic matter, dark matter and dark energy add up to make the universe exactly flat. As the accuracy of experimental tests improve, alternatives to GR that aim to replace dark matter or dark energy will have to explain why.

Results of Testing Theories

PPN Parameters for a Range of Theories

Misner et al. (1973) gives a table for translating parameters from the notation of Ni to that of Will)

General Relativity is now more than 100 years old, during which one alternative theory of gravity after another has failed to agree with ever more accurate observations. One illustrative example is Parameterized post-Newtonian formalism (PPN).

The following table lists PPN values for a large number of theories. If the value in a cell matches that in the column heading then the full formula is too complicated to include here.

	γ	β	ξ	α_1	α_2	α_3	ζ_1	ζ_2	ζ_3	ζ_4
Einstein (1916) GR	1	1	0	0	0	0	0	0	0	0
Scalar-Tensor theories										
Bergmann (1968), Wagoner (1970)	$\dfrac{1+\omega}{2+\omega}$	β	0	0	0	0	0	0	0	0

Nordtvedt (1970), Bekenstein (1977)	$\frac{1+\omega}{2+\omega}$	β	o	o	o	o	o	o	o	o
Brans–Dicke (1961)	.	1	o	o	o	o	o	o	o	o
Vector-Tensor theories										
Hellings-Nordtvedt (1973)	γ	β	o	α_1	α_2	o	o	o	o	o
Will-Nordtvedt (1972)	1	1	o	o	α_2	o	o	o	o	o
Bimetric theories										
Rosen (1975)	1	1	o	o	c_0/c_1-1	o	o	o	o	o
Rastall (1979)	1	1	o	o	α_2	o	o	o	o	o
Lightman-Lee (1973)	γ	β	o		α_2	o	o	o	o	o
Stratified theories										
Lee-Lightman-Ni (1974)	ac_0/c_1	β	ξ	α_1	α_2	o	o	o	o	o
Ni (1973)	ac_0/c_1 z	bc_0	o	α_1	α_2	o	o	o	o	o
Scalar Field theories										
Einstein (1912) {Not GR}	o	o		-4	o	-2	o	-1	o	o†
Whitrow-Morduch (1965)	o	-1		-4	o	o	o	-3	o	o†
Rosen (1971)	λ	$\frac{3}{4}+\frac{\lambda}{4}$		$-4-4\lambda$	o	-4	o	-1	o	o
Papetrou (1954a, 1954b)	1	1		-8	-4	o	o	2	o	o
Ni (1972) (stratified)	-1	1		-8	o	o	o	2	o	o
Yilmaz (1958, 1962)	1	1		-8	o	-4	o	-2	o	-1†
Page-Tupper (1968)	γ	β		$-4-4\gamma$	o	$-2-2\gamma$	o	ζ_2	o	ζ_4
Nordström (1912)	-1	$\frac{1}{2}$		o	o	o	o	o	o	o†
Nordström (1913), Einstein-Fokker (1914)	-1	$\frac{1}{2}$		o	o	o	o	o	o	o
Ni (1972) (flat)	-1	$1-q$		o	o	o	o	ζ_2	o	o†
Whitrow-Morduch (1960)	-1	$1-q$		o	o	o	o	q	o	o†

Littlewood (1953), Bergman(1956)	-1	$\dfrac{1}{2}$		0	0	0	0	-1	0	0†

† The theory is incomplete, and ζ_4 can take one of two values. The value closest to zero is listed.

All experimental tests agree with GR so far, and so PPN analysis immediately eliminates all the scalar field theories in the table.

A full list of PPN parameters is not available for Whitehead (1922), Deser-Laurent (1968), Bollini-Giambiagi-Tiomino (1970), but in these three cases $\beta = \xi =$, which is in strong conflict with GR and experimental results. In particular, these theories predict incorrect amplitudes for the Earth's tides. (A minor modification of Whitehead's theory avoids this problem. However, the modification predicts the Nordtvedt effect, which has been experimentally constrained.)

Theories that Fail Other Tests

The stratified theories of Ni (1973), Lee Lightman and Ni (1974) are non-starters because they all fail to explain the perihelion advance of Mercury.

The bimetric theories of Lightman and Lee (1973), Rosen (1975), Rastall (1979) all fail some of the tests associated with strong gravitational fields.

The scalar-tensor theories include GR as a special case, but only agree with the PPN values of GR when they are equal to GR to within experimental error. As experimental tests get more accurate, the deviation of the scalar-tensor theories from GR is being squashed to zero.

The same is true of vector-tensor theories, the deviation of the vector-tensor theories from GR is being squashed to zero. Further, vector-tensor theories are semi-conservative; they have a nonzero value for α_2 which can have a measurable effect on the Earth's tides.

Non-metric theories, such as Belinfante and Swihart (1957a, 1957b), usually fail to agree with experimental tests of Einstein's equivalence principle.

And that leaves, as a likely valid alternative to GR, nothing except possibly Cartan (1922).

That was the situation until cosmological discoveries pushed the development of modern alternatives.

History of Gravitational Theory

- Gravitation portal

In physics, theories of gravitation postulate mechanisms of interaction governing the movements of bodies with mass. There have been numerous theories of gravitation since ancient times.

Antiquity

In the 4th century BC, the Greek philosopher Aristotle believed that there is no effect or motion without a cause. The cause of the downward motion of heavy bodies, such as the element earth, was related to their nature, which caused them to move downward toward the center of the universe, which was their natural place. Conversely, light bodies such as the element fire, move by their nature upward toward the inner surface of the sphere of the Moon. Thus in Aristotle's system heavy bodies are not attracted to the earth by an external force of gravity, but tend toward the center of the universe because of an inner *gravitas* or heaviness.

In Book VII of his *De Architectura*, the Roman engineer and architect Vitruvius contends that gravity is not dependent on a substance's "weight" but rather on its "nature" (cf. specific gravity).

If the quicksilver is poured into a vessel, and a stone weighing one hundred pounds is laid upon it, the stone swims on the surface, and cannot depress the liquid, nor break through, nor separate it. If we remove the hundred pound weight, and put on a scruple of gold, it will not swim, but will sink to the bottom of its own accord. Hence, it is undeniable that the gravity of a substance depends not on the amount of its weight, but on its nature.

Brahmagupta, the Indian astronomer and mathematician whose work influenced Arab mathematics in the 9th century, held the view that the earth was spherical and that it attracted objects. Al Hamdānī and Al Biruni quote Brahmagupta saying "Disregarding this, we say that the earth on all its sides is the same; all people on the earth stand upright, and all heavy things fall down to the earth by a law of nature, for it is the nature of the earth to attract and to keep things, as it is the nature of water to flow, that of fire to burn, and that of the wind to set in motion. If a thing wants to go deeper down than the earth, let it try. The earth is the only *low* thing, and seeds always return to it, in whatever direction you may throw them away, and never rise upwards from the earth."

Modern Era

During the 17th century, Galileo found that, counter to Aristotle's teachings, all objects accelerated equally when falling.

In the late 17th century, as a result of Robert Hooke's suggestion that there is a gravitational force which depends on the inverse square of the distance, Isaac Newton was able to mathematically derive Kepler's three kinematic laws of planetary motion, including the elliptical orbits for the six then known planets and the Moon:

> "I deduced that the forces which keep the planets in their orbs must be reciprocally as the squares of their distances from the centres about which they revolve, and thereby compared the force requisite to keep the moon in her orb with the force of gravity at the surface of the earth and found them to answer pretty nearly."

> — *Isaac Newton, 1666*

So Newton's original formula was:

$$\text{Force of gravity} \propto \frac{\text{mass of object 1} \times \text{mass of object 2}}{\text{distance from centers}^2}$$

where the symbol $\propto$ means "is proportional to".

To make this into an equal-sided formula or equation, there needed to be a multiplying factor or constant that would give the correct force of gravity no matter the value of the masses or distance between them. This gravitational constant was first measured in 1797 by Henry Cavendish.

In 1907 Albert Einstein, in what was described by him as "*the happiest thought of my life*", realized that an observer who is falling from the roof of a house experiences no gravitational field. In other words, gravitation was exactly equivalent to acceleration. Between 1911 and 1915 this idea, initially stated as the Equivalence principle, was formally developed into Einstein's theory of general relativity.

Newton's Theory of Gravitation

In 1687, English mathematician Sir Isaac Newton published *Principia*, which hypothesizes the inverse-square law of universal gravitation. In his own words, "I deduced that the forces which keep the planets in their orbs must be reciprocally as the squares of their distances from the centers about which they revolve; and thereby compared the force requisite to keep the Moon in her orb with the force of gravity at the surface of the Earth; and found them answer pretty nearly."

Newton's theory enjoyed its greatest success when it was used to predict the existence of Neptune based on motions of Uranus that could not be accounted by the actions of the other planets. Calculations by John Couch Adams and Urbain Le Verrier both predicted the general position of the planet, and Le Verrier's calculations are what led Johann Gottfried Galle to the discovery of Neptune.

Years later, it was another discrepancy in a planet's orbit that showed Newton's theory to be inaccurate. By the end of the 19th century, it was known that the orbit of Mercury could not be accounted for entirely under Newtonian gravity, and all searches for another perturbing body (such as a planet orbiting the Sun even closer than Mercury) have been fruitless. This issue was resolved in 1915 by Albert Einstein's new general theory of relativity, which accounted for the discrepancy in Mercury's orbit.

Paul Dirac developed the hypothesis that gravitation should have slowly and steadily decreased over the course of the history of the universe.

Although Newton's theory has been superseded, most modern non-relativistic gravitational calculations still use it because it is much easier to work with and is sufficiently accurate for most applications.

Mechanical Explanations of Gravitation

The mechanical theories or explanations of the gravitation are attempts to explain the law of gravity by aid of basic mechanical processes, such as pushes, and without the use of any action at a distance. These theories were developed from the 16th until the 19th century in connection with the aether theories.

René Descartes (1644) and Christiaan Huygens (1690) used vortices to explain gravitation. Robert Hooke (1671) and James Challis (1869) assumed, that every body emits waves which lead to an attraction of other bodies. Nicolas Fatio de Duillier (1690) and Georges-Louis Le Sage (1748) proposed a corpuscular model, using some sort of screening or shadowing mechanism. Later a similar model was created by Hendrik Lorentz, who used electromagnetic radiation instead of the corpuscles. Isaac Newton (1675) and Bernhard Riemann (1853) argued that aether streams carry all bodies to each other. Newton (1717) and Leonhard Euler (1760) proposed a model, in which the aether loses density near the masses, leading to a net force directing to the bodies. Lord Kelvin (1871) proposed that every body pulsates, which might be an explanation of gravitation and the electric charges.

However, those models were overthrown because most of them lead to an unacceptable amount of drag, which is not observed. Other models are violating the energy conservation law and are incompatible with modern thermodynamics.

General Relativity

In general relativity, the effects of gravitation are ascribed to spacetime curvature instead of to a force. The starting point for general relativity is the equivalence principle, which equates free fall with inertial motion. The issue that this creates is that free-falling objects can accelerate with respect to each other. In Newtonian physics, no such acceleration can occur unless at least one of the objects is being operated on by a force (and therefore is not moving inertially).

To deal with this difficulty, Einstein proposed that spacetime is curved by matter, and that free-falling objects are moving along locally straight paths in curved spacetime. (This type of path is called a geodesic). More specifically, Einstein and Hilbert discovered the field equations of general relativity, which relate the presence of matter and the curvature of spacetime and are named after Einstein. The Einstein field equations are a set of 10 simultaneous, non-linear, differential equations. The solutions of the field equations are the components of the metric tensor of spacetime. A metric tensor describes the geometry of spacetime. The geodesic paths for a spacetime are calculated from the metric tensor.

Notable solutions of the Einstein field equations include:

- The Schwarzschild solution, which describes spacetime surrounding a spherically symmetric non-rotating uncharged massive object. For compact enough objects, this solution generated a black hole with a central singularity. For radial distances from the center which are much greater than the Schwarzschild radius, the accelerations predicted by the Schwarzschild solution are practically identical to those predicted by Newton's theory of gravity.

- The Reissner–Nordström solution, in which the central object has an electrical charge. For charges with a geometrized length which are less than the geometrized length of the mass of the object, this solution produces black holes with an event horizon surrounding a Cauchy horizon.

- The Kerr solution for rotating massive objects. This solution also produces black holes with multiple horizons.

- The cosmological Robertson–Walker solution, which predicts the expansion of the universe.

General relativity has enjoyed much success because of the way its predictions of phenomena which are not called for by the older theory of gravity have been regularly confirmed. For example:

- General relativity accounts for the anomalous perihelion precession of the planet Mercury.

- The prediction that time runs slower at lower potentials has been confirmed by the Pound–Rebka experiment, the Hafele–Keating experiment, and the GPS.

- The prediction of the deflection of light was first confirmed by Arthur Eddington in 1919, and has more recently been strongly confirmed through the use of a quasar which passes behind the Sun as seen from the Earth.

- The time delay of light passing close to a massive object was first identified by Irwin Shapiro in 1964 in interplanetary spacecraft signals.

- Gravitational radiation has been indirectly confirmed through studies of binary pulsars. In 2016, the LIGO experiments directly detected gravitational radiation from two colliding black holes, making this the first direct observation of both the gravitational radiation as well as black holes.

- The expansion of the universe (predicted by the Robertson–Walker metric) was confirmed by Edwin Hubble in 1929.

Gravity and Quantum Mechanics

Several decades after the discovery of general relativity it was realized that it cannot be the complete theory of gravity because it is incompatible with quantum mechanics. Later it was understood that it is possible to describe gravity in the framework of quantum field theory like the other fundamental forces. In this framework the attractive force of gravity arises due to exchange of virtual gravitons, in the same way as the electromagnetic force arises from exchange of virtual photons. This reproduces general relativity in the classical limit. However, this approach fails at short distances of the order of the Planck length.

It is notable that in general relativity, gravitational radiation, which under the rules of quantum mechanics must be composed of gravitons, is created only in situations where the curvature of spacetime is oscillating, such as is the case with co-orbiting objects. The amount of gravitational radiation emitted by the solar system is far too small to measure.

However, gravitational radiation has been observed both indirectly, as an energy loss over time in binary pulsar systems such as PSR 1913+16, and directly by the LIGO gravitational wave observatory, whose first detection (named GW150914) occurred on 14 September 2015 and matched theoretical predictions of signals due to the inward spiral and merger of a pair of black holes. It is believed that neutron star mergers and black hole formation may also create detectable amounts of gravitational radiation.

References

- Halliday, David; Robert Resnick; Kenneth S. Krane (2001). Physics v. 1. New York: John Wiley & Sons. ISBN 0-471-32057-9.

- Serway, Raymond A.; Jewett, John W. (2004). Physics for Scientists and Engineers (6th ed.). Brooks/Cole.

ISBN 0-534-40842-7.

- Tipler, Paul (2004). Physics for Scientists and Engineers: Mechanics, Oscillations and Waves, Thermodynamics (5th ed.). W. H. Freeman. ISBN 0-7167-0809-4.

- Purrington, Robert D. (2009). The First Professional Scientist: Robert Hooke and the Royal Society of London. Springer. p. 168. ISBN 3-0346-0036-4.Extract of page 168

- Misner, Charles W.; Thorne, Kip S.; Wheeler, John Archibald (1973). Gravitation. New York: W. H.Freeman and Company. ISBN 0-7167-0344-0 Page 1049.

- Bartusiak, Marcia (2000), Einstein's Unfinished Symphony: Listening to the Sounds of Space-Time, Berkley, ISBN 978-0-425-18620-6

- Berry, Michael V. (1989), Principles of Cosmology and Gravitation (2nd ed.), Institute of Physics Publishing, ISBN 0-85274-037-9

- Bertotti, Bruno (2005), "The Cassini Experiment: Investigating the Nature of Gravity", in Renn, Jürgen, One hundred authors for Einstein, Wiley-VCH, pp. 402–405, ISBN 3-527-40574-7

- Blair, David; McNamara, Geoff (1997), Ripples on a Cosmic Sea. The Search for Gravitational Waves, Perseus, ISBN 0-7382-0137-5

- Greene, Brian (1999), The Elegant Universe: Superstrings, Hidden Dimensions, and the Quest for the Ultimate Theory, Vintage, ISBN 0-375-70811-1

3

Key Concepts of General Relativity

The key concepts of general relativity are spacetime, Galilean invariance, Lorentz covariance, cosmological constant and the wormhole. The mathematical model that incorporates space and time into a single intertwined continuum is known as spacetime whereas Galilean invariance considers the laws of motion to be the same in all frames. The key concepts of general relativity are considered in this chapter.

Spacetime

In physics, spacetime is any mathematical model that combines space and time into a single interwoven continuum. Since 300 BCE, the spacetime of our universe has historically been interpreted from a Euclidean space perspective, which regards space as consisting of three dimensions, and time as consisting of one dimension, the "fourth dimension". By combining space and time into a single manifold called Minkowski space in 1908, physicists have significantly simplified a large number of physical theories, as well as described in a more uniform way the workings of the universe at both the supergalactic and subatomic levels.

Explanation

In non-relativistic classical mechanics, the use of Euclidean space instead of spacetime is appropriate, because time is treated as universal with a constant rate of passage that is independent of the state of motion of an observer. In relativistic contexts, time cannot be separated from the three dimensions of space, because the observed rate at which time passes for an object depends on the object's velocity relative to the observer and also on the strength of gravitational fields, which can slow the passage of time for an object as seen by an observer outside the field.

In cosmology, the concept of spacetime combines space and time to a single abstract universe. Mathematically it is a manifold whose points correspond to physical events. In a local coordinate system whose domain is an open set of the spacetime manifold, three *spacelike coordinates* and one *timelike coordinate* typically emerge. Dimensions are independent components of a coordinate grid needed to locate a point in a certain defined "space". For example, on the globe the latitude and longitude are two independent coordinates which together uniquely determine a location. In spacetime, a coordinate grid that spans the 3+1 dimensions locates events (rather than just points in space), i.e., time is added as another dimension to the coordinate grid. This way the coordinates specify *where* and *when* events occur. However, the unified nature of spacetime and the freedom of coordinate choice it allows, imply that to express the temporal coordinate in one coordinate system requires both temporal and spatial coordinates in another coordinate system. Unlike in normal spatial coordinates, there are still restrictions for how measurements can be made spatially and temporally. These restrictions correspond roughly to a particular mathematical model which differs from Euclidean space in its manifest symmetry.

Until the beginning of the 20th century, time was believed to be independent of motion, progressing at a fixed rate in all reference frames; however, following its prediction by special relativity, later experiments confirmed that time slows at higher speeds of the reference frame relative to another reference frame. Such slowing, called time dilation, is explained in special relativity theory. Many experiments have confirmed time dilation, such as the relativistic decay of muons from cosmic ray showers and the slowing of atomic clocks aboard a Space Shuttle relative to synchronized Earth-bound inertial clocks. The duration of time can therefore vary according to events and reference frames.

When dimensions are understood as mere components of the grid system, rather than physical attributes of space, it is easier to understand the alternate dimensional views as being simply the result of coordinate transformations.

The term *spacetime* has taken on a generalized meaning beyond treating spacetime events with the normal 3+1 dimensions. It is really the combination of space and time. Other proposed spacetime theories include additional dimensions—normally spatial but there exist some speculative theories that include additional temporal dimensions and even some that include dimensions that are neither temporal nor spatial (e.g., superspace). How many dimensions are needed to describe the universe is still an open question. Speculative theories such as string theory predict 10 or 26 dimensions (with M-theory predicting 11 dimensions: 10 spatial and 1 temporal), but the existence of more than four dimensions would only appear to make a difference at the subatomic level.

Spacetime in Literature

Incas regarded space and time as a single concept, referred to as *pacha* (Quechua: *pacha*, Aymara: *pacha*). The peoples of the Andes maintain a similar understanding.

The idea of a unified spacetime is stated by Edgar Allan Poe in his essay on cosmology titled *Eureka* (1848) that "Space and duration are one". In 1895, in his novel *The Time Machine*, H. G. Wells wrote, "There is no difference between time and any of the three dimensions of space except that our consciousness moves along it", and that "any real body must have extension in four directions: it must have Length, Breadth, Thickness, and Duration".

Marcel Proust, in his novel *Swann's Way* (published 1913), describes the village church of his childhood's Combray as "a building which occupied, so to speak, four dimensions of space—the name of the fourth being Time".

Mathematical Concept

In Encyclopedie, published in 1754, under the term *dimension* Jean le Rond d'Alembert speculated that duration (time) might be considered a fourth dimension if the idea was not too novel.

Another early venture was by Joseph Louis Lagrange in his *Theory of Analytic Functions* (1797, 1813). He said, "One may view mechanics as a geometry of four dimensions, and mechanical analysis as an extension of geometric analysis".

The ancient idea of the cosmos gradually was described mathematically with differential equations, differential geometry, and abstract algebra. These mathematical articulations blossomed in the nineteenth century as electrical technology stimulated men like Michael Faraday and James Clerk Maxwell to describe the reciprocal relations of electric and magnetic fields. Daniel Siegel phrased Maxwell's role in relativity as follows:

[...] the idea of the propagation of forces at the velocity of light through the electromagnetic field as described by Maxwell's equations—rather than instantaneously at a distance—formed the necessary basis for relativity theory.

Maxwell used vortex models in his papers on On Physical Lines of Force, but ultimately gave up on any substance but the electromagnetic field. Pierre Duhem wrote:

[Maxwell] was not able to create the theory that he envisaged except by giving up the use of any model, and by extending by means of analogy the abstract system of electrodynamics to displacement currents.

In Siegel's estimation, "this very abstract view of the electromagnetic fields, involving no visualizable picture of what is going on out there in the field, is Maxwell's legacy." Describing the behaviour of electric fields and magnetic fields led Maxwell to view the combination as an electromagnetic field. These fields have a value at every point of spacetime. It is the intermingling of electric and magnetic manifestations, described by Maxwell's equations, that give spacetime its structure. In particular, the rate of motion of an observer determines the electric and magnetic profiles of the electromagnetic field. The propagation of the field is determined by the electromagnetic wave equation, which requires spacetime for description.

Spacetime was described as an affine space with quadratic form in Minkowski space of 1908. In his 1914 textbook The Theory of Relativity, Ludwik Silberstein used biquaternions to represent events in Minkowski space. He also exhibited the Lorentz transformations between observers of differing velocities as biquaternion mappings. Biquaternions were described in 1853 by W. R. Hamilton, so while the physical interpretation was new, the mathematics was well known in English literature, making relativity an instance of applied mathematics.

The first inkling of general relativity in spacetime was articulated by W. K. Clifford. Description of the effect of gravitation on space and time was found to be most easily visualized as a "warp" or stretching in the geometrical fabric of space and time, in a smooth and continuous way that changed smoothly from point-to-point along the spacetime fabric. In 1947 James Jeans provided a concise summary of the development of spacetime theory in his book The Growth of Physical Science.

Basic Concepts

The basic elements of spacetime are events. In any given spacetime, an event is a unique position at a unique time. Because events are spacetime points, an example of an event in classical relativistic physics is (x, y, z, t), the location of an elementary (point-like) particle at a particular time. A spacetime itself can be viewed as the union of all events in the same way that a line is the union of all of its points, formally organized into a manifold, a space which can be described at small scales using coordinate systems.

Spacetime is independent of any observer. However, in describing physical phenomena (which occur at certain moments of time in a given region of space), each observer chooses a convenient metrical coordinate system. Events are specified by four real numbers in any such coordinate system. The trajectories of elementary (point-like) particles through space and time are thus a continuum of events called the world line of the particle. Extended or composite objects (consisting of many elementary particles) are thus a union of many world lines twisted together by virtue of their interactions through spacetime into a "world-braid".

However, in physics, it is common to treat an extended object as a "particle" or "field" with its own unique (e.g., center of mass) position at any given time, so that the world line of a particle or light beam is the path that this particle or beam takes in the spacetime and represents the history of the particle or beam. The world line of the orbit of the Earth (in such a description) is depicted in two spatial dimensions x and y (the plane of the Earth's orbit) and a time dimension orthogonal to x and y. The orbit of the Earth is an ellipse in space alone, but its world line is a helix in spacetime.

The unification of space and time is exemplified by the common practice of selecting a metric (the measure that specifies the interval between two events in spacetime) such that all four dimensions are measured in terms of units of distance: representing an event as $(x_0, x_1, x_2, x_3) = (ct, x, y, z)$ (in the Lorentz metric) or $(x_1, x_2, x_3, x_4) = (x, y, z, ict)$ (in the original Minkowski metric) where c is the speed of light. The metrical descriptions of Minkowski Space and spacelike, lightlike, and timelike intervals given below follow this convention, as do the conventional formulations of the Lorentz transformation.

Spacetime Intervals in Flat Space

In a Euclidean space, the separation between two points is measured by the distance between the two points. The distance is purely spatial, and is always positive. In spacetime, the displacement four-vector ΔR is given by the space displacement vector Δr and the time difference Δt between the events. The *spacetime interval*, also called *invariant interval*, between the two events, s^2, is defined as:

$$s^2 = \Delta r^2 - c^2 \Delta t^2 \quad \text{(spacetime interval)},$$

where c is the speed of light. The choice of signs for s^2 above follows the space-like convention $(-+++)$. Spacetime intervals may be classified into three distinct types, based on whether the temporal separation $(c^2 \Delta t^2)$ is greater than, equal to, or smaller than the spatial separation (Δr^2), corresponding to resp. time-like, light-like, or space-like separated intervals.

Certain types of world lines are called geodesics of the spacetime – straight lines in the case of Minkowski space and their closest equivalent in the curved spacetime of general relativity. In the case of purely time-like paths, geodesics are (locally) the paths of greatest separation (spacetime interval) as measured along the path between two events, whereas in Euclidean space and Riemannian manifolds, geodesics are paths of shortest distance between two points. The concept of geodesics becomes central in general relativity, since geodesic motion may be thought of as "pure motion" (inertial motion) in spacetime, that is, free from any external influences.

Time-like Interval

$$c^2 \Delta t^2 > \Delta r^2$$
$$s^2 < 0$$

For two events separated by a time-like interval, enough time passes between them that there could be a cause–effect relationship between the two events. For a particle traveling through space at less than the speed of light, any two events which occur to or by the particle must be separated by a time-like interval. Event pairs with time-like separation define a negative spacetime interval ($s^2 < 0$) and may be said to occur in each other's future or past. There exists a reference frame such that the two events are observed to occur in the same spatial location, but there is no reference frame in which the two events can occur at the same time.

The measure of a time-like spacetime interval is described by the proper time interval, $\Delta \tau$:

$$\Delta \tau = \sqrt{\Delta t^2 - \frac{\Delta r^2}{c^2}} \quad \text{(proper time interval)}.$$

The proper time interval would be measured by an observer with a clock traveling between the two events in an inertial reference frame, when the observer's path intersects each event as that event occurs. (The proper time interval defines a real number, since the interior of the square root is positive.)

Light-like Interval

$$c^2 \Delta t^2 = \Delta r^2$$
$$s^2 = 0$$

In a light-like interval, the spatial distance between two events is exactly balanced by the time between the two events. The events define a spacetime interval of zero ($s^2 = 0$). Light-like intervals are also known as "null" intervals.

Events which occur to or are initiated by a photon along its path (i.e., while traveling at c, the speed of light) all have light-like separation. Given one event, all those events which follow at light-like intervals define the propagation of a light cone, and all the events which preceded from a light-like interval define a second (graphically inverted, which is to say "*pastward*") light cone.

Space-like Interval

$$c^2 \Delta t^2 < \Delta r^2$$
$$s^2 > 0$$

When a space-like interval separates two events, not enough time passes between their occurrences for there to exist a causal relationship crossing the spatial distance between the two events at the speed of light or slower. Generally, the events are considered not to occur in each other's future or past. There exists a reference frame such that the two events are observed to occur at the same time, but there is no reference frame in which the two events can occur in the same spatial location.

For these space-like event pairs with a positive spacetime interval ($s^2 > 0$), the measurement of space-like separation is the proper distance, $\Delta \sigma$:

$$\Delta\sigma = \sqrt{s^2} = \sqrt{\Delta r^2 - c^2 \Delta t^2} \quad \text{(proper distance)}.$$

Like the proper time of time-like intervals, the proper distance of space-like spacetime intervals is a real number value.

Interval as Area

The interval has been presented as the area of an oriented rectangle formed by two events and iso-tropic lines through them. Time-like or space-like separations correspond to oppositely oriented rectangles, one type considered to have rectangles of negative area. The case of two events separated by light corresponds to the rectangle degenerating to the segment between the events and zero area. The transformations leaving interval-length invariant are the area-preserving squeeze mappings.

The parameters traditionally used rely on quadrature of the hyperbola, which is the natural logarithm. This transcendental function is essential in mathematical analysis as its inverse unites circular functions and hyperbolic functions: The exponential function, e^t, t a real number, used in the hyperbola (e^t, e^{-t}), generates hyperbolic sectors and the hyperbolic angle parameter. The functions cosh and sinh, used with rapidity as hyperbolic angle, provide the

common representation of squeeze in the form $\begin{pmatrix} \cosh\phi & \sinh\phi \\ \sinh\phi & \cosh\phi \end{pmatrix}$, or as the split-complex unit $e^{j\phi} = \cosh\phi + j\sinh\phi$.

Mathematics of Spacetimes

For physical reasons, a spacetime continuum is mathematically defined as a four-dimensional, smooth, connected Lorentzian manifold (M, g). This means the smooth Lorentz metric g has signature $(3,1)$. The metric determines the geometry of spacetime, as well as determining the geodesics of particles and light beams. About each point (event) on this manifold, coordinate charts are used to represent observers in reference frames. Usually, Cartesian coordinates (x, y, z, t) are used. Moreover, for simplicity's sake, units of measurement are usually chosen such that the speed of light   is equal to 1.

A reference frame (observer) can be identified with one of these coordinate charts; any such observer can describe any event p. Another reference frame may be identified by a second coordinate chart about p. Two observers (one in each reference frame) may describe the same event p but obtain different descriptions.

Usually, many overlapping coordinate charts are needed to cover a manifold. Given two coordinate charts, one containing p (representing an observer) and another containing q (representing another observer), the intersection of the charts represents the region of spacetime in which both observers can measure physical quantities and hence compare results. The relation between the two sets of measurements is given by a non-singular coordinate transformation on this intersection. The idea of coordinate charts as local observers who can perform measurements in their vicinity also makes good physical sense, as this is how one actually collects physical data—locally.

For example, two observers, one of whom is on Earth, but the other one who is on a fast rocket to Jupiter, may observe a comet crashing into Jupiter (this is the event p). In general, they will disagree about the exact location and timing of this impact, i.e., they will have different 4-tuples (x, y, z, t) (as they are using different coordinate systems). Although their kinematic descriptions will differ, dynamical (physical) laws, such as momentum conservation and the first law of thermodynamics, will still hold. In fact, relativity theory requires more than this in the sense that it stipulates these (and all other physical) laws must take the same form in all coordinate systems. This introduces tensors into relativity, by which all physical quantities are represented.

Geodesics are said to be time-like, null, or space-like if the tangent vector to one point of the geodesic is of this nature. Paths of particles and light beams in spacetime are represented by time-like and null (light-like) geodesics, respectively.

Topology

The assumptions contained in the definition of a spacetime are usually justified by the following considerations.

The connectedness assumption serves two main purposes. First, different observers making measurements (represented by coordinate charts) should be able to compare their observations on the non-empty intersection of the charts. If the connectedness assumption were dropped, this would not be possible. Second, for a manifold, the properties of connectedness and path-connectedness are equivalent, and one requires the existence of paths (in particular, geodesics) in the spacetime to represent the motion of particles and radiation.

Every spacetime is paracompact. This property, allied with the smoothness of the spacetime, gives rise to a smooth linear connection, an important structure in general relativity. Some important theorems on constructing spacetimes from compact and non-compact manifolds include the following:

- A compact manifold can be turned into a spacetime if, and only if, its Euler characteristic is 0. (Proof idea: the existence of a Lorentzian metric is shown to be equivalent to the existence of a nonvanishing vector field.)

- Any non-compact 4-manifold can be turned into a spacetime.

Spacetime Symmetries

Often in relativity, spacetimes that have some form of symmetry are studied. As well as helping to classify spacetimes, these symmetries usually serve as a simplifying assumption in specialized work. Some of the most popular ones include:

- Axisymmetric spacetimes

- Spherically symmetric spacetimes

- Static spacetimes

- Stationary spacetimes

Causal Structure

The causal structure of a spacetime describes causal relationships between pairs of points in the spacetime based on the existence of certain types of curves joining the points.

Spacetime in Special Relativity

The geometry of spacetime in special relativity is described by the Minkowski metric on R^4. This spacetime is called Minkowski space. The Minkowski metric is usually denoted by η and can be written as a four-by-four matrix:

$$\eta_{ab} = \text{diag}(1, -1, -1, -1)$$

where the Landau–Lifshitz time-like convention is being used. A basic assumption of relativity is that coordinate transformations must leave spacetime intervals invariant. Intervals are invariant under Lorentz transformations. This invariance property leads to the use of four-vectors (and other tensors) in describing physics.

Strictly speaking, one can also consider events in Newtonian physics as a single spacetime. This is Galilean–Newtonian relativity, and the coordinate systems are related by Galilean transformations. However, since these preserve spatial and temporal distances independently, such a spacetime can always be decomposed into spatial coordinates plus temporal coordinates, which is not possible for general spacetimes.

Spacetime in General Relativity

In general relativity, it is assumed that spacetime is curved by the presence of matter (energy), this curvature being represented by the Riemann tensor. In special relativity, the Riemann tensor is identically zero, and so this concept of "non-curvedness" is sometimes expressed by the statement *Minkowski spacetime is flat.*

The earlier discussed notions of time-like, light-like and space-like intervals in special relativity can similarly be used to classify one-dimensional curves through curved spacetime. A time-like curve can be understood as one where the interval between any two infinitesimally close events on the curve is time-like, and likewise for light-like and space-like curves. Technically the three types of curves are usually defined in terms of whether the tangent vector at each point on the curve is time-like, light-like or space-like. The world line of a slower-than-light object will always be a time-like curve, the world line of a massless particle such as a photon will be a light-like curve, and a space-like curve could be the world line of a hypothetical tachyon. In the local neighborhood of any event, time-like curves that pass through the event will remain inside that event's past and future light cones, light-like curves that pass through the event will be on the surface of the light cones, and space-like curves that pass through the event will be outside the light cones. One can also define the notion of a three-dimensional "spacelike hypersurface", a continuous three-dimensional "slice" through the four-dimensional property with the property that every curve that is contained entirely within this hypersurface is a space-like curve.

Many spacetime continua have physical interpretations which most physicists would consider bizarre or unsettling. For example, a compact spacetime has closed timelike curves, which violate our usual ideas of causality (that is, future events could affect past ones). For this reason, mathematical physicists usually consider only restricted subsets of all the possible spacetimes. One way to do this is to study "realistic" solutions of the equations of general relativity. Another way is to add some additional "physically reasonable" but still fairly general geometric restrictions and try to prove interesting things about the resulting spacetimes. The latter approach has led to some important results, most notably the Penrose–Hawking singularity theorems.

Quantized Spacetime

In general relativity, spacetime is assumed to be smooth and continuous—and not just in the mathematical sense. In the theory of quantum mechanics, there is an inherent discreteness present in physics. In attempting to reconcile these two theories, it is sometimes postulated that spacetime should be quantized at the very smallest scales. Current theory is focused on the nature of spacetime at the Planck scale. Causal sets, loop quantum gravity, string theory, causal dynamical triangulation, and black hole thermodynamics all predict a quantized spacetime with agreement on the order of magnitude. Loop quantum gravity makes precise predictions about the geometry of spacetime at the Planck scale.

Spin networks provide a language to describe quantum geometry of space. Spin foam does the same job on spacetime. A spin network is a one-dimensional graph, together with labels on its vertices and edges which encodes aspects of a spatial geometry.

Galilean Invariance

Galilean invariance or Galilean relativity states that the laws of motion are the same in all inertial frames. Galileo Galilei first described this principle in 1632 in his *Dialogue Concerning the Two Chief World Systems* using the example of a ship travelling at constant velocity, without rocking, on a smooth sea; any observer doing experiments below the deck would not be able to tell whether the ship was moving or stationary.

Formulation

Specifically, the term *Galilean invariance* today usually refers to this principle as applied to Newtonian mechanics, that is, Newton's laws hold in all frames related to one another by a Galilean transformation. In other words, all frames related to one another by such a transformation is inertial (meaning, Newton's equation of motion is valid in this frame). In this context it is sometimes called *Newtonian relativity*.

Among the axioms from Newton's theory are:

1. There exists an *absolute space*, in which Newton's laws are true. An inertial frame is a reference frame in relative uniform motion to absolute space.

2. All inertial frames share a *universal time*.

Galilean relativity can be shown as follows. Consider two inertial frames S and S'. A physical event in S will have position coordinates $r = (x, y, z)$ and time t; similarly for S'. By the second axiom above, one can synchronize the clock in the two frames and assume $t = t'$. Suppose S' is in relative uniform motion to S with velocity v. Consider a point object whose position is given by $r'(t) = r(t)$ in S. We see that

$$r'(t) = r(t) - vt.$$

The velocity of the particle is given by the time derivative of the position:

$$u'(t) = \frac{d}{dt} r'(t) = \frac{d}{dt} r(t) - v = u(t) - v.$$

Another differentiation gives the acceleration in the two frames:

$$a'(t) = \frac{d}{dt} u'(t) = \frac{d}{dt} u(t) - 0 = a(t).$$

It is this simple but crucial result that implies Galilean relativity. Assuming that mass is invariant in all inertial frames, the above equation shows Newton's laws of mechanics, if valid in one frame, must hold for all frames. But it is assumed to hold in absolute space, therefore Galilean relativity holds.

Newton's Theory Versus Special Relativity

A comparison can be made between Newtonian relativity and special relativity.

Some of the assumptions and properties of Newton's theory are:

1. The existence of infinitely many inertial frames. Each frame is of infinite size (the entire universe may be covered by many linearly equivalent frames). Any two frames may be in relative uniform motion. (The relativistic nature of mechanics derived above shows that the absolute space assumption is not necessary.)

2. The inertial frames may move in *all* possible relative forms of uniform motion.

3. There is a universal, or absolute, notion of time.

4. Two inertial frames are related by a Galilean transformation.

5. In all inertial frames, Newton's laws, and gravity, hold.

In comparison, the corresponding statements from special relativity are as follows:

1. The existence, as well, of infinitely many non-inertial frames, each of which referenced to (and physically determined by) a unique set of spacetime coordinates. Each frame may be of infinite size, but its definition is always determined locally by contextual physical conditions. Any two frames may be in relative non-uniform motion (as long as it is assumed that this condition of relative motion implies a relativistic dynamical effect -and later, mechanical effect in general relativity- between both frames).

2. Rather than freely allowing all conditions of relative uniform motion between frames of reference, the relative velocity between two inertial frames becomes bounded above by the speed of light.

3. Instead of universal time, each inertial frame possesses its own notion of time.

4. The Galilean transformations are replaced by Lorentz transformations.

5. In all inertial frames, *all* laws of physics are the same.

Notice both theories assume the existence of inertial frames. In practice, the size of the frames in which they remain valid differ greatly, depending on gravitational tidal forces.

In the appropriate context, a *local Newtonian inertial frame*, where Newton's theory remains a good model, extends to, roughly, 10^7 light years.

In special relativity, one considers *Einstein's cabins*, cabins that fall freely in a gravitational field. According to Einstein's thought experiment, a man in such a cabin experiences (to a good approximation) no gravity and therefore the cabin is an approximate inertial frame. However, one has to assume that the size of the cabin is sufficiently small so that the gravitational field is approximately parallel in its interior. This can greatly reduce the sizes of such approximate frames, in comparison to Newtonian frames. For example, an artificial satellite orbiting around earth can be viewed as a cabin. However, reasonably sensitive instruments would detect "microgravity" in such a situation because the "lines of force" of the Earth's gravitational field converge.

In general, the convergence of gravitational fields in the universe dictates the scale at which one might consider such (local) inertial frames. For example, a spaceship falling into a black hole or neutron star would (at a certain distance) be subjected to tidal forces so strong that it would be crushed. In comparison, however, such forces might only be uncomfortable for the astronauts inside (compressing their joints, making it difficult to extend their limbs in any direction perpendicular to the gravity field of the star). Reducing the scale further, the forces at that distance might have almost no effects at all on a mouse. This illustrates the idea that all freely falling frames are locally inertial (acceleration and gravity-free) if the scale is chosen correctly.

Electromagnetism

Maxwell's equations governing electromagnetism possess a different symmetry, Lorentz invariance, under which lengths and times are affected by a change in velocity, which is then described mathematically by a Lorentz transformation.

Albert Einstein's central insight in formulating special relativity was that, for full consistency with electromagnetism, mechanics must also be revised such that Lorentz invariance replaces Galilean invariance. At the low relative velocities characteristic of everyday life, Lorentz invariance and Galilean invariance are nearly the same, but for relative velocities close to that of light they are very different.

Work, Kinetic Energy, and Momentum

Because the distance covered while applying a force to an object depends on the inertial frame of reference, so does the work done. Due to Newton's law of reciprocal actions there is a reaction force; it does work depending on the inertial frame of reference in an opposite way. The total work done is independent of the inertial frame of reference.

Correspondingly the kinetic energy of an object, and even the change in this energy due to a change in velocity, depends on the inertial frame of reference. The total kinetic energy of an isolated system also depends on the inertial frame of reference: it is the sum of the total kinetic energy in a center of momentum frame and the kinetic energy the total mass would have if it were concentrated in the center of mass. Due to the conservation of momentum the latter does not change with time, so changes with time of the total kinetic energy do not depend on the inertial frame of reference.

By contrast, while the momentum of an object also depends on the inertial frame of reference, its change due to a change in velocity does not.

Lorentz Covariance

In physics, Lorentz symmetry, named for Hendrik Lorentz, is "the feature of nature that says experimental results are independent of the orientation or the boost velocity of the laboratory through space". In everyday language, it means that the laws of physics stay the same for all observers that are moving with respect to one another with a uniform velocity. Lorentz covariance, a related concept, is a key property of spacetime following from the special theory of relativity. Lorentz covariance has two distinct, but closely related meanings:

1. A physical quantity is said to be Lorentz covariant if it transforms under a given representation of the Lorentz group. According to the representation theory of the Lorentz group, these quantities are built out of scalars, four-vectors, four-tensors, and spinors. In particular, a Lorentz covariant scalar (e.g., the space-time interval) remains the same under Lorentz transformations and is said to be a *Lorentz invariant* (i.e., they transform under the trivial representation).

2. An equation is said to be Lorentz covariant if it can be written in terms of Lorentz covariant quantities (confusingly, some use the term *invariant* here). The key property of such equations is that if they hold in one inertial frame, then they hold in any inertial frame; this follows from the result that if all the components of a tensor vanish in one frame, they vanish in every frame. This condition is a requirement according to the principle of relativity; i.e., all non-gravitational laws must make the same predictions for identical experiments taking place at the same spacetime event in two different inertial frames of reference.

This usage of the term *covariant* should not be confused with the related concept of a *covariant vector*. On manifolds, the words *covariant* and *contravariant* refer to how objects transform under general coordinate transformations. Confusingly, both covariant and contravariant four-vectors can be Lorentz covariant quantities.

Local Lorentz covariance, which follows from general relativity, refers to Lorentz covariance applying only *locally* in an infinitesimal region of spacetime at every point. There is a generalization of this concept to cover Poincaré covariance and Poincaré invariance.

Examples

In general, the nature of a Lorentz tensor can be identified by its tensor order, which is the number of free indices it has. No indices implies it is a scalar, one implies that it is a vector, etc. Furthermore, any number of new scalars, vectors ,etc. can be made by contracting or creating an outer product of any kinds of tensors together, but many of these may not have any real physical meaning. Some of those tensors that do have a physical interpretation are listed (by no means exhaustively) below.

Please note, the metric sign convention such that $\eta = \text{diag}(1, -1, -1, -1)$ is used throughout the article.

Scalars

Spacetime interval

$$\Delta s^2 = \Delta x^a \Delta x^b \eta_{ab} = c^2 \Delta t^2 - \Delta x^2 - \Delta y^2 - \Delta z^2$$

Proper time (for timelike intervals)

$$\Delta \tau = \sqrt{\frac{\Delta s^2}{c^2}}, \Delta s^2 > 0$$

Proper distance (for spacelike intervals)

$$L = \sqrt{-\Delta s^2}, \Delta s^2 < 0$$

Rest mass

$$m_0^2 c^2 = P^a P^b \eta_{ab} = \frac{E^2}{c^2} - p_x^2 - p_y^2 - p_z^2$$

Electromagnetism invariants

$$F_{ab} F^{ab} = 2\left(B^2 - \frac{E^2}{c^2} \right)$$

$$G_{cd} F^{cd} = \frac{1}{2} \epsilon_{abcd} F^{ab} F^{cd} = -\frac{4}{c} \left(\vec{B} \cdot \vec{E} \right)$$

D'Alembertian/wave operator

$$\Box = \eta^{\mu\nu} \partial_\mu \partial_\nu = \frac{1}{c^2} \frac{\partial^2}{\partial t^2} - \frac{\partial^2}{\partial x^2} - \frac{\partial^2}{\partial y^2} - \frac{\partial^2}{\partial z^2}$$

Four-vectors

4-displacement

$$\Delta X^a = \left(c\Delta t, \overrightarrow{\Delta x} \right) = (c\Delta t, \Delta x, \Delta y, \Delta z)$$

4-position

$$X^a = \left(ct, \vec{x}\right) = (ct, x, y, z)$$

4-gradient

which is the 4D partial derivative:

$$\partial^a = \left(\frac{\partial_t}{c}, -\vec{\nabla}\right) = \left(\frac{1}{c}\frac{\partial}{\partial t}, -\frac{\partial}{\partial x}, -\frac{\partial}{\partial y}, -\frac{\partial}{\partial z}\right)$$

4-velocity

$$U^a = \gamma\left(c, \vec{u}\right) = \gamma\left(c, \frac{dx}{dt}, \frac{dy}{dt}, \frac{dz}{dt}\right)$$

where $U^a = \dfrac{dX^a}{d\tau}$

4-momentum

$$P^a = \left(mc, \vec{p}\right) = \left(\frac{E}{c}, \vec{p}\right) = \left(\frac{E}{c}, p_x, p_y, p_z\right)$$

where $P^a = m_o U^a$

4-current

$$J^a = \left(c\rho, \vec{j}\right) = \left(c\rho, j_x, j_y, j_z\right)$$

where $J^a = \rho_o U^a$

Four-tensors

Kronecker delta

$$\delta_b^a = \begin{cases} 1 & \text{if } a = b, \\ 0 & \text{if } a \neq b. \end{cases}$$

Minkowski metric (the metric of flat space according to general relativity)

$$\eta_{ab} = \eta^{ab} = \begin{cases} 1 & \text{if } a = b = 0, \\ -1 & \text{if } a = b = 1, 2, 3, \\ 0 & \text{if } a \neq b. \end{cases}$$

Levi-Civita symbol

$$\epsilon_{abcd} = -\epsilon^{abcd} = \begin{cases} +1 & \text{if } \{abcd\} \text{ is an even permutation of } \{0123\}, \\ -1 & \text{if } \{abcd\} \text{ is an odd permutation of } \{0123\}, \\ 0 & \text{otherwise.} \end{cases}$$

Electromagnetic field tensor (using a metric signature of $+---$)

$$F_{ab} = \begin{bmatrix} 0 & \frac{1}{c}E_x & \frac{1}{c}E_y & \frac{1}{c}E_z \\ -\frac{1}{c}E_x & 0 & -B_z & B_y \\ -\frac{1}{c}E_y & B_z & 0 & -B_x \\ -\frac{1}{c}E_z & -B_y & B_x & 0 \end{bmatrix}$$

electromagnetic field tensor

Dual

$$G_{cd} = \frac{1}{2}\epsilon_{abcd}F^{ab} = \begin{bmatrix} 0 & B_x & B_y & B_z \\ -B_x & 0 & \frac{1}{c}E_z & -\frac{1}{c}E_y \\ -B_y & -\frac{1}{c}E_z & 0 & \frac{1}{c}E_x \\ -B_z & \frac{1}{c}E_y & -\frac{1}{c}E_x & 0 \end{bmatrix}$$

Lorentz Violating Models

In standard field theory, there are very strict and severe constraints on marginal and relevant Lorentz violating operators within both QED and the Standard Model. Irrelevant Lorentz violating operators may be suppressed by a high cutoff scale, but they typically induce marginal and relevant Lorentz violating operators via radiative corrections. So, we also have very strict and severe constraints on irrelevant Lorentz violating operators.

Since some approaches to quantum gravity lead to violations of Lorentz invariance, these studies are part of Phenomenological Quantum Gravity.

Lorentz violating models typically fall into four classes:

- The laws of physics are exactly Lorentz covariant but this symmetry is spontaneously broken. In special relativistic theories, this leads to phonons, which are the Goldstone bosons. The phonons travel at *less* than the speed of light.

- Similar to the approximate Lorentz symmetry of phonons in a lattice (where the speed of sound plays the role of the critical speed), the Lorentz symmetry of special relativity (with the speed of light as the critical speed in vacuum) is only a low-energy limit of the laws of physics, which involve new phenomena at some fundamental scale. Bare conventional "elementary" particles are not point-like field-theoretical objects at very small distance scales, and a nonzero fundamental length must be taken into account. Lorentz symmetry violation is governed by an energy-dependent parameter which tends to zero as momentum decreases. Such patterns require the existence of a privileged local inertial frame (the "vacuum rest frame"). They can be tested, at least partially, by ultra-high energy cosmic ray experiments like the Pierre Auger Observatory.

- The laws of physics are symmetric under a deformation of the Lorentz or more generally, the Poincaré group, and this deformed symmetry is exact and unbroken. This deformed symmetry is also typically a quantum group symmetry, which is a generalization of a group symmetry. Deformed special relativity is an example of this class of models. It is not accurate to call such models Lorentz-violating as much as Lorentz deformed any more than special relativity can be called a violation of Galilean symmetry rather than a deformation of it. The deformation is scale dependent, meaning that at length scales much larger than the Planck scale, the symmetry looks pretty much like the Poincaré group. Ultra-high energy cosmic ray experiments cannot test such models.

- Very special relativity forms a class of its own; if charge-parity (CP) is an exact symmetry, a subgroup of the Lorentz group is sufficient to give us all the standard predictions. This is, however, not the case.

Models belonging to the first two classes can be consistent with experiment if Lorentz breaking happens at Planck scale or beyond it, or even before it in suitable preonic models, and if Lorentz symmetry violation is governed by a suitable energy-dependent parameter. One then has a class of models which deviate from Poincaré symmetry near the Planck scale but still flows towards an exact Poincaré group at very large length scales. This is also true for the third class, which is furthermore protected from radiative corrections as one still has an exact (quantum) symmetry.

Even though there is no evidence of the violation of Lorentz invariance, several experimental searches for such violations have been performed during recent years. A detailed summary of the results of these searches is given in the Data Tables for Lorentz and CPT Violation.

Cosmological Constant

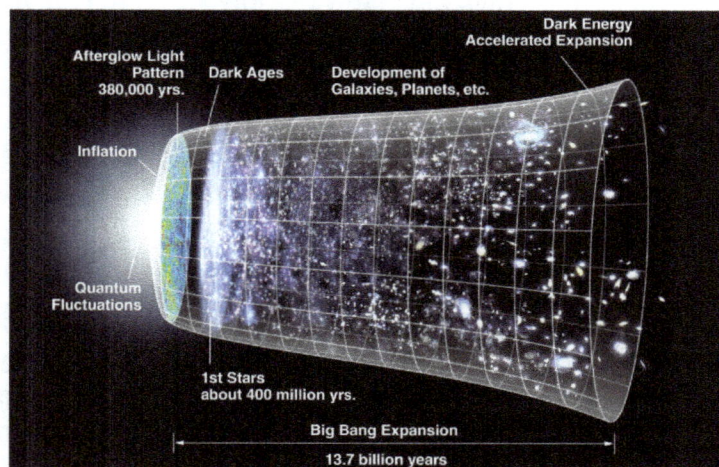

Sketch of the timeline of the universe in the ΛCDM model. The accelerated expansion in the last third of the timeline represents the dark-energy dominated era.

TODAY

13.7 BILLION YEARS AGO
(Universe 380,000 years old)

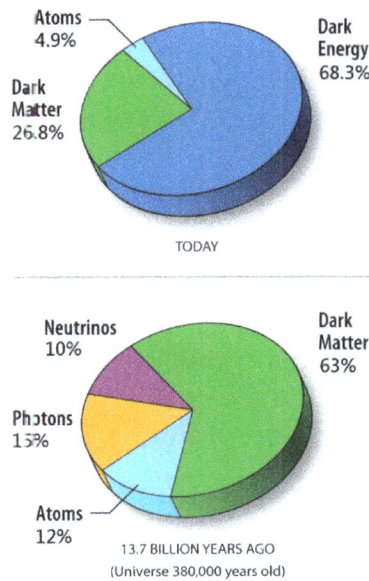

Estimated ratios of dark matter and dark energy (which may be the cosmological constant) in the universe. According to current theories of physics, dark energy now dominates as the largest source of energy of the universe, in contrast to earlier epochs when it was insignificant.

In cosmology, the cosmological constant (usually denoted by the Greek capital letter lambda: Λ) is the value of the energy density of the vacuum of space. It was originally introduced by Albert Einstein in 1917 as an addition to his theory of general relativity to "hold back gravity" and achieve a static universe, which was the accepted view at the time. Einstein abandoned the concept after Hubble's 1929 discovery that all galaxies outside the Local Group (the group that contains the Milky Way Galaxy) are moving away from each other, implying an overall expanding universe. From 1929 until the early 1990s, most cosmology researchers assumed the cosmological constant to be zero.

Since the 1990s, several developments in observational cosmology, especially the discovery of the accelerating universe from distant supernovae in 1998 (in addition to independent evidence from the cosmic microwave background and large galaxy redshift surveys), have shown that around 68% of the mass–energy density of the universe can be attributed to dark energy. While dark energy is poorly understood at a fundamental level, the main required properties of dark energy are that it functions as a type of anti-gravity, it dilutes much more slowly than matter as the universe expands, and it clusters much more weakly than matter, or perhaps not at all. The cosmological constant is the simplest possible form of dark energy since it is constant in both space and time, and this leads to the current standard model of cosmology known as the Lambda-CDM model, which provides a good fit to many cosmological observations as of 2016.

Equation

The cosmological constant Ë appears in Einstein's field equation in the form of

$$R_{\mu\nu} - \frac{1}{2} R g_{\mu\nu} + \Lambda g_{\mu\nu} = \frac{8\pi G}{c^4} T_{\mu\nu},$$

where R and g describe the structure of spacetime, T pertains to matter and energy affecting that structure, and G and c are conversion factors that arise from using traditional units of measurement. When Λ is zero, this reduces to the original field equation of general relativity. When T is zero, the field equation describes empty space (the vacuum).

The cosmological constant has the same effect as an intrinsic energy density of the vacuum, ρ_{vac} (and an associated pressure). In this context, it is commonly moved onto the right-hand side of the equation, and defined with a proportionality factor of 8π: $\Lambda = 8\pi\rho_{vac}$, where unit conventions of general relativity are used (otherwise factors of G and c would also appear, i.e. $\Lambda = 8\pi (G/c^2)\rho_{vac} = \kappa \rho_{vac}$, where κ is Einstein's constant). It is common to quote values of energy density directly, though still using the name "cosmological constant", with convention $8\pi G = 1$. (In fact, the true dimension of Λ is a length^{-2} and it has the value of $1.19 \cdot 10^{-52}$ m^{-2} or in reduced Planck units : $\sim 3 \cdot 10^{-122}$, calculated with the best present (2015) values of $\Omega_\Lambda = 0.6911 \pm 0.0062$ and $H_0 = 67.74 \pm 0.46$ km/s / Mpc = $2.195 \pm 0.015\ 10^{-18}$ s^{-1}).

A positive vacuum energy density resulting from a cosmological constant implies a negative pressure, and vice versa. If the energy density is positive, the associated negative pressure will drive an accelerated expansion of the universe, as observed.

Ω_Λ (Omega Lambda)

Instead of the cosmological constant itself, cosmologists often refer to the ratio between the energy density due to the cosmological constant and the critical density of the universe, the tipping point for a sufficient density to stop the universe from expanding forever. This ratio is usually denoted Ω_Λ, and is estimated to be 0.6911 ± 0.0062, according to results published by the Planck Collaboration in 2015.

In a flat universe Ω_Λ is the fraction of the energy of the universe due to the cosmological constant, i.e., what we would intuitively call the fraction of the universe that is made up of dark energy. Note that this value changes over time: the critical density changes with cosmological time, but the energy density due to the cosmological constant remains unchanged throughout the history of the universe: the amount of dark energy increases as the universe grows, while the amount of matter does not.

Equation of State

Another ratio that is used by scientists is the equation of state, usually denoted w, which is the ratio of pressure that dark energy puts on the universe to the energy per unit volume. This ratio is $w = -1$ for a true cosmological constant, and is generally different for alternative time-varying forms of vacuum energy such as quintessence.

History

Einstein included the cosmological constant as a term in his field equations for general relativity because he was dissatisfied that otherwise his equations did not allow, apparently, for a static universe: gravity would cause a universe that was initially at dynamic equilibrium to contract. To counteract this possibility, Einstein added the cosmological constant. However, soon after Einstein developed his static theory, observations by Edwin Hubble indicated that the universe appears to be expanding; this was consistent with a cosmological solution to the *original* general relativity equations that had been found by the mathematician Friedmann, working on the Einstein equations of general relativity. Einstein later reputedly referred to his failure to accept the validation of his equations—when they had predicted the expansion of the universe in theory, before it was demonstrated in observation of the cosmological red shift—as the "biggest blunder" of his life.

In fact, adding the cosmological constant to Einstein's equations does not lead to a static universe at equilibrium because the equilibrium is unstable: if the universe expands slightly, then the expansion releases vacuum energy, which causes yet more expansion. Likewise, a universe that contracts slightly will continue contracting.

However, the cosmological constant remained a subject of theoretical and empirical interest. Empirically, the onslaught of cosmological data in the past decades strongly suggests that our universe has a positive cosmological constant. The explanation of this small but positive value is an outstanding theoretical challenge.

Finally, it should be noted that some early generalizations of Einstein's gravitational theory, known as classical unified field theories, either introduced a cosmological constant on theoretical grounds or found that it arose naturally from the mathematics. For example, Sir Arthur Stanley Eddington claimed that the cosmological constant version of the vacuum field equation expressed the "epistemological" property that the universe is "self-gauging", and Erwin Schrödinger's pure-affine theory using a simple variational principle produced the field equation with a cosmological term.

Positive Value

Observations announced in 1998 of distance–redshift relation for Type Ia supernovae indicated that the expansion of the universe is accelerating. When combined with measurements of the cosmic microwave background radiation these implied a value of $\Omega_\Lambda \approx 0.7$, a result which has been supported and refined by more recent measurements. There are other possible causes of an accelerating universe, such as quintessence, but the cosmological constant is in most respects the simplest solution. Thus, the current standard model of cosmology, the Lambda-CDM model, includes the cosmological constant, which is measured to be on the order of 10^{-52} m^{-2}, in metric units. Multiplied by other constants that appear in the equations, it is often expressed as 10^{-52} m^{-2}, 10^{-35} s^{-2}, 10^{-47} GeV^4, 10^{-29} g/cm^3. In terms of Planck units, and as a natural dimensionless value, the cosmological constant, Λ, is on the order of 10^{-122}. Modern calculations considering the vacuum energy of all known scalar and vector fields leads to 10^{-54} orders of magnitude smaller than the prediction.

As was only recently seen, by works of 't Hooft, Susskind and others, a positive cosmological constant has surprising consequences, such as a finite maximum entropy of the observable universe.

Predictions

Quantum Field Theory

A major outstanding problem is that most quantum field theories predict a huge value for the quantum vacuum. A common assumption is that the quantum vacuum is equivalent to the cosmological constant. Although no theory exists that supports this assumption, arguments can be made in its favor.

Such arguments are usually based on dimensional analysis and effective field theory. If the universe is described by an effective local quantum field theory down to the Planck scale, then we would expect a cosmological constant of the order of M_{pl}^4. As noted above, the measured cosmological constant is smaller than this by a factor of 10^{-120}. This discrepancy has been called "the worst theoretical prediction in the history of physics!".

Some supersymmetric theories require a cosmological constant that is exactly zero, which further complicates things. This is the *cosmological constant problem*, the worst problem of fine-tuning in physics: there is no known natural way to derive the tiny cosmological constant used in cosmology from particle physics.

Anthropic Principle

One possible explanation for the small but non-zero value was noted by Steven Weinberg in 1987 following the anthropic principle. Weinberg explains that if the vacuum energy took different values in different domains of the universe, then observers would necessarily measure values similar to that which is observed: the formation of life-supporting structures would be suppressed in domains where the vacuum energy is much larger. Specifically, if the vacuum energy is negative and its absolute value is substantially larger than it appears to be in the observed universe (say, a factor of 10 larger), holding all other variables (e.g. matter density) constant, that would mean that the universe is closed; furthermore, its lifetime would be shorter than the age of our universe, possibly too short for intelligent life to form. On the other hand, a universe with a large positive cosmological constant would expand too fast, preventing galaxy formation. According to Weinberg, domains where the vacuum energy is compatible with life would be comparatively rare. Using this argument, Weinberg predicted that the cosmological constant would have a value of less than a hundred times the currently accepted value. In 1992, Weinberg refined this prediction of the cosmological constant to 5 to 10 times the matter density.

This argument depends on a lack of a variation of the distribution (spatial or otherwise) in the vacuum energy density, as would be expected if dark energy were the cosmological constant. There is no evidence that the vacuum energy does vary, but it may be the case if, for example, the vacuum energy is (even in part) the potential of a scalar field such as the residual inflaton. Another theoretical approach that deals with the issue is that of multiverse theories, which predict a large number of "parallel" universes with different laws of physics and/or values of fundamental constants. Again, the anthropic principle states that we can only live in one of the universes that is compatible with some form of intelligent life. Critics claim that these theories, when used as an explanation for fine-tuning, commit the inverse gambler's fallacy.

In 1995, Weinberg's argument was refined by Alexander Vilenkin to predict a value for the cosmological constant that was only ten times the matter density, i.e. about three times the current value since determined.

BKL Singularity

A Belinsky-Khalatnikov-Lifshitz (BKL) singularity is a model of the dynamic evolution of the Universe near the initial singularity, described by an anisotropic, homogeneous, chaotic solution to Einstein's field equations of gravitation. According to this model, the Universe is oscillating around a gravitational singularity in which time and space become equal to zero. This singularity is physically real in the sense that it is a necessary property of the solution, and will appear also in the exact solution of those equations. The singularity is not artificially created by the assumptions and simplifications made by the other special solutions such as the Friedmann–Lemaître–Robertson–Walker, quasi-isotropic, and Kasner solutions.

The Mixmaster universe is a solution to general relativity that exhibits properties similar to those discussed by BKL.

Existence of Time Singularity

The basis of modern cosmology are the special solutions of the Einstein field equations found by Alexander Friedmann in 1922–1924. The Universe is assumed homogeneous (space has the same metric properties (measures) in all points) and is isotropic (space has the same measures in all directions). Friedmann's solutions allow two possible geometries for space: closed model with a ball-like, outwards-bowed space (positive curvature) and open model with a saddle-like, inwards-bowed space (negative curvature). In both models, the Universe is not standing still, it is constantly either expanding (becoming larger) or contracting (shrinking, becoming smaller). This was confirmed by Edwin Hubble who established the Hubble redshift of receding galaxies. The present consensus is that the isotropic model, in general, gives an adequate description of the present state of the Universe.

Another important property of the isotropic model is the inevitable existence of a time singularity: time flow is not continuous, but stops or reverses after time reaches some (very large or very small) value. Between singularities, time flows in one direction, away from the singularity (arrow of time). In the open model, there is one time singularity so time is limited at one end but unlimited at the other, while in the closed model there are two singularities that limit time at both ends (the Big Bang and Big Crunch).

The adequacy of the isotropic model in describing the present state of the Universe by itself is not a reason to expect that it is adequate for describing the early stages of Universe evolution. At the same time, it is obvious that in the real world homogeneity is, at best, only an approximation. Even if one can speak about a homogeneous distribution of matter density at distances that are large compared to the intergalactic space, this homogeneity vanishes at smaller scales. On the other hand, the homogeneity assumption goes very far in a mathematical aspect: it makes the solution highly symmetric which can give the solution specific properties that disappear when considering a more general case.

One of the principal problems studied by the Landau group (to which BKL belong) was whether relativistic cosmological models necessarily contain a time singularity or whether the time singularity is an artifact of the assumptions used to simplify these models. The independence of the singularity on symmetry assumptions would mean that time singularities exist not only in the special, but also in the general solutions of the Einstein equations. A criterion for generality of solutions is the number of independent space coordinate functions that they contain. These include only the "physically independent" functions whose number cannot be reduced by any choice of reference frame. In the general solution, the number of such functions must be enough to fully define the initial conditions (distribution and movement of matter, distribution of gravitational field) at some moment of time chosen as initial. This number is four for an empty (vacuum) space, and eight for a matter and/or radiation-filled space.

For a system of non-linear differential equations, such as the Einstein equations, a general solution is not unambiguously defined. In principle, there may be multiple general integrals, and each of those may contain only a finite subset of all possible initial conditions. Each of those integrals may contain all required independent functions which, however, may be subject to some conditions (e.g., some inequalities). Existence of a general solution with a singularity, therefore, does not preclude the existence of other additional general solutions that do not contain a singularity. For example, there is no reason to doubt the existence of a general solution without a singularity that describes an isolated body with a relatively small mass.

It is impossible to find a general integral for all space and for all time. However, this is not necessary for resolving the problem: it is sufficient to study the solution near the singularity. This would also resolve another aspect of the problem: the characteristics of spacetime metric evolution in the general solution when it reaches the physical singularity, understood as a point where matter density and invariants of the Riemann curvature tensor become infinite. The BKL paper concerns only the cosmological aspect. This means, that the subject is a time singularity in the whole spacetime and not in some limited region as in a gravitational collapse of a finite body.

Previous work by the Landau group (reviewed in) led to the conclusion that the general solution does not contain a physical singularity. This search for a broader class of solutions with a singularity has been done, essentially, by a trial-and-error method, since a systematic approach to the study of the Einstein equations is lacking. A negative result, obtained in this way, is not convincing by itself; a solution with the necessary degree of generality would invalidate it, and at the same time would confirm any positive results related to the specific solution.

It is reasonable to suggest that if a singularity is present in the general solution, there must be some indications that are based only on the most general properties of the Einstein equations, although those indications by themselves might be insufficient for characterizing the singularity. At that time, the only known indication was related to the form of the Einstein equations written in a synchronous frame, that is, in a frame in which the proper time $x^o = t$ is synchronized throughout the whole space; in this frame the space distance element dl is separate from the time interval dt. The Einstein equation

$$R_0^0 = T_0^0 - \tfrac{1}{2}T$$

(eq. 1)

written in synchronous frame gives a result in which the metric determinant g inevitably becomes zero in a finite time irrespective of any assumptions about matter distribution.

This indication, however, was dropped after it became clear that it is linked with a specific geometric property of the synchronous frame: the crossing of time line coordinates. This crossing takes place on some encircling hypersurfaces which are four-dimensional analogs of the caustic surfaces in geometrical optics; g becomes zero exactly at this crossing. Therefore, although this singularity is general, it is fictitious, and not a physical one; it disappears when the reference frame is changed. This, apparently, removed the incentive among the researchers for further investigations along these lines.

However, the interest in this problem waxed again in the 1960s after Penrose published his theorems that linked the existence of a singularity of unknown character with some very general assumptions that did not have anything in common with a choice of reference frame. Other similar theorems were found later on by Hawking and Geroch. This revived interest in the search for singular solutions.

Generalized Kasner Solution

Further generalization of solutions depended on some solution classes found previously. The Friedmann solution, for example, is a special case of a solution class that contains three physically arbitrary coordinate functions. In this class the space is anisotropic; however, its compression when approaching the singularity has "quasi-isotropic" character: the linear distances in all directions diminish as the same power of time. Like the fully homogeneous and isotropic case, this class of solutions exist only for a matter-filled space.

Much more general solutions are obtained by a generalization of an exact particular solution derived by Edward Kasner for a field in vacuum, in which the space is homogeneous and has Euclidean metric that depends on time according to the Kasner metric

$$dl^2 = t^{2p_1} dx^2 + t^{2p_2} dy^2 + t^{2p_3} dz^2 \qquad \text{(eq. 2)}$$

Here, p_1, p_2, p_3 are any 3 numbers that are related by

$$p_1 + p_2 + p_3 = p_1^2 + p_2^2 + p_3^2 = 1. \qquad \text{(eq. 3)}$$

Because of these relationships, only 1 of the 3 numbers is independent. All 3 numbers are never the same; 2 numbers are the same only in the sets of values and (0, 0, 1). In all other cases the numbers are different, one number is negative and the other two are positive. If the numbers are arranged in increasing order, $p_1 < p_2 < p_3$, they change in the ranges

$$-\tfrac{1}{3} \leq p_1 \leq 0,$$
$$0 \leq p_2 \leq \tfrac{2}{3}, \qquad \text{(eq. 4)}$$
$$\frac{2}{3} \leq p_3 \leq 1.$$

The numbers p_1, p_2, p_3 can be written parametrically as

$$p_1(u) = \frac{-u}{1+u+u^2}, \; p_2(u) = \frac{1+u}{1+u+u^2}, \; p_3(u) = \frac{u(1+u)}{1+u+u^2} \qquad \text{(eq. 5)}$$

All different values of p_1, p_2, p_3 ordered as above are obtained by changing the value of the parameter u in the range $u \geq 1$. The values $u < 1$ are brought into this range according to

$$p_1\left(\frac{1}{u}\right) = p_1(u),\ p_2\left(\frac{1}{u}\right) = p_3(u),\ p_3\left(\frac{1}{u}\right) = p_2(u)$$

(eq. 6)

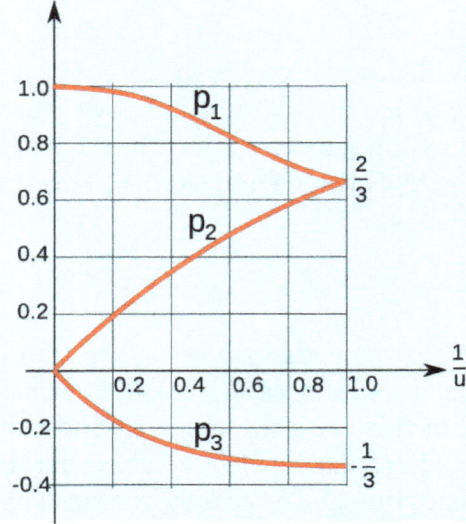

Figure is a plot of p_1, p_2, p_3 with an argument $1/u$. The numbers $p_1(u)$ and $p_3(u)$ are monotonously increasing while $p_2(u)$ is monotonously decreasing function of the parameter u.

In the generalized solution, the form corresponding to eq. 2 applies only to the asymptotic metric (the metric close to the singularity $t = 0$), respectively, to the major terms of its series expansion by powers of t. In the synchronous reference frame it is written in the form of eq. 1 with a space distance element

$$dl^2 = \left(a^2 l_\alpha l_\beta + b^2 m_\alpha m_\beta + c^2 n_\alpha n_\beta\right) dx^\alpha dx^\beta$$

(eq. 7)

where

$$a = t^{p_l},\ b = t^{p_m},\ c = t^{p_n}$$

(eq. 8)

The three-dimensional vectors l, m, n define the directions at which space distance changes with time by the power laws eq. 8. These vectors, as well as the numbers p_l, p_m, p_n which, as before, are related by eq. 3, are functions of the space coordinates. The powers p_l, p_m, p_n are not arranged in increasing order, reserving the symbols p_1, p_2, p_3 for the numbers in eq. 5 that remain arranged in increasing order. The determinant of the metric of eq. 7 is

$$-g = a^2 b^2 c^2 v^2 = t^2 v^2$$

(eq. 9)

where $v = l[mn]$. It is convenient to introduce the following quantities

$$\lambda = \frac{l\ \text{rot}\ l}{v},\ \mu = \frac{m\ \text{rot}\ m}{v},\ v = \frac{n\ \text{rot}\ n}{v}.$$

(eq. 10)

The space metric in eq. 7 is anisotropic because the powers of t in eq. 8 cannot have the same values. On approaching the singularity at $t = 0$, the linear distances in each space element decrease in two directions and increase in the third direction. The volume of the element decreases in proportion to t.

The Einstein equations in vacuum in synchronous reference frame are

$$R_0^0 = -\frac{1}{2}\frac{\partial \varkappa_\alpha^\alpha}{\partial t} - \frac{1}{4}\varkappa_\alpha^\beta \varkappa_\beta^\alpha = 0, \qquad (eq.\ 11)$$

$$R_\alpha^\beta = -\left(\frac{1}{2}\sqrt{-g}\right)\frac{\partial}{\partial t}\left(\sqrt{-g}\varkappa_\alpha^\beta\right) - P_\alpha^\beta = 0, \qquad (eq.\ 12)$$

$$R_\alpha^0 = \frac{1}{2}\left(\varkappa_{\alpha;\beta}^\beta - \varkappa_{\beta;\alpha}^\beta\right) = 0, \qquad (eq.\ 13)$$

where $\varkappa_\alpha^\beta$ is the 3-dimensional tensor $\varkappa_\alpha^\beta = \dfrac{\partial \gamma_\alpha^\beta}{\partial t}$, and $P_{\alpha\beta}$ is the 3-dimensional Ricci tensor, which is expressed by the 3-dimensional metric tensor $\gamma_{\alpha\beta}$ in the same way as R_{ik} is expressed by g_{ik}; $P_{\alpha\beta}$ contains only the space (but not the time) derivatives of $\gamma_{\alpha\beta}$.

The Kasner metric is introduced in the Einstein equations by substituting the respective metric tensor $\gamma_{\alpha\beta}$ from eq. 7 without defining *a priori* the dependence of a, b, c from t:

$$\varkappa_\alpha^\beta = \left(\frac{2\dot{a}}{a}\right)l_\alpha l^\beta + \left(\frac{2\dot{b}}{b}\right)m_\alpha m^\beta + \left(\frac{2\dot{c}}{c}\right)n_\alpha n^\beta$$

where the dot above a symbol designates differentiation with respect to time. The Einstein equation eq. 11 takes the form

$$-R_0^0 = \frac{\ddot{a}}{a} + \frac{\ddot{b}}{b} + \frac{\ddot{c}}{c} = 0. \qquad (eq.\ 14)$$

All its terms are to a second order for the large (at $t \to 0$) quantity $1/t$. In the Einstein equations eq. 12, terms of such order appear only from terms that are time-differentiated. If the components of $P_{\alpha\beta}$ do not include terms of order higher than 2, then

$$-R_l^l = \frac{(\dot{a}bc).}{abc} = 0, \quad -R_m^m = \frac{(a\dot{b}c).}{abc} = 0, \quad -R_n^n = \frac{(ab\dot{c}).}{abc} = 0 \qquad (eq.\ 15)$$

where indices l, m, n designate tensor components in the directions l, m, n. These equations together with eq. 14 give the expressions eq. 8 with powers that satisfy eq. 3.

However, the presence of 1 negative power among the 3 powers p_l, p_m, p_n results in appearance of terms from $P_{\alpha\beta}$ with an order greater than t^{-2}. If the negative power is p_l ($p_l = p_1 < 0$), then $P_{\alpha\beta}$ contains the coordinate function λ and eq. 12 become

$$-R_l^l = \frac{(\dot{abc})^{\cdot}}{abc} + \frac{\lambda^2 a^2}{2b^2 c^2} = 0,$$

$$-R_m^m = \frac{(a\dot{b}c)^{\cdot}}{abc} - \frac{\lambda^2 a^2}{2b^2 c^2} = 0,$$

$$-R_n^n = \frac{(ab\dot{c})^{\cdot}}{abc} - \frac{\lambda^2 a^2}{2b^2 c^2} = 0.$$

(*eq. 16*)

Here, the second terms are of order $t^{-2(p_m + p_n - p_l)}$ whereby $p_m + p_n - p_l = 1 + 2\,|p_l| > 1$. To remove these terms and restore the metric eq. 7, it is necessary to impose on the coordinate functions the condition $\lambda = 0$.

The remaining 3 Einstein equations eq. 13 contain only first order time derivatives of the metric tensor. They give 3 time-independent relations that must be imposed as necessary conditions on the coordinate functions in eq. 7. This, together with the condition $\lambda = 0$, makes 4 conditions. These conditions bind 10 different coordinate functions: 3 components of each of the vectors l, m, n, and one function in the powers of t (any one of the functions p_l, p_m, p_n, which are bound by the conditions eq. 3). When calculating the number of physically arbitrary functions, it must be taken into account that the synchronous system used here allows time-independent arbitrary transformations of the 3 space coordinates. Therefore, the final solution contains overall 10 − 4 − 3 = 3 physically arbitrary functions which is 1 less than what is needed for the general solution in vacuum.

The degree of generality reached at this point is not lessened by introducing matter; matter is written into the metric eq. 7 and contributes 4 new coordinate functions necessary to describe the initial distribution of its density and the 3 components of its velocity. This makes possible to determine matter evolution merely from the laws of its movement in an *a priori* given gravitational field. These movement laws are the hydrodynamic equations

$$\frac{1}{\sqrt{-g}} \frac{\partial}{\partial x^i}\left(\sqrt{-g}\,\sigma u^i\right) = 0,$$

(*eq. 17*)

$$(p+\varepsilon)u^k \left\{ \frac{\partial u_i}{\partial x^k} - \frac{1}{2}u^l \frac{\partial g_{kl}}{\partial x^i} \right\} = -\frac{\partial p}{\partial x^i} - u_i u^k \frac{\partial p}{\partial x^k},$$

(*eq. 18*)

where u^i is the 4-dimensional velocity, ε and σ are the densities of energy and entropy of matter. For the ultrarelativistic equation of state $p = \varepsilon/3$ the entropy $\sigma \sim \varepsilon^{1/4}$. The major terms in eq. 17 and eq. 18 are those that contain time derivatives. From eq. 17 and the space components of eq. 18 one has

$$\frac{\partial}{\partial t}\left(\sqrt{-g}\,u_0 \varepsilon^{\frac{3}{4}}\right) = 0,\; 4\varepsilon \cdot \frac{\partial u_\alpha}{\partial t} + u_\alpha \cdot \frac{\partial \varepsilon}{\partial t} = 0,$$

resulting in

$$abcu_0\varepsilon^{\frac{3}{4}} = \text{const}, \, u_\alpha\varepsilon^{\frac{1}{4}} = \text{const},$$ (eq. 19)

where 'const' are time-independent quantities. Additionally, from the identity $u_i u^i = 1$ one has (because all covariant components of u_α are to the same order)

$$u_0^2 \approx u_n u^n = \frac{u_n^2}{c^2},$$

where u_n is the velocity component along the direction of n that is connected with the highest (positive) power of t (supposing that $p_n = p_3$). From the above relations, it follows that

$$\varepsilon \sim \frac{1}{a^2 b^2}, u_\alpha \sim \sqrt{ab}$$ (eq. 20)

or

$$\varepsilon \sim t^{-2(p_1+p_2)} = t^{-2(1-p_3)}, u_\alpha \sim t^{\frac{(1-p_3)}{2}}.$$ (eq. 21)

The above equations can be used to confirm that the components of the matter stress-energy-momentum tensor standing in the right hand side of the equations

$$R_0^0 = T_0^0 - \frac{1}{2}T, \, R_\alpha^\beta = T_\alpha^\beta - \frac{1}{2}\delta_\alpha^\beta T,$$

are, indeed, to a lower order by $1/t$ than the major terms in their left hand sides. In the equations the presence of matter results only in the change of relations imposed on their constituent coordinate functions.

The fact that ε becomes infinite by the law eq. 21 confirms that in the solution to eq. 7 one deals with a physical singularity at any values of the powers p_1, p_2, p_3 excepting only (0, 0, 1). For these last values, the singularity is non-physical and can be removed by a change of reference frame.

The fictional singularity corresponding to the powers (0, 0, 1) arises as a result of time line coordinates crossing over some 2-dimensional "focal surface". As pointed out in, a synchronous reference frame can always be chosen in such a way that this inevitable time line crossing occurs exactly on such surface (instead of a 3-dimensional caustic surface). Therefore, a solution with such simultaneous for the whole space fictional singularity must exist with a full set of arbitrary functions needed for the general solution. Close to the point $t = 0$ it allows a regular expansion by whole powers of t.

Oscillating Mode Towards the Singularity

The four conditions that had to be imposed on the coordinate functions in the solution eq. 7 are of different types: three conditions that arise from the equations $R_\alpha^0 = 0$ are "natural"; they are a consequence of the structure of Einstein equations. However, the additional condition $\lambda = 0$ that causes the loss of one derivative function, is of entirely different type.

The general solution by definition is completely stable; otherwise the Universe would not exist. Any perturbation is equivalent to a change in the initial conditions in some moment of time; since the general solution allows arbitrary initial conditions, the perturbation is not able to change its character. In other words, the existence of the limiting condition $\lambda = 0$ for the solution of eq. 7 means instability caused by perturbations that break this condition. The action of such perturbation must bring the model to another mode which thereby will be most general. Such perturbation cannot be considered as small: a transition to a new mode exceeds the range of very small perturbations.

The analysis of the behavior of the model under perturbative action, performed by BKL, delineates a complex oscillatory mode on approaching the singularity. They could not give all details of this mode in the broad frame of the general case. However, BKL explained the most important properties and character of the solution on specific models that allow far-reaching analytical study.

These models are based on a homogeneous space metric of a particular type. Supposing a homogeneity of space without any additional symmetry leaves a great freedom in choosing the metric. All possible homogeneous (but anisotropic) spaces are classified, according to Bianchi, in 9 classes. BKL investigate only spaces of Bianchi Types VIII and IX.

If the metric has the form of eq. 7, for each type of homogeneous spaces exists some functional relation between the reference vectors l, m, n and the space coordinates. The specific form of this relation is not important. The important fact is that for Type VIII and IX spaces, the quantities λ, μ, ν eq. 10 are constants while all "mixed" products l rot m, l rot n, m rot l, *etc.* are zeros. For Type IX spaces, the quantities λ, μ, ν have the same sign and one can write $\lambda = \mu = \nu = 1$ (the simultaneous sign change of the 3 constants does not change anything). For Type VIII spaces, 2 constants have a sign that is opposite to the sign of the third constant; one can write, for example, $\lambda = -1$, $\mu = \nu = 1$.

The study of the effect of the perturbation on the "Kasner mode" is thus confined to a study on the effect of the λ-containing terms in the Einstein equations. Type VIII and IX spaces are the most suitable models exactly in this connection. Since all 3 quantities λ, μ, ν differ from zero, the condition $\lambda = 0$ does not hold irrespective of which direction l, m, n has negative power law time dependence.

The Einstein equations for the Type VIII and Type IX space models are

$$-R_l^l = \frac{(\dot{a}bc)^{\cdot}}{abc} + \frac{1}{2}\left(a^2b^2c^2\right)\left[\lambda^2 a^4 - \left(\mu b^2 - \nu c^2\right)^2\right] = 0,$$

$$-R_m^m = \frac{(a\dot{b}c)^{\cdot}}{abc} + \frac{1}{2}\left(a^2b^2c^2\right)\left[\mu^2 b^4 - \left(\lambda a^2 - \nu c^2\right)^2\right] = 0, \qquad \textit{(eq. 22)}$$

$$-R_n^n = \frac{(ab\dot{c})^{\cdot}}{abc} + \frac{1}{2}\left(a^2b^2c^2\right)\left[\nu^2 c^4 - \left(\lambda a^2 - \mu b^2\right)^2\right] = 0,$$

$$-R_0^0 = \frac{\ddot{a}}{a} + \frac{\ddot{b}}{b} + \frac{\ddot{c}}{c} = 0 \qquad \textit{(eq. 23)}$$

(the remaining components R_l^0, R_m^0, R_n^0, R_l^m, R_l^n, R_m^n are identically zeros). These equations contain only functions of time; this is a condition that has to be fulfilled in all homogeneous spaces.

Here, the eq. 22 and eq. 23 are exact and their validity does not depend on how near one is to the singularity at $t = 0$.

The time derivatives in eq. 22 and eq. 23 take a simpler form if a, b, c are substituted by their logarithms α, β, γ:

$$a = e^{\alpha}, b = e^{\beta}, c = e^{\gamma}, \qquad\qquad\qquad\qquad\qquad (eq.\ 24)$$

substituting the variable t for τ according to:

$$dt = abc\ d\tau. \qquad\qquad\qquad\qquad\qquad (eq.\ 25)$$

Then:

$$
\begin{aligned}
2\alpha_{\tau\tau} &= \left(\mu b^2 - v c^2\right)^2 - \lambda^2 a^4 = 0, \\
2\beta_{\tau\tau} &= \left(\lambda a^2 - v c^2\right)^2 - \mu^2 b^4 = 0, \\
2\gamma_{\tau\tau} &= \left(\lambda a^2 - \mu b^2\right)^2 - v^2 c^4 = 0,
\end{aligned}
\qquad\qquad (eq.\ 26)
$$

$$\frac{1}{2}\left(\alpha + \beta + \gamma\right)_{\tau\tau} = \alpha_{\tau}\beta_{\tau} + \alpha_{\tau}\gamma_{\tau} + \beta_{\tau}\gamma_{\tau}. \qquad\qquad\qquad (eq.\ 27)$$

Adding together equations eq. 26 and substituting in the left hand side the sum $(\alpha + \beta + \gamma)_{\tau\tau}$ according to eq. 27, one obtains an equation containing only first derivatives which is the first integral of the system eq. 26:

$$\alpha_{\tau}\beta_{\tau} + \alpha_{\tau}\gamma_{\tau} + \beta_{\tau}\gamma_{\tau} = \frac{1}{4}\left(\lambda^2 a^4 + \mu^2 b^4 + v^2 c^4 - 2\lambda\mu a^2 b^2 - 2\lambda v a^2 c^2 - 2\mu v b^2 c^2\right). \qquad (eq.\ 28)$$

This equation plays the role of a binding condition imposed on the initial state of eq. 26. The Kasner mode eq. 8 is a solution of eq. 26 when ignoring all terms in the right hand sides. But such situation cannot go on (at $t \to 0$) indefinitely because among those terms there are always some that grow. Thus, if the negative power is in the function $a(t)$ $(p_l = p_1)$ then the perturbation of the Kasner mode will arise by the terms $\lambda^2 a^4$; the rest of the terms will decrease with decreasing t. If only the growing terms are left in the right hand sides of eq. 26, one obtains the system:

$$\alpha_{\tau\tau} = -\frac{1}{2}\lambda^2 e^{4\alpha}, \beta_{\tau\tau} = \gamma_{\tau\tau} = \frac{1}{2}\lambda^2 e^{4\alpha} \qquad\qquad\qquad (eq.\ 29)$$

(compare eq. 16; below it is substituted $\lambda^2 = 1$). The solution of these equations must describe the metric evolution from the initial state, in which it is described by eq. 8 with a given set of powers (with $p_l < 0$); let $p_l = p_1, p_m = p_2, p_n = p_3$ so that

$$a \sim t^{p_1}, b \sim t^{p_2}, c \sim t^{p_3}. \qquad\qquad\qquad\qquad (eq.\ 30)$$

Then

$$abc = \Lambda t, \tau = \Lambda^{-1} \ln t + \text{const} \qquad (eq.\ 31)$$

where Λ is constant. Initial conditions for eq. 29 are redefined as

$$\alpha_\tau = \Lambda p_1, \beta_\tau = \Lambda p_2, \gamma_\tau = \Lambda p_3 \text{ at } \tau \to \infty \qquad (eq.\ 32)$$

Equations eq. 29 are easily integrated; the solution that satisfies the condition eq. 32 is

$$\begin{cases} a^2 = \dfrac{2\,|\,p_1\,|\,\Lambda}{\text{ch}(2\,|\,p_1\,|\,\Lambda\tau)}, \\ b^2 = b_0^2 e^{2\Lambda(p_2-|p_1|)\tau}\,\text{ch}(2\,|\,p_1\,|\,\Lambda\tau), \\ c^2 = c_0^2 e^{2\Lambda(p_2-|p_1|)\tau}\,\text{ch}(2\,|\,p_1\,|\,\Lambda\tau), \end{cases} \qquad (eq.\ 33)$$

where b_0 and c_0 are two more constants.

It can easily be seen that the asymptotic of functions eq. 33 at $t \to 0$ is eq. 30. The asymptotic expressions of these functions and the function $t(\tau)$ at $\tau \to -\infty$ is

$$a \sim e^{-\Lambda p_1 \tau}, b \sim e^{\Lambda(p_2+2p_1)\tau}, c \sim e^{\Lambda(p_3+2p_1)\tau}, t \sim e^{\Lambda(1+2p_1)\tau}.$$

Expressing a, b, c as functions of t, one has

$$a \sim t^{p'_l}, b \sim t^{p'_m}, c \sim t^{p'_n} \qquad (eq.\ 34)$$

where

$$p'_l = \frac{|\,p_1\,|}{1-2\,|\,p_1\,|}, p'_m = -\frac{2\,|\,p_1\,|}{1-2\,|\,p_1\,|}\frac{p_2}{}, p'_n = \frac{p_3}{1-2\,|\,p_1\,|}\frac{2\,|\,p_1\,|}{}. \qquad (eq.\ 35)$$

Then

$$abc = \Lambda' t, \Lambda' = (1-2\,|\,p_1\,|)\Lambda. \qquad (eq.\ 36)$$

The above shows that perturbation acts in such a way that it changes one Kasner mode with another Kasner mode, and in this process the negative power of t flips from direction l to direction m: if before it was $p_l < 0$, now it is $p'_m < 0$. During this change the function $a(t)$ passes through a maximum and $b(t)$ passes through a minimum; b, which before was decreasing, now increases: a from increasing becomes decreasing; and the decreasing $c(t)$ decreases further. The perturbation itself ($\lambda^2 a^{4\alpha}$ in eq. 29), which before was increasing, now begins to decrease and die away. Further evolution similarly causes an increase in the perturbation from the terms with μ^2 (instead of λ^2) in eq. 26, next change of the Kasner mode, and so on.

It is convenient to write the power substitution rule eq. 35 with the help of the parametrization eq. 5:

$$\text{if} \qquad p_l = p_1(u) \qquad p_m = p_2(u) \qquad p_n = p_3(u)$$
$$\text{then} \qquad p_l' = p_2(u-1) \qquad p_m' = p_1(u-1) \qquad p_n' = p_3(u-1) \qquad \textbf{(eq. 37)}$$

The greater of the two positive powers remains positive.

BKL call this flip of negative power between directions a *Kasner epoch*. The key to understanding the character of metric evolution on approaching singularity is exactly this process of Kasner epoch alternation with flipping of powers p_l, p_m, p_n by the rule eq. 37.

The successive alternations eq. 37 with flipping of the negative power p_1 between directions l and m (Kasner epochs) continues by depletion of the whole part of the initial u until the moment at which $u < 1$. The value $u < 1$ transforms into $u > 1$ according to eq. 6; in this moment the negative power is p_l or p_m while p_n becomes the lesser of two positive numbers ($p_n = p_2$). The next series of Kasner epochs then flips the negative power between directions n and l or between n and m. At an arbitrary (irrational) initial value of u this process of alternation continues unlimited.

In the exact solution of the Einstein equations, the powers p_l, p_m, p_n lose their original, precise, sense. This circumstance introduces some "fuzziness" in the determination of these numbers (and together with them, to the parameter u) which, although small, makes meaningless the analysis of any definite (for example, rational) values of u. Therefore, only these laws that concern arbitrary irrational values of u have any particular meaning.

The larger periods in which the scales of space distances along two axes oscillate while distances along the third axis decrease monotonously, are called *eras*; volumes decrease by a law close to ~ t. On transition from one era to the next, the direction in which distances decrease monotonously, flips from one axis to another. The order of these transitions acquires the asymptotic character of a random process. The same random order is also characteristic for the alternation of the lengths of successive eras (by era length, BKL understand the number of Kasner epoch that an era contains, and not a time interval).

The era series become denser on approaching $t = 0$. However, the natural variable for describing the time course of this evolution is not the world time t, but its logarithm, ln t, by which the whole process of reaching the singularity is extended to $-\infty$.

According to eq. 33, one of the functions a, b, c, that passes through a maximum during a transition between Kasner epochs, at the peak of its maximum is

$$a_{\max} = \sqrt{2\Lambda \,|\, p_1(u)\,|} \qquad \textbf{(eq. 38)}$$

where it is supposed that $a_{\max}$ is large compared to b_0 and c_0; in eq. 38 u is the value of the parameter in the Kasner epoch before transition. It can be seen from here that the peaks of consecutive maxima during each era are gradually lowered. Indeed, in the next Kasner epoch this parameter has the value $u' = u - 1$, and Λ is substituted according to eq. 36 with $\Lambda' = \Lambda(1 - 2|p_1(u)|)$. Therefore, the ratio of 2 consecutive maxima is

$$\frac{a'_{max}}{a_{max}} = \left[\frac{p_1(u-1)}{p_1(u)} (1 - 2 \mid p_1(u) \mid) \right]^{\frac{1}{2}};$$

and finally

$$\frac{a'_{max}}{a_{max}} = \sqrt{\frac{u-1}{u}} \equiv \sqrt{\frac{u'}{u}}. \qquad \qquad (eq.\ 39)$$

The above are solutions to Einstein equations in vacuum. As for the pure Kasner mode, matter does not change the qualitative properties of this solution and can be written into it disregarding its reaction on the field.

However, if one does this for the model under discussion, understood as an exact solution of the Einstein equations, the resulting picture of matter evolution would not have a general character and would be specific for the high symmetry imminent to the present model. Mathematically, this specificity is related to the fact that for the homogeneous space geometry discussed here, the Ricci tensor components R_α^0 are identically zeros and therefore the Einstein equations would not allow movement of matter (which gives non-zero stress energy-momentum tensor components T_α^0).

This difficulty is avoided if one includes in the model only the major terms of the limiting (at $t \to 0$) metric and writes into it a matter with arbitrary initial distribution of densities and velocities. Then the course of evolution of matter is determined by its general laws of movement eq. 17 and eq. 18 that result in eq. 21. During each Kasner epoch, density increases by the law

$$\varepsilon = t^{-2(1-p_3)}, \qquad \qquad (eq.\ 40)$$

where p_3 is, as above, the greatest of the numbers p_1, p_2, p_3. Matter density increases monotonously during all evolution towards the singularity.

To each era (s-th era) correspond a series of values of the parameter u starting from the greatest, $u_{max}^{(s)}$, and through the values $u_{max}^{(s)} - 1$, $u_{max}^{(s)} - 2$, ..., reaching to the smallest, $u_{min^{(s)}} < 1$. Then

$$u_{min}^{(s)} = x^{(s)}, \ u_{max}^{(s)} = k^{(s)} + x^{(s)}, \qquad \qquad (eq.\ 41)$$

that is, $k^{(s)} = [u_{max^{(s)}}]$ where the brackets mean the whole part of the value. The number $k^{(s)}$ is the era length, measured by the number of Kasner epochs that the era contains. For the next era

$$u_{max^{(s+1)}} = \frac{1}{x^{(s)}}, \ k^{(s+1)} = \left[\frac{1}{x^{(s)}} \right]. \qquad \qquad (eq.\ 42)$$

In the limiteless series of numbers u, composed by these rules, there are infinitesimally small (but never zero) values $x^{(s)}$ and correspondingly infinitely large lengths $k^{(s)}$.

Metric Evolution

Very large u values correspond to Kasner powers

$$p_1 \approx -\frac{1}{u}, \ p_2 \approx \frac{1}{u}, \ p_2 \approx 1 - \frac{1}{u^2}, \tag{eq. 43}$$

which are close to the values (0, 0, 1). Two values that are close to zero, are also close to each other, and therefore the changes in two out of the three types of "perturbations" (the terms with λ, μ and ν in the right hand sides of eq. 26) are also very similar. If in the beginning of such long era these terms are very close in absolute values in the moment of transition between two Kasner epochs (or made artificially such by assigning initial conditions) then they will remain close during the greatest part of the length of the whole era. In this case (BKL call this the case of *small oscillations*), analysis based on the action of one type of perturbations becomes incorrect; one must take into account the simultaneous effect of two perturbation types.

Two Perturbations

Consider a long era, during which 2 out of the 3 functions *a*, *b*, *c* (let them be *a* and *b*) undergo small oscillations while the third function (*c*) decreases monotonously. The latter function quickly becomes small; consider the solution just in the region where one can ignore *c* in comparison to *a* and *b*. The calculations are first done for the Type IX space model by substituting accordingly $\lambda = \mu = \nu = 1$.

After ignoring function *c*, the first 2 equations eq. 26 give

$$\alpha_{\tau\tau} + \beta_{\tau\tau} = 0, \tag{eq. 44}$$

$$\alpha_{\tau\tau} - \beta_{\tau\tau} = e^{4\beta} - e^{4\alpha}, \tag{eq. 45}$$

and as a third equation, eq. 28 can be used, which takes the form

$$\gamma_{\tau\tau}\left(\alpha_{\tau\tau} + \beta_{\tau\tau}\right) = -\alpha_\tau \beta_\tau + \frac{1}{4}\left(e^{2\alpha} - e^{2\beta}\right)^2. \tag{eq. 46}$$

The solution of eq. 44 is written in the form

$$\alpha + \beta = \left(\frac{2a_0^2}{\xi_0}\right)(\tau - \tau_0) + 2\ln a_0,$$

where α_0, ξ_0 are positive constants, and τ_0 is the upper limit of the era for the variable τ. It is convenient to introduce further a new variable (instead of τ)

$$\xi = \xi_0 \exp\left[\frac{2a_0^2}{\xi_0}(\tau - \tau_0)\right]. \tag{eq. 47}$$

Then

$$\alpha + \beta = \ln\left(\frac{\xi}{\xi_0}\right) + 2\ln a_0. \tag{eq. 48}$$

Equations eq. 45 and eq. 46 are transformed by introducing the variable χ = α − β:

$$\chi_{\xi\xi} = \frac{\chi_\xi}{\xi} + \frac{1}{2}\,\mathrm{sh}\,2\chi = 0,$$

(eq. 49)

$$\gamma_\xi = -\frac{1}{4}\xi + \frac{1}{8}\xi\left(2\chi_\xi^2 + \mathrm{ch}\,2\chi - 1\right).$$

(eq. 50)

Decrease of τ from τ_0 to $-\infty$ corresponds to a decrease of ξ from ξ_0 to 0. The long era with close a and b (that is, with small χ), considered here, is obtained if ξ_0 is a very large quantity. Indeed, at large ξ the solution of eq. 49 in the first approximation by $1/\xi$ is

$$\chi = \alpha - \beta = \left(\frac{2A}{\sqrt{\xi}}\right)\sin\left(\xi - \xi_0\right),$$

(eq. 51)

where A is constant; the multiplier $\frac{1}{\sqrt{\xi}}$ makes χ a small quantity so it can be substituted in eq. 49 by $\mathrm{sh}\,2\chi \approx 2\chi$.

From eq. 50 one obtains

$$\gamma_\xi = \frac{1}{4}\xi\left(2\chi_\xi^2 + \chi^2\right) = A^2, \gamma = A^2\left(\xi - \xi_0\right) + \mathrm{const.}$$

After determining α and β from eq. 48 and eq. 51 and expanding e^α and e^β in series according to the above approximation, one obtains finally:

$$\begin{cases} a \\ b \end{cases} = a_0\sqrt{\frac{\xi}{\xi_0}}\left[1 \pm \frac{A}{\sqrt{\xi}}\sin\left(\xi - \xi_0\right)\right],$$

(eq. 52)

$$c = c_0 e^{-A^2(\xi_0 - \xi)}.$$

(eq. 53)

The relation between the variable ξ and time t is obtained by integration of the definition $dt = abc\,d\tau$ which gives

$$\frac{t}{t_0} = e^{-A^2(\xi_0 - \xi)}.$$

(eq. 54)

The constant c_0 (the value of c at $\xi = \xi_0$) should be now $c_0 \ll a_0$.

Let us now consider the domain $\xi \ll 1$. Here the major terms in the solution of eq. 49 are:

$$\chi = \alpha - \beta = k \ln \xi + \mathrm{const},$$

where k is a constant in the range $-1 < k < 1$; this condition ensures that the last term in eq. 49 is small ($\mathrm{sh}\,2\chi$ contains ξ^{2k} and ξ^{-2k}). Then, after determining α, β, and t, one obtains

$$a \sim \xi^{\frac{1+k}{2}}, b \sim \xi^{\frac{1-k}{2}}, c \sim \xi^{-\frac{1-k^2}{4}}, t \sim \xi^{\frac{3+k^2}{4}}.$$

(eq. 55)

This is again a Kasner mode with the negative t power coming into the function $c(t)$.

These results picture an evolution that is qualitatively similar to that, described above. During a long period of time that corresponds to a large decreasing ξ value, the two functions a and b oscil-

late, remaining close in magnitude $\dfrac{a-b}{a} \sim \dfrac{1}{\sqrt{\xi}}$; in the same time, both functions a and b slowly (

$\sim \sqrt{\xi}$) decrease. The period of oscillations is constant by the variable ξ : $\Delta\xi = 2\pi$ (or, which is the same, with a constant period by logarithmic time: $\Delta \ln t = 2\pi A^2$). The third function, c, decreases monotonously by a law close to $c = c_c t/t_0$.

This evolution continues until $\xi \sim 1$ and formulas eq. 52 and eq. 53 are no longer applicable. Its time duration corresponds to change of t from t_0 to the value t_1, related to ξ_0 according to

$$A^2 \xi_0 = \ln \frac{t_0}{t_1}. \qquad\qquad (eq.\ 56)$$

The relationship between ξ and t during this time can be presented in the form

$$\frac{\xi}{\xi_0} = \frac{\ln \frac{t}{t_1}}{\ln \frac{t_0}{t_1}}. \qquad\qquad (eq.\ 57)$$

After that, as seen from eq. 55, the decreasing function c starts to increase while functions a and b start to decrease. This Kasner epoch continues until terms c^2/a^2b^2 in eq. 22 become $\sim t^2$ and a next series of oscillations begins.

The law for density change during the long era under discussion is obtained by substitution of eq. 52 in eq. 20:

$$\varepsilon \sim \left(\frac{\xi_0}{\xi}\right)^2. \qquad\qquad (eq.\ 58)$$

When ξ changes from ξ_0 to $\xi \sim 1$, the density increases ξ_0^2 times.

It must be stressed that although the function $c(t)$ changes by a law, close to $c \sim t$, the metric eq. 52 does not correspond to a Kasner metric with powers (0, 0, 1). The latter corresponds to an exact solution (found by Taub) which is allowed by eqs. 26'–'27 and in which

$$a^2 = b^2 = \frac{p}{2}\frac{\operatorname{ch}(2p\tau + \delta_1)}{\operatorname{ch}^2(p\tau + \delta_2)}, c^2 = \frac{2p}{\operatorname{ch}(2p\tau + \delta_1)}, \qquad\qquad (eq.\ 59)$$

where p, δ_1, δ_2 are constant. In the asymptotic region $\tau \to -\infty$, one can obtain from here $a = b =$ const, $c =$ const.t after the substitution $e^{p\tau} = t$. In this metric, the singularity at $t = 0$ is non-physical.

Let us now describe the analogous study of the Type VIII model, substituting in eqs. eqs. 26'–'28 $\lambda = -1$, $\mu = \nu = 1$.

If during the long era, the monotonically decreasing function is a, nothing changes in the foregoing analysis: ignoring a^2 on the right side of equations 26 and 28, goes back to the same equations 49 and 50 (with altered notation). Some changes occur, however, if the monotonically decreasing function is b or c; let it be c.

As before, one has equation 49 with the same symbols, and, therefore, the former expressions eq. 52 for the functions $a(\xi)$ and $b(\xi)$, but equation 50 is replaced by

$$\gamma_\xi = -\frac{1}{4}\xi + \frac{1}{8}\xi\left(2\chi_\xi^2 + \mathrm{ch}2\chi + 1\right). \qquad \text{(eq. 60)}$$

The major term at large ξ now becomes

$$\gamma_\xi \approx \frac{1}{8}\xi \cdot 2, \quad \gamma \approx \frac{1}{8}\left(\xi^2 - \xi_0^2\right),$$

so that

$$\frac{c}{c_0} = \frac{t}{t_0} = e^{-\frac{1}{8}\left(\xi_0^2 - \xi^2\right)}. \qquad \text{(eq. 61)}$$

The value of c as a function of time t is, as before $c = c_0 t/t_0$, but the time dependence of ξ changes. The length of a long era depends on ξ_0 according to

$$\xi_0 = \sqrt{8\ln\frac{t}{t_0}}. \qquad \text{(eq. 62)}$$

On the other hand, the value ξ_0 determines the number of oscillations of the functions a and b during an era (equal to $\xi_0/2\pi$). Given the length of an era in logarithmic time (i.e., with given ratio t_0/t_1) the number of oscillations for Type VIII will be, generally speaking, less than for Type IX. For the period of oscillations one gets now $\Delta \ln t = \pi\xi/2$; contrary to Type IX, the period is not constant throughout the long era, and slowly decreases along with ξ.

The Small-time Domain

As shown above, long eras violate the "regular" course of evolution; this fact makes it difficult to study the evolution of time intervals, encompassing several eras. It can shown, however, that such "abnormal" cases appear in the spontaneous evolution of the model to a singular point in the asymptotically small times t at sufficiently large distances from a start point with arbitrary initial conditions. Even in long eras both oscillatory functions during transitions between Kasner epochs remain so different that the transition occurs under the influence of only one perturbation. All results in this section relate equally to models of the types VIII and IX.

During each Kasner epoch $abc = \Lambda t$, i. e. $\alpha + \beta + \gamma = \ln\Lambda + \ln t$. In transitions between epochs the constant $\ln\Lambda$ changes to the first order (cf. eq. 36). However, asymptotically to very large $|\ln t|$ values one can ignore not only these changes, but also the constant $\ln\Lambda$ itself. In other words, this approximation corresponds to ignoring all values whose ratio to $|\ln t|$ converges to zero at $t \to 0$. Then

$$\alpha + \beta + \gamma = -\Omega, \qquad \text{(eq. 63)}$$

where Ω is the "logarithmic time"

$$\Omega = -\ln t. \qquad \text{(eq. 64)}$$

In this approximation, the process of epoch transitions can be regarded as a series of brief time flashes. The constant in the right hand side of condition eq. 38 $\alpha_{max} = \frac{1}{2} \ln (2|p_1|\Lambda)$ that defines the periods of transition can also be ignored, *i. e.* this condition becomes $\alpha = 0$ (or similar conditions for β or γ if the initial negative power is related to the functions b or c). Thus, α_{max}, β_{max}, and γ_{max} become zeros meaning that α, β, and γ will run only through negative values which are related in each moment by the relationship eq. 64.

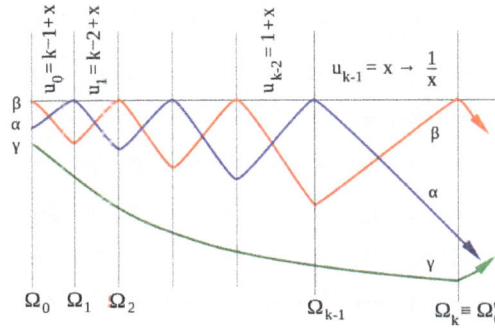

Considering such instant change of epochs, the transition periods are ignored as small in comparison to the epoch length; this condition is actually fulfilled. Replacement of α, β, and γ maxima with zeros requires that quantities $\ln (|p_1|\Lambda)$ be small in comparison with the amplitudes of oscillations of the respective functions. As mentioned above, during transitions between eras $|p_1|$ values can become very small while their magnitude and probability for occurrence are not related to the oscillation amplitudes in the respective moment. Therefore, in principle, it is possible to reach so small $|p_1|$ values that the above condition (zero maxima) is violated. Such drastic drop of α_{max} can lead to various special situations in which the transition between Kasner epochs by the rule eq. 37 becomes incorrect (including the situations described above). These "dangerous" situations could break the laws used for the statistical analysis below. As mentioned, however, the probability for such deviations converges asymptotically to zero; this issue will be discussed below.

Consider an era that contains k Kasner epochs with a parameter u running through the values

$$u_n = k + x - 1 - n, \quad n = 0, 1, \cdots, k - 1, \qquad \textbf{\textit{(eq. 65)}}$$

and let α and β are the oscillating functions during this era.

Initial moments of Kasner epochs with parameters u_n are Ω_n. In each initial moment, one of the values α or β is zero, while the other has a minimum. Values α or β in consecutive minima, that is, in moments Ω_n are

$$\alpha_n = -\delta_n \Omega_n \qquad \textbf{\textit{(eq. 66)}}$$

(not distinguishing minima α and β). Values δ_n that measure those minima in respective Ω_n units can run between 0 and 1. Function γ monotonously decreases during this era; according to eq. 63 its value in moment Ω_n is

$$\gamma_n = -\Omega_n(1-\delta_n).$$

(eq. 67)

During the epoch starting at moment Ω_n and ending at moment Ω_{n+1} one of the functions α or β increases from $-\delta_n\Omega_n$ to zero while the other decreases from 0 to $-\delta_{n+1}\Omega_{n+1}$ by linear laws, respectively:

$$\text{const}+\mid p_1(u_n)\mid\Omega \text{ and } \text{const} - p_2(u_n)\Omega$$

resulting in the recurrent relationship

$$\delta_{n+1}\Omega_{n+1} = \frac{1+u_n}{u_n}\delta_n\Omega_n = \frac{1+u_0}{u_n}\delta_0\Omega_0$$

(eq. 68)

and for the logarithmic epoch length

$$\Delta_{n+1} \equiv \Omega_{n+1} - \Omega_n = \frac{f(u_n)}{u_n}\delta_n\Omega_n = \frac{f(u_n)(1+u_{n-1})}{f(u_{n-1})u_n}\Delta_n,$$

(eq. 69)

where, for short, $f(u) = 1 + u + u^2$. The sum of n epoch lengths is obtained by the formula

$$\Omega_n - \Omega_0 = \left[n(n-1)+\frac{nf(u_{n-1})}{u_{n-1}}\right]\delta_0\Omega_0.$$

(eq. 70)

It can be seen from eq. 68 that $|\alpha_{n+1}| > |\alpha_n|$, i.e., the oscillation amplitudes of functions α and β increase during the whole era although the factors δ_n may be small. If the minimum at the beginning of an era is deep, the next minima will not become shallower; in other words, the residue $|\alpha - \beta|$ at the moment of transition between Kasner epochs remains large. This assertion does not depend upon era length k because transitions between epochs are determined by the common rule eq. 37 also for long eras.

The last oscillation amplitude of functions α or β in a given era is related to the amplitude of the first oscillation by the relationship $|\alpha_{k-1}| = |\alpha_0| (k + x) / (1 + x)$. Even at k's as small as several units x can be ignored in comparison to k so that the increase of α and β oscillation amplitudes becomes proportional to the era length. For functions $a = e^\alpha$ and $b = e^\beta$ this means that if the amplitude of their oscillations in the beginning of an era was A_0, at the end of this era the amplitude will become $A_0^{k/(1+x)}$.

The length of Kasner epochs (in logarithmic time) also increases inside a given era; it is easy to calculate from eq. 69 that $\Delta_{n+1} > \Delta_n$. The total era length is

$$\Omega_0' - \Omega_0 \equiv \Omega_k - \Omega_0 = k\left(k+x+\frac{1}{x}\right)\delta_0\Omega_0$$

(eq. 71)

(the term with $1/x$ arises from the last, k-th, epoch whose length is great at small x; cf. Fig. 2). Moment Ω_n when the k-th epoch of a given era ends is at the same time moment Ω_0' of the beginning of the next era.

In the first Kasner epoch of the new era function γ is the first to rise from the minimal value $\gamma_k = -\Omega_k(1 - \delta_k)$ that it reached in the previous era; this value plays the role of a starting amplitude $\delta_0'\Omega_0'$ for the new series of oscillations. It is easily obtained that:

$$\delta'_0 \Omega'_0 = \left(\delta_0^{-1} + k^2 + kx - 1 \right) \delta_0 \Omega_0.$$ (eq. 72)

It is obvious that $\delta'_0 \Omega'_0 > \delta_0 \Omega_0$. Even at not very great k the amplitude increase is very significant: function $c = e^y$ begins to oscillate from amplitude $A'_0 \sim A_0^{k^2}$. The issue about the abovementioned "dangerous" cases of drastic lowering of the upper oscillation limit is left aside for now.

According to eq. 40 the increase in matter density during the first $(k - 1)$ epochs is given by the formula

$$\ln \left(\frac{\varepsilon_{n+1}}{\varepsilon_n} \right) = 2 \left[1 - p_3(u_n) \right] \Delta_{n+1}.$$

For the last k epoch of a given era, it should be taken into account that at $u = x < 1$ the greatest power is $p_2(x)$ (not $p_3(x)$). Therefore, for the density increase over the whole era one obtains

$$\ln \left(\frac{\varepsilon_k}{\varepsilon_0} \right) \equiv \ln \left(\frac{\varepsilon'_0}{\varepsilon_0} \right) = 2(k - 1 + x) \delta_0 \Omega_0.$$ (eq. 73)

Therefore, even at not very great k values, $\varepsilon_{0'} / \varepsilon_0 \sim A_0^{2k}$. During the next era (with a length k') density will increase faster because of the increased starting amplitude A_0': $\varepsilon_{0''} / \varepsilon_{0'} \sim A_{0'}^{2k''} \sim A_0^{2k^2 k'}$, etc. These formulae illustrate the steep increase in matter density.

Statistical Analysis Near the Singularity

The sequencing order of era lengths $k^{(s)}$, measured by the number of Kasner epochs contained in them, exhibits the character of a random process. The source of this stochasticity is the rule eqs. 41'–'42 according to which the transition from one era to the next is determined from an infinite numerical sequence of u values.

In the statistical description of this sequence, instead of a fixed initial value $u_{max} = k^{(0)} + x^{(0)}$, BKL consider values of $x^{(0)}$ that are distributed in the interval from 0 to 1 by some probabilistic distributional law. Then the values of $x^{(s)}$ that finish each (s-th) number series will also be distributed according to some laws. It can be shown that with growing s these distributions converge to a definite static (s-independent) distribution of probabilities $w(x)$ in which the initial conditions are completely "forgotten":

$$w(x) = \frac{1}{(1 + x) \ln 2}.$$ (eq. 74)

This allows to find the distribution of probabilities for length k:

$$W(k) = \left(\ln 2 \right)^{-1} \ln \left[\frac{(k + 1)^2}{k(k + 2)} \right].$$ (eq. 75)

The above formulae are the basis on which the statistical properties of the model evolution are studied.

This study is complicated by the slow decrease of the distribution function eq. 75 at large k:

$$W(k) \approx \frac{1}{k^2 \ln 2}.$$

(*eq. 76*)

The mean value $\bar{k}$, calculated from this distribution, diverges logarithmically. For a sequence, cut off at a very large, but still finite number N, one has $\bar{k} \sim \ln N$. The usefulness of the mean in this case is very limited because of its instability: because of the slow decrease of $W(k)$, fluctuations in k diverge faster than its mean. A more adequate characteristic of this sequence is the probability that a randomly chosen number from it belongs to a series of length K where K is large. This probability is $\ln K/\ln N$. It is small if $1 \ll K \ll N$. In this respect one can say that a randomly chosen number from the given sequence belongs to the long series with a high probability.

The recurrent formulae defining transitions between eras are re-written and detailed below. Index s numbers the successive eras (not the Kasner epochs in a given era!), beginning from some era (s = 0) defined as initial. $\Omega^{(s)}$ and $\varepsilon^{(s)}$ are, respectively, the initial moment and initial matter density in the s-th era; $\delta_s \Omega_s$ is the initial oscillation amplitude of that pair of functions α, β, γ, which oscillates in the given era: $k^{(s)}$ is the length of s-th era, and $x^{(s)}$ determines the length of the next era according to $k^{(s+1)} = [1/x^{(s)}]$. According to eqs. 71'–'73

$$\Omega^{(s+1)} / \Omega^{(s)} = 1 + \delta^{(s)} k^{(s)} \left(k^{(s)} + x^{(s)} + 1/x^{(s)} \right) \equiv \varepsilon^{\xi_s},$$

(*eq. 77*)

$$\delta^{(s+1)} = 1 - \frac{\left(k^{(s)} / x^{(s)} + 1 \right) \delta^{(s)}}{1 + \delta^{(s)} k^{(s)} \left(1 + x^{(s)} + 1/x^{(s)} \right)},$$

(*eq. 78*)

$$\ln \left(\frac{\varepsilon^{(s+1)}}{\varepsilon^{(s)}} \right) = 2 \left(k^{(s)} + x^{(s)} - 1 \right) \delta^{(s)} \Omega^{(s)}$$

(*eq. 79*)

(ξ_s is introduced in eq. 77 to be used further on).

The values of $\delta^{(s)}$ (ranging from 0 to 1) have their own static statistical distribution. It satisfies an integral equation expressing the fact that $\delta^{(s)}$ and $\delta^{(s+1)}$ which are related through eq. 78 have an identical distribution; this equation can be solved numerically (cf.). Since eq. 78 does not contain a singularity, the distribution is perfectly stable; the mean values of δ or its powers calculated through it are definite finite numbers. In particular, the mean value of δ is

The statistical relation between large time intervals Ω and the number of eras s contained in them is found by repeated application of eq. 77:

$$\frac{\Omega^{(s)}}{\Omega^{(0)}} = \exp \left(\sum_{p=0}^{s-1} \xi_p \right).$$

(*eq. 80*)

Direct averaging of this equation, however, does not make sense: because of the slow decrease of function $W(k)$ mean values of $\exp(\xi_s)$ are unstable in the above sense. This instability is removed by taking logarithm: the "double-logarithmic" time interval

$$\tau_s \equiv \ln \left(\frac{\Omega^{(s)}}{\Omega^{(0)}} \right) = \sum_{p=0}^{s-1} \xi_p$$

(*eq. 81*)

is expressed by the sum of values ξ_p which have a stable statistical distribution. The mean values of ξ_s and their powers (calculated from the distributions of values x, k and δ) are finite; numeric calculation gives

Averaging eq. 81 at a given s obtains

$$\bar{\tau}_s = 2.1s, \qquad\qquad (eq.\ 82)$$

which determines the mean double-logarithmic time interval containing s successive eras.

In order to calculate the mean square of fluctuations of this value one writes

$$\overline{(\tau_s - \bar{\tau}_s)^2} = \sum_{p,q=0}^{s-1} \left(\overline{\xi_p \xi_q} - \bar{\xi}_p \bar{\xi}_q \right) = s \sum_{p=0}^{s-1} \left(\overline{\xi_0 \xi_p} - \bar{\xi}^2 \right).$$

In the last equation, it is taken into account that in the static limit the statistical correlation between $\xi^{(s)}$ and $\xi'^{(s)}$ depends only on the difference $|s - s'|$. Due to the existing recurrent relationship between $x^{(s)}$, $k^{(s)}$, $\delta^{(s)}$ and $x^{(s+1)}$, $k^{(s+1)}$, $\delta^{(s+1)}$ this correlation is, strictly speaking, different from zero. It, however, quickly decreases with increasing $|s - s'|$ and numeric calculation shows that even at $|s - s'| = 1$, $\overline{\xi_{s+1}\xi_s} - \bar{\xi}^2 = -0.4$. Leaving the first two terms in the sum by p, one obtains

$$\left[\overline{(\tau_s - \bar{\tau}_s)^2} \right]^{\frac{1}{2}} = 1.4\sqrt{s}, \qquad\qquad (eq.\ 83)$$

At $s \to \infty$ the relative fluctuation (i.e., the ratio between the mean squared fluctuations eq. 83 and the mean value eq. 82), therefore, approaches zero as $s^{-1/2}$. In other words, the statistical relationship eq. 82 at large s becomes close to certainty. This is a corollary that according to eq. 81 τ_s can be presented as a sum of a large number of quasi-independent additives (i.e., it has the same origin as the certainty of the values of additive thermodynamic properties of macroscopic bodies). Therefore, the probabilities of various τ_s values (at given s) have a Gaussian distribution:

$$\rho(\tau_s) \propto \exp \left\{ -\frac{(\tau_s - 2.1s)^2}{4s} \right\}. \qquad\qquad (eq.\ 84)$$

Certainty of relationship eq. 82 allows its reversal, i.e., express it as a dependence of the mean number of eras $\bar{s}_\tau$ contained in a given interval of double-logarithmic time τ:

$$\bar{s}_\tau = 0.47\tau. \qquad\qquad (eq.\ 85)$$

The respective statistical distribution is given by the same Gaussian distribution in which the random variable is now s_τ at a given τ:

$$\rho(s_\tau) \propto \exp \left\{ -(s_\tau - 0.47\tau)^2 / 0.43\tau \right\}. \qquad\qquad (eq.\ 86)$$

Respective to matter density, eq. 79 can be re-written with account of eq. 80 in the form

$$\ln \ln \frac{\varepsilon^{(s+1)}}{\varepsilon^{(s)}} = \eta_s + \sum_{p=0}^{s-1} \xi_p, \quad \eta_s = \ln \left[2\delta^{(s)} \left(k^{(s)} + x^{(s)} - 1 \right) \Omega^{(0)} \right]$$

and then, for the complete energy change during s eras,

$$\ln \ln \frac{\varepsilon^{(s)}}{\varepsilon^{(0)}} = \ln \sum_{p=0}^{s-1} \exp \left\{ \sum_{q=0}^{p} \xi_q + \eta_p \right\}. \qquad \textit{(eq. 87)}$$

The term with the sum by p gives the main contribution to this expression because it contains an exponent with a large power. Leaving only this term and averaging eq. 87, one gets in its right hand side the expression $s\bar{\xi}$ which coincides with eq. 82; all other terms in the sum (also terms with η_s in their powers) lead only to corrections of a relative order $1/s$. Therefore,

$$\overline{\ln \ln \left(\frac{\varepsilon^{(s)}}{\varepsilon^{(0)}} \right)} = \overline{\ln \left(\frac{\Omega^{(s)}}{\Omega^{(0)}} \right)}. \qquad \textit{(eq. 88)}$$

Thanks to the above established almost certain character of the relation between τ_s and s eq. 88 can be written as

$$\overline{\ln \ln \left(\varepsilon_\tau / \varepsilon^{(0)} \right)} = \tau \quad \text{or} \quad \overline{\ln \ln \left(\varepsilon^{(s)} / \varepsilon^{(0)} \right)} = 2.1s,$$

which determines the value of the double logarithm of density increase averaged by given double-logarithmic time intervals τ or by a given number of eras s.

These stable statistical relationships exist specifically for double-logarithmic time intervals and for the density increase. For other characteristics, e.g., $\ln (\varepsilon^{(s)}/\varepsilon^{(0)})$ the relative fluctuation increase by a power law with the increase of the averaging range thereby devoiding the term mean value of its sense of stability.

As shown below, in the limiting asymptotic case the abovementioned "dangerous" cases that disturb the regular course of evolution expressed by the recurrent relationships eqs. 77'–'79, do not occur in reality.

Dangerous are cases when at the end of an era the value of the parameter $u = x$ (and with it also $|p_1| \approx x$). A criterion for selection of such cases is the inequality

$$x^{(s)} \exp \left| \alpha^{(s)} \right| < 1, \qquad \textit{(eq. 89)}$$

where $| \alpha^{(s)} |$ is the initial minima depth of the functions that oscillate in era s (it would have been better to take the final amplitude, but that would only strengthen the selection criterion).

The value of $x^{(0)}$ in the first era is determined by the initial conditions. Dangerous are values in the interval $\delta x^{(0)} \sim \exp (- |\alpha^{(0)}|)$, and also in intervals that could result in dangerous cases in the next eras. In order that $x^{(s)}$ comes into the dangerous interval $\delta x^{(s)} \sim \exp (- | \alpha^{(s)} |)$, the initial value $x^{(0)}$ should lie into an interval of a width $\delta x^{(0)} \sim \delta x^{(s)} / k^{(1)^\wedge 2} \dots k^{(s)^\wedge 2}$. Therefore, from a unit interval of all possible values of $x^{(0)}$, dangerous cases will appear in parts λ of this interval:

$$\lambda = \exp \left(\left| -\alpha^{(s)} \right| \right) + \sum_{s=1}^{\infty} \sum_{k} \frac{\exp \left(\left| -\alpha^{(s)} \right| \right)}{k^{(1)^2} k^{(2)^2} \cdots k^{(s)^2}} \qquad \textit{(eq. 90)}$$

(the inner sum is taken by all values $k^{(1)}, k^{(2)}, \dots , k^{(s)}$ from 1 to ∞). It is easy to show that this series converges to the value $\lambda \ll 1$ whose order of magnitude is determined by the first term in eq. 90.

This can be shown by a strong majoration of the series for which one substitutes $| \alpha^{(s)} | = (s + 1) | \alpha^{(0)} |$, regardless of the lengths of eras $k^{(1)}$, $k^{(2)}$, ... (In fact $| \alpha^{(s)} |$ increase much faster; even in the most unfavorable case $k^{(1)} = k^{(2)} = ... = 1$ values of $| \alpha^{(s)} |$ increase as $q^s | \alpha^{(0)} |$ with $q > 1$.) Noting that

$$\sum_k 1 / k^{(1)^2} k^{(2)^2} \cdots k^{(s)^2} = \left(\pi^2 / 6 \right)^s$$

one obtains

$$\lambda = \exp\left(-\left|\alpha^{(0)}\right|\right) \sum_{s=0}^{\infty} \left[\left(\pi^2 / 6 \right) \exp\left(-\left|\alpha^{(0)}\right|\right) \right]^s \approx \exp\left(-\left|\alpha^{(0)}\right|\right).$$

If the initial value of $x^{(0)}$ lies outside the dangerous region λ there will be no dangerous cases. If it lies inside this region dangerous cases occur, but upon their completion the model resumes a "regular" evolution with a new initial value which only occasionally (with a probability λ) may come into the dangerous interval. Repeated dangerous cases occur with probabilities λ^2, λ^3, ... , asymptotically converging to zero.

General Solution with Small Oscillations

In the above models, metric evolution near the singularity is studied on the example of homogeneous space metrics. It is clear from the characteristic of this evolution that the analytic construction of the general solution for a singularity of such type should be made separately for each of the basic evolution components: for the Kasner epochs, for the process of transitions between epochs caused by "perturbations", for long eras with two perturbations acting simultaneously. During a Kasner epoch (i.e. at small perturbations), the metric is given by eq. 7 without the condition $\lambda = 0$.

BKL further developed a matter distribution-independent model (homogeneous or non-homogeneous) for long era with small oscillations. The time dependence of this solution turns out to be very similar to that in the particular case of homogeneous models; the latter can be obtained from the distribution-independent model by a special choice of the arbitrary functions contained in it.

It is convenient, however, to construct the general solution in a system of coordinates somewhat different from synchronous reference frame: $g_{0\alpha} = 0$ as in the synchronous frame, but instead of $g_{00} = 1$ it is now $g_{00} = -g_{33}$. Defining again the space metric tensor $\gamma_{\alpha\beta} = -g_{\alpha\beta}$ one has, therefore

$$g_{00} = \gamma_{33}, \quad g_{0\alpha} = 0. \tag{eq. 91}$$

The special space coordinate is written as $x^3 = z$ and the time coordinate is written as $x^0 = \xi$ (as different from proper time t); it will be shown that ξ corresponds to the same variable defined in homogeneous models. Differentiation by ξ and z is designated, respectively, by dot and prime. Latin indices a, b, c take values 1, 2, corresponding to space coordinates x^1, x^2 which will be also written as x, y. Therefore, the metric is

$$ds^2 = \gamma_{33} \left(d\xi^2 - dz^2 \right) - \gamma_{ab} dx^a dx^b - 2\gamma_{a3} dx^a dz. \tag{eq. 92}$$

N BThe required solution should satisfy the inequalities

$$\gamma_{33} \ll \gamma_{ab}, \tag{eq. 93}$$

$$\gamma_{a3}^2 \ll \gamma_{aa}\gamma_{33} \tag{eq. 94}$$

(these conditions specify that one of the functions a^2, b^2, c^2 is small compared to the other two which was also the case with homogeneous models).

Inequality eq. 94 means that components γ_{a3} are small in the sense that at any ratio of the shifts dx^a and dz, terms with products $dx^a dz$ can be omitted in the square of the spatial length element dl^2. Therefore, the first approximation to a solution is a metric eq. 92 with $\gamma_{a3} = 0$:

$$ds^2 = \gamma_{33}\left(d\xi^2 - dz^2\right) - \gamma_{ab}dx^a dx^b. \tag{eq. 95}$$

One can be easily convinced by calculating the Ricci tensor components R_0^0, R_3^0, R_3^3, R_a^b using metric eq. 95 and the condition eq. 93 that all terms containing derivatives by coordinates x^a are small compared to terms with derivatives by ξ and z (their ratio is $\sim \gamma_{33} / \gamma_{ab}$). In other words, to obtain the equations of the main approximation, γ_{33} and γ_{ab} in eq. 95 should be differentiated as if they do not depend on x^a. Designating

$$\gamma_{33} = e^{\psi}, \quad \dot{\gamma}_{ab} = \varkappa_{ab}, \quad \gamma'_{ab} = \lambda_{ab}, \quad |\gamma_{ab}| = G^2, \tag{eq. 96}$$

one obtains the following equations:

$$2e^{\psi} R_a^b = G^{-1}\left(G\lambda_a^b\right)' - G^{-1}\left(G\varkappa_a^b\right)\cdot = 0, \tag{eq. 97}$$

$$2e^{\psi} R_3^0 = \frac{1}{2}\varkappa\psi' + \frac{1}{2}\lambda\dot{\psi} - \varkappa' - \frac{1}{2}\varkappa_a^b\lambda_b^a = 0, \tag{eq. 98}$$

$$2e^{\psi}\left(R_0^0 - R_3^3\right) = \lambda\psi' + \varkappa\dot{\psi} - \dot{\varkappa} - \lambda' - \frac{1}{2}\varkappa_a^b\varkappa_b^a - \frac{1}{2}\lambda_a^b\lambda_b^a = 0. \tag{eq. 99}$$

Index raising and lowering is done here with the help of γ_{ab}. The quantities $\dot{\upsilon}$ and λ are the contractions $\dot{\upsilon}_a^a$ and λ_a^a whereby

$$\varkappa = 2\dot{G}/G, \quad \lambda = 2G'/G. \tag{eq. 100}$$

As to the Ricci tensor components R_a^0, R_a^3, by this calculation they are identically zero. In the next approximation (i.e., with account to small γ_{a3} and derivatives by x, y), they determine the quantities γ_{a3} by already known γ_{33} and γ_{ab}.

Contraction of eq. 97 gives $G'' + \ddot{G} = 0$, and, hence,

$$G = f_1(x, y, \xi + z) + f_2(x, y, \xi - z). \tag{eq. 101}$$

Different cases are possible depending on the G variable. In the above case $g^{00} = \gamma^{33} \gg \gamma^{ab}$ and

$N \approx g^{00}\left(\dot{G}\right)^2 - \gamma^{33}\left(G'\right)^2 = 4\gamma^{33}\dot{f}_1\dot{f}_2..$ The case $N > 0$ (quantity N is time-like) leads to time singularities of interest. Substituting in **eq. 101** $f_1 = 1/2 (\xi + z) \sin y, f_2 = 1/2 (\xi - z) \sin y$ results in G of type

$$G = \xi \sin y. \qquad\qquad (eq.\ 102)$$

This choice does not diminish the generality of conclusions; it can be shown that generality is possible (in the first approximation) just on account of the remaining permissible transformations of variables. At $N < 0$ (quantity N is space-like) one can substitute $G = z$ which generalizes the well-known Einstein–Rosen metric. At $N = 0$ one arrives at the Robinson–Bondi wave metric that depends only on $\xi + z$ or only on $\xi - z$ (cf.). The factor $\sin y$ in eq. 102 is put for convenient comparison with homogeneous models. Taking into account eq. 102, equations 97'–'99 become

$$\dot{\varkappa}_a^b + \xi^{-1}\varkappa_a^b - \lambda_a^{b'} = 0, \qquad\qquad (eq.\ 103)$$

$$\dot{\psi} = -\xi^{-1} + \frac{1}{4}\xi\left(\varkappa_a^b\varkappa_t^a + \lambda_a^b\lambda_b^a\right). \qquad\qquad (eq.\ 104)$$

$$\psi' = \frac{1}{2}\xi_a^b\lambda_b^a. \qquad\qquad (eq.\ 105)$$

The principal equations are eq. 103 defining the γ_{ab} components; then, function ψ is found by a simple integration of eqs. 104'–'105.

The variable ξ runs through the values from 0 to ∞. The solution of eq. 103 is considered at two boundaries, $\xi \gg 1$ and $\ll 1$. At large ξ values, one can look for a solution that takes the form of a $1 / \sqrt{\xi}$ decomposition:

$$\gamma_{ab} = \xi\left[a_{ab}(x,y,z) + O(1/\sqrt{\xi})\right], \qquad\qquad (eq.\ 106)$$

whereby

$$|a_{ab}| = \sin^2 y \qquad\qquad (eq.\ 107)$$

(equation 107 needs condition 102 to be true). Substituting eq. 103 in eq. 106, one obtains in the first order

$$\left(a^{ac'}a_{bc}\right)' = 0, \qquad\qquad (eq.\ 108)$$

where quantities a^{ac} constitute a matrix that is inverse to matrix a_{ac}. The solution of eq. 108 has the form

$$a_{ab} = l_a l_b e^{-2\rho z} + m_a m_b e^{2\rho z}, \qquad\qquad (eq.\ 109)$$

$$l_1 m_2 + l_2 m_1 = \sin y, \qquad\qquad (eq.\ 110)$$

where l_a, m_a, ρ, are arbitrary functions of coordinates x, y bound by condition eq. 110 derived from eq. 107.

To find higher terms of this decomposition, it is convenient to write the matrix of required quantities γ_{ab} in the form

$$\gamma_{ab} = \xi\left(\tilde{L}e^H L\right)_{ab}, \qquad\qquad (eq.\ 111)$$

$$L = \begin{bmatrix} l_1 e^{-\rho z} & l_2 e^{-\rho z} \\ m_1 e^{\rho z} & m_2 e^{\rho z} \end{bmatrix},$$

$$(eq.\ 112)$$

where the symbol ~ means matrix transposition. Matrix H is symmetric and its trace is zero. Presentation eq. 111 ensures symmetry of γ_{ab} and fulfillment of condition eq. 102. If exp H is substituted with 1, one obtains from eq. 111 $\gamma_{ab} = \xi a_{ab}$ with a_{ab} from eq. 109. In other words, the first term of γ_{ab} decomposition corresponds to $H = 0$; higher terms are obtained by powers decomposition of matrix H whose components are considered small.

The independent components of matrix H are written as σ and φ so that

$$H = \begin{bmatrix} \sigma & \varphi \\ \varphi & -\sigma \end{bmatrix}.$$

$$(eq.\ 113)$$

Substituting eq. 111 in eq. 103 and leaving only terms linear by H, one derives for σ and φ

$$\ddot{\sigma} + \xi^{-1}\dot{\sigma} - \sigma'' = 0,$$

$$\ddot{\varphi} + \xi^{-1}\dot{\varphi} - \varphi'' + 4\rho^2\varphi = 0.$$

$$(eq.\ 114)$$

If one tries to find a solution to these equations as Fourier series by the z coordinate, then for the series coefficients, as functions of ξ, one obtains Bessel equations. The major asymptotic terms of the solution at large ξ are

$$\sigma = \frac{1}{\sqrt{\xi}} \sum_{n=-\infty}^{\infty} \left(A_{1n} e^{in\omega\xi} + B_{1n} e^{-in\omega\xi} \right) e^{in\omega z},$$

$$\varphi = \frac{1}{\sqrt{\xi}} \sum_{n=-\infty}^{\infty} \left(A_{2n} e^{in\omega\xi} + B_{2n} e^{-in\omega\xi} \right) e^{in\omega z},$$

$$\omega_n^2 = n^2\omega^2 + 4\rho^2.$$

$$(eq.\ 115)$$

Coefficients A and B are arbitrary complex functions of coordinates x, y and satisfy the necessary conditions for real σ and φ; the base frequency ω is an arbitrary real function of x, y. Now from eqs. 104'–'105 it is easy to obtain the first term of the function ψ:

$$\psi = \rho^2 \xi^2$$

$$(eq.\ 116)$$

(this term vanishes if $\rho = 0$; in this case the major term is the one linear for ξ from the decomposition: $\psi = \xi q\,(x,\,y)$ where q is a positive function).

Therefore, at large ξ values, the components of the metric tensor γ_{ab} oscillate upon decreasing ξ on the background of a slow decrease caused by the decreasing ξ factor in eq. 111. The component $\gamma_{33} = e^{\psi}$ decreases quickly by a law close to exp $(\rho^2\xi^2)$; this makes it possible for condition eq. 93.

Next BKL consider the case $\xi \ll 1$. The first approximation to a solution of eq. 103 is found by the assumption (confirmed by the result) that in these equations terms with derivatives by coordinates can be left out:

$$\ddot{\varkappa}_a^b + \xi^{-1}\varkappa_a^b = 0. \qquad\qquad (eq.\ 117)$$

This equation together with the condition eq. 102 gives

$$\gamma_{ab} = \lambda_a\lambda_b\xi^{2s_1} + \mu_a\mu_b\xi^{2s_2}, \qquad\qquad (eq.\ 118)$$

where λ_a, μ_a, s_1, s_2 are arbitrary functions of all 3 coordinates x, y, z, which are related with other conditions

$$\lambda_1\mu_2 - \lambda_2\mu_1 = \sin y, \quad s_1 + s_2 = 1. \qquad\qquad (eq.\ 119)$$

Equations 104'–'105 give now

$$\gamma_{33} = e^{\psi} \sim \xi^{-(1-s_1^2-s_2^2)}. \qquad\qquad (eq.\ 120)$$

The derivatives $\lambda_a^{b'}$, calculated by eq. 118, contain terms $\sim \xi^{4s_1-2}$ and $\sim \xi^{4s_2-2}$ while terms left in eq. 117 are $\sim \xi^{-2}$. Therefore, application of eq. 103 instead of eq. 117 is permitted on conditions $s_1 > 0$, $s_2 > 0$; hence $1 - s_1^2 - s_2^2 > 0$.

Thus, at small ξ oscillations of functions γ_{ab} cease while function γ_{33} begins to increase at decreasing ξ. This is a Kasner mode and when γ_{33} is compared to γ_{ab}, the above approximation is not applicable.

In order to check the compatibility of this analysis, BKL studied the equations $R_\alpha^0 = 0$, $R_\alpha^3 = 0$, and, calculating from them the components γ_{a3}, confirmed that the inequality eq. 94 takes place. This study showed that in both asymptotic regions the components γ_{a3} were $\sim \gamma_{33}$. Therefore, correctness of inequality eq. 93 immediately implies correctness of inequality eq. 94.

This solution contains, as it should be for the general case of a field in vacuum, four arbitrary functions of the three space coordinates x, y, z. In the region $\xi \ll 1$ these functions are, e.g., λ_1, λ_2, μ_1, s_1. In the region $\xi \gg 1$ the four functions are defined by the Fourier series by coordinate z from eq. 115 with coefficients that are functions of x, y; although Fourier series decomposition (or integral?) characterizes a special class of functions, this class is large enough to encompass any finite subset of the set of all possible initial conditions.

The solution contains also a number of other arbitrary functions of the coordinates x, y. Such *two-dimensional* arbitrary functions appear, generally speaking, because the relationships between three-dimensional functions in the solutions of the Einstein equations are differential (and not algebraic), leaving aside the deeper problem about the geometric meaning of these functions. BKL did not calculate the number of independent two-dimensional functions because in this case it is hard to make unambiguous conclusions since the three-dimensional functions are defined by a set of two-dimensional functions (cf. for more details).

Finally, BKL go on to show that the general solution contains the particular solution obtained above for homogeneous models.

Substituting the basis vectors for Bianchi Type IX homogeneous space in eq. 7 the space-time metric of this model takes the form

$$ds_{IX}^2 = dt^2 - \left[\left(a^2 \sin^2 z + b^2 \cos^2 z\right)\sin^2 y + c^2 \cos^2 y\right]dx^2 - \left[a^2 \cos^2 z + b^2 \sin^2 z\right]dy^2 - c^2 dz^2 +$$

$$\left(b^2 - a^2\right)\sin 2z \sin y\, dxdy - 2c^2 \cos y\, dxdz. \qquad \text{(eq. 121)}$$

When $c^2 \ll a^2$, b^2, one can ignore c^2 everywhere except in the term $c^2\, dz^2$. To move from the synchronous frame used in eq. 121 to a frame with conditions eq. 91, the transformation $dt = c\, d\xi/2$ and substitution $z \to z/2$ are done. Assuming also that $\chi \equiv \ln(a/b) \ll 1$, one obtains from eq. 121 in the first approximation:

$$ds_{IX}^2 = \tfrac{1}{4}c^2\left(d\xi^2 - dz^2\right) - ab\left\{\sin^2 y\left(1 - \chi \cos z\right)dx^2 + \left(1 + \chi \cos z\right) + 2\chi \sin z \sin y\, dx\, dy\right\}. \qquad \text{(eq. 122)}$$

Similarly, with the basis vectors of Bianchi Type VIII homogeneous space, one obtains

$$ds_{VIII}^2 = \tfrac{1}{4}c^2\left(d\xi^2 - dz^2\right) - ab\left\{\sin^2 y\left(\text{ch}\, z - \chi\right)dx^2 + \left(\text{ch}\, z + \chi\right) + 2\,\text{sh}\, z \sin y\, dx\, dy\right\}. \qquad \text{(eq. 123)}$$

According to the analysis of homogeneous spaces above, in both cases $ab = \xi$ (simplifying $= \xi_0$) and χ is from eq. 51; function $c\,(\xi)$ is given by formulae eq. 53 and eq. 61, respectively, for models of Types IX and VIII.

Identical metric for Type VIII is obtained from eqs. 112, 115, 116 choosing two-dimensional vectors l_a and m_a in the form

$$l_1 = m_1 = \frac{1}{\sqrt{2}}\sin y, \qquad l_2 = m_2 = \frac{1}{\sqrt{2}} \qquad \text{(eq. 124)}$$

and substituting

$$\rho = \tfrac{1}{2}, \quad A_{20}^* = B_{20} = iAe^{i\xi_0}, \quad A_{1n} = A_{2n} = B_{1n} = B_{2n} = 0 \quad (n \neq 0). \qquad \text{(eq. 125)}$$

To obtain the metric for Type IX, one should substitute

$$\rho = 0, \omega = 1,$$

$$A_{11} = -B_{11}^* = A_{1,-1}^* = -B_{1,-1} = -\frac{1}{2}Ae^{-i\xi_0},$$

$$A_{21} = B_{21}^* = A_{2,-1}^* = B_{2,-1} = -\frac{1}{2}iAe^{-i\xi_0}, \qquad \text{(eq. 126)}$$

$$A_{1n} = A_{2n} = B_{1n} = B_{2n} = 0 \quad (n \neq \pm 0)$$

(for calculation of $c\,(\xi)$ the approximation in eq. 116 is not sufficient and the term in ψ linear by ξ is calculated)

This analysis was done for empty space. Including matter does not make the solution less general and does not change its qualitative characteristics.

Conclusions

BKL describe singularities in the cosmologic solution of Einstein equations that have a complicated oscillatory character. Although these singularities have been studied primarily on spatially homogeneous models, there are convincing reasons to assume that singularities in the general solution of Einstein equations have the same characteristics; this circumstance makes the BKL model important for cosmology.

A basis for such statement is the fact that the oscillatory mode in the approach to singularity is caused by the single perturbation that also causes instability in the generalized Kasner solution. A confirmation of the generality of the model is the analytic construction for long era with small oscillations. Although this latter behavior is not a necessary element of metric evolution close to the singularity, it has all principal qualitative properties: metric oscillation in two spatial dimensions and monotonous change in the third dimension with a certain perturbation of this mode at the end of some time interval. However, the transitions between Kasner epochs in the general case of non-homogeneous spatial metric have not been elucidated in details.

The problem connected with the possible limitations upon space geometry caused by the singularity was left aside for further study. It is clear from the outset, however, that the original BKL model is applicable to both finite or infinite space; this is evidenced by the existence of oscillatory singularity models for both closed and open spacetimes.

The oscillatory mode of the approach to singularity gives a new aspect to the term 'finiteness of time'. Between any finite moment of the world time t and the moment $t = 0$ there is an infinite number of oscillations. In this sense, the process acquires an infinite character. Instead of time t, a more adequate variable for its description is $\ln t$ by which the process is extended to $-\infty$.

BKL consider metric evolution in the direction of decreasing time. The Einstein equations are symmetric in respect to the time sign so that a metric evolution in the direction of increasing time is equally possible. However, these two cases are fundamentally different because past and future are not equivalent in the physical sense. Future singularity can be physically meaningful only if it is possible at arbitrary initial conditions existing in a previous moment. Matter distribution and fields in some moment in the evolution of Universe do not necessarily correspond to the specific conditions required for the existence of a given special solution to the Einstein equations.

The choice of solutions corresponding to the real world is related to profound physical requirements which is impossible to find using only the existing relativity theory and which can be found as a result of future synthesis of physical theories. Thus, it may turn out that this choice singles out some special (e.g., isotropic) type of singularity. Nevertheless, it is more natural to assume that because of its general character, the oscillatory mode should be the main characteristic of the initial evolutionary stages.

In this respect, of considerable interest is the property of the model, shown by Misner, related to propagation of light signals. In the isotropic model, a "light horizon" exists, meaning that for each moment of time, there is some longest distance, at which exchange of light signals and, thus, a causal connection, is impossible: the signal cannot reach such distances for the time since the singularity $t = 0$.

Signal propagation is determined by the equation $ds = 0$. In the isotropic model near the singularity $t = 0$ the interval element is $ds^2 = dt^2 - 2t$, where is a time-independent spatial differential form. Substituting $t = \eta^2/2$ yields

$$ds^2 = \eta^2 \left(d\eta^2 - d\bar{l}^2 \right).$$ **(eq. 127)**

The "distance" $\Delta \bar{l}$ reached by the signal is

$$\Delta \bar{l} = \Delta \eta.$$ **(eq. 128)**

Since η, like t, runs through values starting from 0, up to the "moment" η signals can propagate only at the distance $\Delta \bar{l} \leq \eta$ which fixes the farthest distance to the horizon.

The existence of a light horizon in the isotropic model poses a problem in the understanding of the origin of the presently observed isotropy in the relic radiation. According to the isotropic model, the observed isotropy means isotropic properties of radiation that comes to the observer from such regions of space that can not be causally connected with each other. The situation in the oscillatory evolution model near the singularity can be different.

For example, in the homogeneous model for Type IX space, a signal is propagated in a direction in which for a long era, scales change by a law close to $\sim t$. The square of the distance element in this direction is $dl^2 = t^2 \bar{l}^2$, and the respective element of the four-dimensional interval is $ds^2 = dt^2 - t^2 \bar{l}^2$. The substitution $t = e^{\eta}$ puts this in the form

$$ds^2 = e^{2\eta} \left(d\eta^2 - d\bar{l}^2 \right),$$ **(eq. 129)**

and for the signal propagation one has equation of the type eq. 128 again. The important difference is that the variable η runs now through values starting from $-\infty$ (if metric eq. 129 is valid for all t starting from $t = 0$).

Therefore, for each given "moment" η are found intermediate intervals $\Delta\eta$ sufficient for the signal to cover each finite distance.

In this way, during a long era a light horizon is opened in a given space direction. Although the duration of each long era is still finite, during the course of the world evolution eras change an infinite number of times in different space directions. This circumstance makes one expect that in this model a causal connection between events in the whole space is possible. Because of this property, Misner named this model «mixmaster universe» by a brand name of a dough-blending machine.

As time passes and one goes away from the singularity, the effect of matter on metric evolution, which was insignificant at the early stages of evolution, gradually increases and eventually becomes dominant. It can be expected that this effect will lead to a gradual "isotropisation" of space as a result of which its characteristics come closer to the Friedman model which adequately describes the present state of the Universe.

Finally, BKL pose the problem about the feasibility of considering a "singular state" of a world with infinitely dense matter on the basis of the existing relativity theory. The physical application of the Einstein equations in their present form in these conditions can be made clear only in the process of a future synthesis of physical theories and in this sense the problem can not be solved at present.

It is important that the gravitational theory itself does not lose its logical cohesion (i.e., does not lead to internal controversies) at whatever matter densities. In other words, this theory is not limited by the conditions that it imposes, which could make logically inadmissible and controversial its application at very large densities; limitations could, in principle, appear only as a result of factors that are "external" to the gravitational theory. This circumstance makes the study of singularities in cosmological models formally acceptable and necessary in the frame of existing theory.

Relativistic Angular Momentum

In physics, relativistic angular momentum refers to the mathematical formalisms and physical concepts that define angular momentum in special relativity (SR) and general relativity (GR). The relativistic quantity is subtly different from the three-dimensional quantity in classical mechanics.

Angular momentum is a dynamical quantity derived from position and momentum, and is important; angular momentum is a measure of an object's "amount of rotational motion" and resistance to stop rotating. Also, in the same way momentum conservation corresponds to translational symmetry, angular momentum conservation corresponds to rotational symmetry – the connection between symmetries and conservation laws is made by Noether's theorem. While these concepts were originally discovered in classical mechanics – they are also true and significant in special and general relativity. In terms of abstract algebra; the invariance of angular momentum, four-momentum, and other symmetries in spacetime, are described by the Lorentz group, or more generally the Poincaré group.

Physical quantities which remain separate in classical physics are *naturally combined* in SR and GR by enforcing the postulates of relativity. Most notably; space and time coordinates combine into the four-position, and energy and momentum combine into the four-momentum. The components of these four-vectors depend on the frame of reference used, and change under Lorentz transformations to other inertial frames or accelerated frames.

Relativistic angular momentum is less obvious. The classical definition of angular momentum is the cross product of position x with momentum p to obtain a pseudovector x×p, or alternatively as the exterior product to obtain a second order antisymmetric tensor x∧p. What does this combine with, if anything? There is another vector quantity not often discussed – it is the time-varying moment of mass (*not* the moment of inertia) related to the boost of the centre of mass of the system, and this combines with the classical angular momentum to form an antisymmetric tensor of second order. For rotating mass–energy distributions (such as gyroscopes, planets, stars, and black holes) instead of point-like particles, the angular momentum tensor is expressed in terms of the stress–energy tensor of the rotating object.

In special relativity alone, in the rest frame of a spinning object; there is an intrinsic angular momentum analogous to the "spin" in quantum mechanics and relativistic quantum mechanics, although for an extended body rather than a point particle. In relativistic quantum mechanics, elementary particles have *spin* and this is an additional contribution to the *orbital* angular momentum operator, yielding the *total* angular momentum tensor operator. In any case, the intrinsic "spin" addition to the orbital angular momentum of an object can be expressed in terms of the Pauli–Lubanski pseudovector.

Definitions

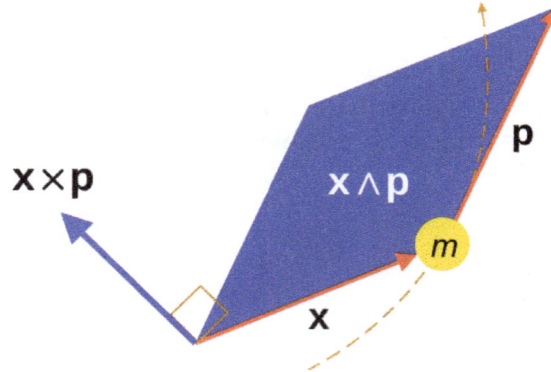

The 3-angular momentum as a bivector (plane element) and axial vector, of a particle of mass m with instantaneous 3-position **x** and 3-momentum **p**.

Orbital 3d Angular Momentum

For reference and background, two closely related forms of angular momentum are given.

In classical mechanics, the orbital angular momentum of a particle with instantaneous three-dimensional position vector x = (x, y, z) and momentum vector p = (p_x, p_y, p_z), is defined as the *axial vector*

$$\mathbf{L} = \mathbf{x} \times \mathbf{p}$$

which has three components, that are systematically given by cyclic permutations of Cartesian directions (e.g. change x to y, y to z, z to x, repeat)

$$L_x = yp_z - zp_y,$$
$$L_y = zp_x - xp_z,$$
$$L_z = xp_y - yp_x.$$

A related definition is to conceive orbital angular momentum as a *plane element*. This can be achieved by replacing the cross product by the exterior product in the language of exterior algebra, and angular momentum becomes a contravariant second order antisymmetric tensor

$$\mathbf{L} = \mathbf{x} \wedge \mathbf{p}$$

or writing x = $(x_1, x_2, x_3) = (x, y, z)$ and momentum vector p = $(p_1, p_2, p_3) = (p_x, p_y, p_z)$, the components can be compactly abbreviated in tensor index notation

$$L^{ij} = x^i p^j - x^j p^i$$

where the indices i and j take the values 1, 2, 3. On the other hand, the components can be systematically displayed fully in a 3×3 antisymmetric matrix

$$\mathbf{L} = \begin{pmatrix} L^{11} & L^{12} & L^{13} \\ L^{21} & L^{22} & L^{23} \\ L^{31} & L^{32} & L^{33} \end{pmatrix} = \begin{pmatrix} 0 & L_{xy} & L_{xz} \\ L_{yx} & 0 & L_{yz} \\ L_{zx} & L_{zy} & 0 \end{pmatrix} = \begin{pmatrix} 0 & L_{xy} & -L_{zx} \\ -L_{xy} & 0 & L_{yz} \\ L_{zx} & -L_{yz} & 0 \end{pmatrix}$$

$$= \begin{pmatrix} 0 & xp_y - yp_x & -(zp_x - xp_z) \\ -(xp_y - yp_x) & 0 & yp_z - zp_y \\ zp_x - xp_z & -(yp_z - zp_y) & 0 \end{pmatrix}$$

This quantity is additive, and for an isolated system, the total angular momentum of a system is conserved.

Dynamic Mass Moment

In classical mechanics, the three-dimensional quantity for a particle of mass m moving with velocity u

$$\mathbf{N} = m(\mathbf{x} - t\mathbf{u}) = m\mathbf{x} - t\mathbf{p}$$

has the dimensions of *mass moment* – length multiplied by mass. It is related to the boost (relative velocity) of the centre of mass (COM) of the particle or system of particles, as measured in the lab frame. There is no universal symbol, nor even a universal name, for this quantity. Different authors may denote it by other symbols if any (for example µ), may designate other names, and may define N to be the negative of what is used here. The above form has the advantage that it resembles the familiar Galilean transformation for position, which in turn is the non-relativistic boost transformation between inertial frames.

This vector is also additive: for a system of particles, the vector sum is the resultant

$$\sum_n \mathbf{N}_n = \sum_n m_n(\mathbf{x}_n - t\mathbf{u}_n) = \left(\mathbf{x}_{\text{COM}} \sum_n m_n - t \sum_n m_n \mathbf{u}_n \right)$$

where the system's centre of mass is

$$\mathbf{x}_{\text{COM}} = \frac{\sum_n m_n \mathbf{x}_n}{\sum_n m_n}$$

For an isolated system, N is conserved in time, which can be seen by differentiating with respect to time. The angular momentum L is a pseudovector, but N is an "ordinary" (polar) vector, and is therefore invariant under rotations.

The resultant N_{total} for a multiparticle system has the physical visualization that, whatever the complicated motion of all the particles are, they move in such a way that the system's COM moves in a straight line. This does not necessarily mean all particles "follow" the COM, nor that all particles all move in almost the same direction simultaneously, only that the motion of all the particles are constrained in relation to the centre of mass.

In special relativity, if the particle moves with velocity u relative to the lab frame, then

$$E = \gamma(\mathbf{u})m_0 c^2, \quad \mathbf{p} = \gamma(\mathbf{u})m_0\mathbf{u}$$

where

$$\gamma(\mathbf{u}) = \frac{1}{\sqrt{1 - \dfrac{\mathbf{u} \cdot \mathbf{u}}{c^2}}}$$

is the Lorentz factor and m_0 the rest mass of the particle. Some authors use relativistic mass

$$m = \gamma(\mathbf{u})m_0$$

or proper velocity

$$\mathbf{w} = \gamma(\mathbf{u})\mathbf{u}$$

The corresponding relativistic mass moment in terms of m_0, m, u, p, E, in the same lab frame is

$$\mathbf{N} = m\mathbf{x} - \mathbf{p}t = \frac{E}{c^2}\mathbf{x} - \mathbf{p}t = \gamma(\mathbf{u})m_0(\mathbf{x} - \mathbf{u}t)$$

defined here so that the relativistic equation in terms of the relativistic mass, and classical definition, have the same form. The Cartesian components are

$$N_x = mx - p_x t = \frac{E}{c^2}x - p_x t = \gamma(u)m_0(x - u_x t)$$

$$N_y = my - p_y t = \frac{E}{c^2}y - p_y t = \gamma(u)m_0(y - u_y t)$$

$$N_z = mz - p_z t = \frac{E}{c^2}z - p_z t = \gamma(u)m_0(z - u_z t)$$

Expressing N in terms of relativistic mass-energy and momentum, rather than rest mass and velocity, avoids extra Lorentz factors. However, relativistic mass is discouraged by some authors since it can be a misleading quantity to apply in certain equations.

Special Relativity

Coordinate Transformations for a Boost in the X Direction

Consider a coordinate frame F′ which moves with velocity v = $(v, 0, 0)$ relative to another frame F, along the direction of the coincident xx' axes. The origins of the two coordinate frames coincide at times $t = t' = 0$. The mass–energy $E = mc^2$ and momentum components p = (p_x, p_y, p_z) of an object, as well as position coordinates x = (x, y, z) and time t in frame F are transformed to $E' = m'c^2$, p′ = (p_x', p_y', p_z'), x′ = (x', y', z'), and t' in F′ according to the Lorentz transformations

$$t' = \gamma(v)\left(t - \frac{vx}{}\right), \quad E' = \gamma(v)(E - vp)$$

$$x' = \gamma(v)(x - vt), \quad p_{x'} = \gamma(v)\left(p_x - \frac{vE}{}\right)$$

$$y' = y \quad p_{y'} = p_y$$

$$z' = z \quad p_{z'} = p_z$$

The Lorentz factor here applies to the velocity v, the relative velocity between the frames. This is not necessarily the same as the velocity u of an object.

For the orbital 3-angular momentum L as a pseudovector, we have

$$L_{x'} = y'p_{z'} - z'p_{y'} = L_x$$

$$L_{y'} = z'p_{x'} - x'p_{z'} = \gamma(v)(L_y - vN_z)$$

$$L_{z'} = x'p_{y'} - y'p_{x'} = \gamma(v)(L_z - vN_y)$$

In the second terms of L_y' and L_z', the y and z components of the cross product v×N can be inferred by recognizing cyclic permutations of $v_x = v$ and $v_y = v_z = 0$ with the components of N,

$$-vN_z = v_z N_x - v_x N_z = (\mathbf{v} \times \mathbf{N})_y$$

$$vN_y = v_x N_y - v_y N_x = (\mathbf{v} \times \mathbf{N})_z$$

Now, L_x is parallel to the relative velocity v, and the other components L_y and L_z are perpendicular to v. The parallel-perpendicular correspondence can be facilitated by splitting the entire 3-angular momentum pseudovector into components parallel (∥) and perpendicular (⊥) to v, in each frame,

$$\mathbf{L} = \mathbf{L}_{\parallel} + \mathbf{L}_{\perp}, \quad \mathbf{L}' = \mathbf{L}_{\parallel'} + \mathbf{L}_{\perp}'.$$

Then the component equations can be collected into the pseudovector equations

$$\mathbf{L}_{\parallel}' = \mathbf{L}_{\parallel}$$

$$\mathbf{L}_{\perp}' = \gamma(\mathbf{v})(\mathbf{L}_{\perp} + \mathbf{v} \times \mathbf{N})$$

Therefore, component of angular momentum along the direction of motion does not change, while the components perpendicular do change. By contrast to the transformations of space and time, time and the spatial coordinates change along the direction of motion, while those perpendicular do not.

These transformations are true for *all* v, not just for motion along the *xx'* axes.

Considering L as a tensor, we get a similar result

$$\mathbf{L}'_\perp = \gamma(\mathbf{v})\left(\mathbf{L}_\perp + \mathbf{v} \wedge \mathbf{N}\right)$$

where
$$v_z N_x - v_x N_z = \left(\mathbf{v} \wedge \mathbf{N}\right)_{zx}$$
$$v_x N_y - v_y N_x = \left(\mathbf{v} \wedge \mathbf{N}\right)_{xy}$$

The boost of the dynamic mass moment along the x direction is
$$N'_x = m'x' - p'_x t' = N_x$$
$$N'_y = m'y' - p'_y t' = \gamma(v)\left(N_y + \frac{vL_z}{c^2}\right)$$
$$N'_z = m'z' - p'_z t' = \gamma(v)\left(N_z - \frac{vL_y}{c^2}\right)$$

Collecting parallel and perpendicular components as before

$$\mathbf{N}'_\| = \mathbf{N}_\|$$
$$\mathbf{N}'_\perp = \gamma(\mathbf{v})\left(\mathbf{N}_\perp - \frac{1}{c^2}\mathbf{v} \times \mathbf{L}\right)$$

Again, the components parallel to the direction of relative motion do not change, those perpendicular do change.

Vector Transformations for a Boost in Any Direction

So far these are only the parallel and perpendicular decompositions of the vectors. The transformations on the full vectors can be constructed from them as follows (throughout here L is a pseudovector for concreteness and compatibility with vector algebra).

Introduce a unit vector in the direction of v, given by n = v/v. The parallel components are given by the vector projection of L or N into n

$$\mathbf{L}_\| = (\mathbf{L} \cdot \mathbf{n})\mathbf{n}, \quad \mathbf{N}_\| = (\mathbf{N} \cdot \mathbf{n})\mathbf{n}$$

while the perpendicular component by vector rejection of L or N from n

$$\mathbf{L}_\perp = \mathbf{L} - (\mathbf{L} \cdot \mathbf{n})\mathbf{n}, \quad \mathbf{N}_\perp = \mathbf{N} - (\mathbf{N} \cdot \mathbf{n})\mathbf{n}$$

and the transformations are

$$\mathbf{L}' = \gamma(\mathbf{v})(\mathbf{L} + v\mathbf{n} \times \mathbf{N}) - (\gamma(\mathbf{v}) - 1)(\mathbf{L} \cdot \mathbf{n})\mathbf{n}$$
$$\mathbf{N}' = \gamma(\mathbf{v})\left(\mathbf{N} - \frac{v}{c^2}\mathbf{n} \times \mathbf{L}\right) - (\gamma(\mathbf{v}) - 1)(\mathbf{N} \cdot \mathbf{n})\mathbf{n}$$

or reinstating $v = v n$,

$$\mathbf{L'} = \gamma(\mathbf{v})(\mathbf{L} + \mathbf{v} \times \mathbf{N}) - (\gamma(\mathbf{v}) - 1)\frac{(\mathbf{L} \cdot \mathbf{v})\mathbf{v}}{v^2}$$

$$\mathbf{N'} = \gamma(\mathbf{v})\left(\mathbf{N} - \frac{1}{c^2}\mathbf{v} \times \mathbf{L}\right) - (\gamma(\mathbf{v}) - 1)\frac{(\mathbf{N} \cdot \mathbf{v})\mathbf{v}}{v^2}$$

These are very similar to the Lorentz transformations of the electric field E and magnetic field B.

Alternatively, starting from the vector Lorentz transformations of time, space, energy, and momentum, for a boost with velocity v,

$$t' = \gamma(\mathbf{v})\left(t - \frac{\mathbf{v} \cdot \mathbf{r}}{c^2}\right),$$

$$\mathbf{r'} = \mathbf{r} + \frac{\gamma(\mathbf{v}) - 1}{v^2}(\mathbf{r} \cdot \mathbf{v})\mathbf{v} - \gamma(\mathbf{v})t\mathbf{v},$$

$$\mathbf{p'} = \mathbf{p} + \frac{\gamma(\mathbf{v}) - 1}{v^2}(\mathbf{p} \cdot \mathbf{v})\mathbf{v} - \gamma(\mathbf{v})\frac{E}{c^2}\mathbf{v},$$

$$E' = \gamma(\mathbf{v})(E - \mathbf{v} \cdot \mathbf{p}),$$

inserting these into the definitions

$$\mathbf{L'} = \mathbf{r'} \times \mathbf{p'}, \quad \mathbf{N'} = \frac{E'}{c^2}\mathbf{r'} - t'\mathbf{p'}$$

gives the transformations.

4d Angular Momentum as a Bivector

In relativistic mechanics, the COM boost and orbital 3-angular momentum of a rotating object are combined into a four-dimensional bivector in terms of the 4-position X and the 4-momentum P of the object

$$\mathbf{M} = \mathbf{X} \wedge \mathbf{P}$$

In components

$$M^{\alpha\beta} = X^\alpha P^\beta - X^\beta P^\alpha$$

which are six independent quantities altogether. Since the components of X and P are frame-dependent, so is M. Three components

$$M^{ij} = x^i p^j - x^j p^i = L^{ij}$$

are those of the familiar classical 3-orbital angular momentum, and the other three

$$M^{0i} = x^0 p^i - x^i p^0 = c\left(tp^i - x^i \frac{E}{c^2} \right) = -cN^i$$

are the relativistic mass moment, multiplied by $-c$. The tensor is antisymmetric;

$$M^{\alpha\beta} = -M^{\beta\alpha}$$

The components of the tensor can be systematically displayed as a matrix

$$
\mathbf{M} = \begin{pmatrix}
M^{00} & M^{01} & M^{02} & M^{03} \\
M^{10} & M^{11} & M^{12} & M^{13} \\
M^{20} & M^{21} & M^{22} & M^{23} \\
M^{30} & M^{31} & M^{32} & M^{33}
\end{pmatrix}
$$
$$
= \begin{pmatrix}
0 & -N^1 c & -N^2 c & -N^3 c \\
N^1 c & 0 & L^{12} & -L^{31} \\
N^2 c & -L^{12} & 0 & L^{23} \\
N^3 c & L^{31} & -L^{23} & 0
\end{pmatrix}
$$
$$
= \begin{pmatrix}
0 & -\mathbf{N}c \\
\mathbf{N}^{\mathrm{T}}c & \mathbf{x} \wedge \mathbf{p}
\end{pmatrix}
$$

in which the last array is a block matrix formed by treating N as a row vector which matrix transposes to the column vector $\mathbf{N}^{\mathrm{T}}$, and x∧p as a 3×3 antisymmetric matrix.

Again, this tensor is additive: the total angular momentum of a system is the sum of the angular momentum tensors for each constituent of the system:

$$\mathbf{M}_{\text{total}} = \sum_n \mathbf{M}_n = \sum_n \mathbf{X}_n \wedge \mathbf{P}_n .$$

Each of the six components forms a conserved quantity when aggregated with the corresponding components for other objects and fields.

The angular momentum tensor M is indeed a tensor, the components change according to a Lorentz transformation matrix Λ, as illustrated in the usual way by tensor index notation

$$
\begin{aligned}
M'^{\alpha\beta} &= X'^{\alpha} P'^{\beta} - X'^{\beta} P'^{\alpha} \\
&= \Lambda^{\alpha}{}_{\gamma} X^{\gamma} \Lambda^{\beta}{}_{\delta} P^{\delta} - \Lambda^{\beta}{}_{\delta} X^{\delta} \Lambda^{\alpha}{}_{\gamma} P^{\gamma} \\
&= \Lambda^{\alpha}{}_{\gamma} \Lambda^{\beta}{}_{\delta} \left(X^{\gamma} P^{\delta} - X^{\delta} P^{\gamma} \right) \\
&= \Lambda^{\alpha}{}_{\gamma} \Lambda^{\beta}{}_{\delta} M^{\gamma\delta}
\end{aligned}
$$

where, for a boost (without rotations) with normalized velocity $\beta = v/c$, the Lorentz transformation matrix elements are

$$\Lambda^0{}_0 = \gamma$$

$$\Lambda^i{}_0 = \Lambda^0{}_i = -\gamma\beta^i$$

$$\Lambda^i{}_j = \delta^i{}_j + \frac{\gamma-1}{\beta^2}\beta^i\beta_j$$

and the covariant β_i and contravariant β^i components of β are the same since these are just parameters.

In other words, one can Lorentz-transform the four position and four momentum separately, and then antisymmetrize those newly found components to obtain the angular momentum tensor in the new frame.

Rigid Body Rotation

For a particle moving in a curve, the cross product of its angular velocity ω (a pseudovector) and position x give its tangential velocity

$$\mathbf{u} = \mathbf{\omega} \times \mathbf{x}$$

which cannot exceed a magnitude of c, since in SR the translational velocity of any massive object cannot exceed the speed of light c. Mathematically this constraint is $0 \le |u| < c$, the vertical bars denote the magnitude of the vector. If the angle between ω and x is θ (assumed to be nonzero, otherwise u would be zero corresponding to no motion at all), then $|u| = |\omega||x|\sin\theta$ and the angular velocity is restricted by

$$0 \le |\omega| < \frac{c}{|\mathbf{x}|\sin\theta}$$

The maximum angular velocity of any massive object therefore depends on the size of the object. For a given $|x|$, the maximum upper limit occurs when ω and x are perpendicular, so that $\theta = \pi/2$ and $\sin\theta = 1$.

For a rotating rigid body rotating with an angular velocity ω, the u is tangential velocity at a point x inside the object. For every point in the object, there is a maximum angular velocity.

The angular velocity (pseudovector) is related to the angular momentum (pseudovector) through the moment of inertia tensor I

$$\mathbf{L} = \mathbf{I} \cdot \mathbf{\omega} \quad \rightleftharpoons \quad L_i = I_{ij}\omega_j$$

(the dot $\cdot$ denotes tensor contraction on one index). The relativistic angular momentum is also limited by the size of the object.

Spin in Special Relativity

Four-spin

A particle may have a "built-in" angular momentum independent of its motion, called spin and denoted s. It is a 3d pseudovector like orbital angular momentum L.

The spin has a corresponding spin magnetic moment, so if the particle is subject to interactions (like electromagnetic fields or spin-orbit coupling), the direction of the particle's spin vector will change, but its magnitude will be constant.

The extension to special relativity is straightforward. For some lab frame F, let F′ be the rest frame of the particle and suppose the particle moves with constant 3-velocity u. Then F′ is boosted with the same velocity and the Lorentz transformations apply as usual; it is more convenient to use $\beta = u/c$. As a four-vector in special relativity, the four-spin S generally takes the usual form of a four-vector with a timelike component s_t and spatial components s, in the lab frame

$$\mathbf{S} \equiv \left(S^0, S^1, S^2, S^3 \right) = (s_t, s_x, s_y, s_z)$$

although in the rest frame of the particle, it is defined so the timelike component is zero and the spatial components are those of particle's actual spin vector, in the notation here s′, so in the particle's frame

$$\mathbf{S}' \equiv \left(S'^0, S'^1, S'^2, S'^3 \right) = \left(0, s'_x, s'_y, s'_z \right)$$

Equating norms leads to the invariant relation

$$s_t^2 - \mathbf{s} \cdot \mathbf{s} = -\mathbf{s}' \cdot \mathbf{s}'$$

so if the magnitude of spin is given in the rest frame of the particle and lab frame of an observer, the magnitude of the timelike component s_t is given in the lab frame also.

The covariant constraint on the spin is orthogonality to the velocity vector,

$$U_\alpha S^\alpha = 0$$

In 3-vector notation for explicitness, the transformations are

$$s_t = \beta \cdot \mathbf{s}$$

$$\mathbf{s}' = \mathbf{s} + \frac{\gamma^2}{\gamma + 1} \beta \left(\beta \cdot \mathbf{s} \right) - \gamma \beta s_t$$

The inverse relations

$$s_t = \gamma \beta \cdot \mathbf{s}'$$

$$\mathbf{s} = \mathbf{s}' + \frac{\gamma^2}{\gamma + 1} \beta \left(\beta \cdot \mathbf{s}' \right)$$

are the components of spin the lab frame, calculated from those in the particle's rest frame. Although the spin of the particle is constant for a given particle, it appears to be different in the lab frame.

The Pauli-Lubanski Pseudovector

The Pauli-Lubanski pseudovector

$$S_\rho = \frac{1}{2} \varepsilon_{\lambda\mu\nu\rho} U^\lambda J^{\mu\nu},$$

applies to both massive and massless particles.

Spin–orbital Decomposition

In general, the total angular momentum tensor splits into an orbital component and a spin component,

$$J^{\mu\nu} = M^{\mu\nu} + S^{\mu\nu}$$

This applies to a particle, a mass-energy-momentum distribution, or field.

Angular Momentum of a Mass-energy-momentum Distribution

Angular Momentum from the Mass-energy-momentum Tensor

The following is a summary from MTW. Throughout for simplicity, Cartesian coordinates are assumed. In special and general relativity, a distribution of mass-energy-momentum, e.g. a fluid, or a star, is described by the stress–energy tensor $T^{\beta\gamma}$ (a second order tensor field depending on space and time). Since T^{00} is the energy density, T^{j0} for $j = 1, 2, 3$ is the jth component of the object's 3d momentum per unit volume, and T^{ij} form components of the stress tensor including shear and normal stresses, the orbital angular momentum density about the position 4-vector X^β is given by a 3rd order tensor

$$\mathcal{M}^{\alpha\beta\gamma} = \left(X^\alpha - \bar{X}^\alpha \right) T^{\beta\gamma} - \left(X^\beta - \bar{X}^\beta \right) T^{\alpha\gamma}$$

This is antisymmetric in α and β. In special and general relativity, T is a symmetric tensor, but in other contexts (e.g., quantum field theory), it may not be.

Let Ω be a region of 4d spacetime. The boundary is a 3d spacetime hypersurface ("spacetime surface volume" as opposed to "spatial surface area"), denoted $\partial\Omega$ where "∂" means "boundary". Integrating the angular momentum density over a 3d spacetime hypersurface yields the angular momentum tensor about X,

$$M^{\alpha\beta}\left(\bar{X}\right) = \oint_{\partial\Omega} \mathcal{M}^{\alpha\beta\gamma} d\Sigma_\gamma$$

where $d\Sigma_\gamma$ is the volume 1-form playing the role of a unit vector normal to a 2d surface in ordinary 3d Euclidean space. The integral is taken over the coordinates X, not X. The integral within a spacelike surface of constant time is

$$M^{ij} = \oint_{\partial\Omega} \mathcal{M}^{ij0} d\Sigma_0 = \oint_{\partial\Omega} \left[\left(X^i - Y^i \right) T^{j0} - \left(X^j - Y^j \right) T^{i0} \right] dx\,dy\,dz$$

which collectively form the angular momentum tensor.

Angular Momentum about the Centre of Mass

There is an intrinsic angular momentum in the centre-of-mass frame, in other words, the angular momentum about any event

$$\mathbf{X}_{\text{COM}} = \left(X^0_{\text{COM}}, X^1_{\text{COM}}, X^2_{\text{COM}}, X^3_{\text{COM}} \right)$$

on the wordline of the object's center of mass. Since T^{00} is the energy density of the object, the spatial coordinates of the center of mass are given by

$$X^i_{\text{COM}} = \frac{1}{m_0} \int_{\partial\Omega} X^i T^{00} dx dy dz$$

Setting $Y = \mathrm{X}_{\text{COM}}$ obtains the orbital angular momentum density about the centre-of-mass of the object.

Angular Momentum Conservation

The conservation of energy–momentum is given in differential form by the continuity equation

$$\partial_\gamma T^{\beta\gamma} = 0$$

where ∂_γ is the four gradient. (In non-Cartesian coordinates and general relativity this would be replaced by the covariant derivative). The total angular momentum conservation is given by another continuity equation

$$\partial_\gamma \mathcal{J}^{\alpha\beta\gamma} = 0$$

The integral equations use Gauss' theorem in spacetime

$$\int_V \partial_\gamma T^{\beta\gamma} c dt dx dy dz = \oint_{\partial V} T^{\beta\gamma} d^3\Sigma_\gamma = 0$$

$$\int_V \partial_\gamma \mathcal{J}^{\alpha\beta\gamma} c dt dx dy dz = \oint_{\partial V} \mathcal{J}^{\alpha\beta\gamma} d^3\Sigma_\gamma = 0$$

Torque in Special Relativity

The torque acting on a point-like particle is defined as the derivative of the angular momentum tensor given above with respect to proper time:

$$\tau = \frac{d\mathbf{M}}{d\tau} = \mathbf{X} \wedge \mathbf{F}$$

or in tensor components:

$$\Gamma_{\alpha\beta} = X_\alpha F_\beta - X_\beta F_\alpha$$

where F is the 4d force acting on the particle at the event X. As with angular momentum, torque is additive, so for an extended object one sums or integrates over the distribution of mass.

Angular Momentum as the Generator of Spacetime Boosts and Rotations

The angular momentum tensor is the generator of boosts and rotations for the Lorentz group. Lorentz boosts can be parametrized by rapidity, and a 3d unit vector n pointing in the direction of the boost, which combine into the "rapidity vector"

$$\zeta = \zeta n = n \tanh^{-1} \beta$$

where $\beta = v/c$ is the speed of the relative motion divided by the speed of light. Spatial rotations can be parametrized by the axis–angle representation, the angle θ and a unit vector a pointing in the direction of the axis, which combine into an "axis-angle vector"

$$\theta = \theta a$$

Each unit vector only has two independent components, the third is determined from the unit magnitude. Altogether there are six parameters of the Lorentz group; three for rotations and three for boosts. The (homogeneous) Lorentz group is 6-dimensional.

The boost generators K and rotation generators J can be combined into one generator for Lorentz transformations; M the antisymmetric angular momentum tensor, with components

$$M^{0i} = -M^{i0} = K_i, \quad M^{ij} = \varepsilon_{ijk} J_k.$$

and correspondingly, the boost and rotation parameters are collected into another antisymmetric four-dimensional matrix ω, with entries:

$$\omega_{0i} = -\omega_{i0} = \zeta_i, \quad \omega_{ij} = \varepsilon_{ijk} \theta_k,$$

where the summation convention over the repeated indices i, j, k has been used to prevent clumsy summation signs. The general Lorentz transformation is then given by the Matrix exponential

$$\Lambda(\zeta,\theta) = \exp\left(\frac{1}{2}\omega_{\alpha\beta} M^{\alpha\beta}\right) = \exp(\zeta \cdot K + \cdot \theta J)$$

and the summation convention has been applied to the repeated matrix indices α and β.

The general Lorentz transformation Λ is the transformation law for any four vector A = (A_0, A_1, A_2, A_3), giving the components of this same 4-vector in another inertial frame of reference

$$A' = \Lambda(\zeta,\theta)A$$

The angular momentum tensor forms 6 of the 10 generators of the Poincaré group, the other four are the components of the four-momentum for spacetime translations.

Angular Momentum in General Relativity

The angular momentum of test particles in a gently curved background is more complicated in GR but can be generalized in a straightforward manner. If the Lagrangian is expressed with respect to angular variables as the generalized coordinates, then the angular momenta are the functional derivatives of the Lagrangian with respect to the angular velocities. Referred to Cartesian coordinates, these are typically given by the off-diagonal shear terms of the spacelike part of the stress–energy tensor. If the spacetime supports a Killing vector field tangent to a circle, then the angular momentum about the axis is conserved.

One also wishes to study the effect of a compact, rotating mass on its surrounding spacetime. The prototype solution is of the Kerr metric, which describes the spacetime around an axially symmetric black hole. It is obviously impossible to draw a point on the event horizon of a Kerr black hole and watch it circle around. However, the solution does support a constant of the system that acts mathematically similar to an angular momentum.

Wormhole

A wormhole or "Einstein-Rosen bridge" is a hypothetical topological feature that would fundamentally be a shortcut connecting two separate points in spacetime. A wormhole may connect extremely long distances such as a billion light years or more, short distances such as a few feet, different universes, and different points in time. A wormhole is much like a tunnel with two ends, each at separate points in spacetime.

For a simplified notion of a wormhole, space can be visualized as a two-dimensional (2D) surface. In this case, a wormhole would appear as a hole in that surface, lead into a 3D tube (the inside surface of a cylinder), then re-emerge at another location on the 2D surface with a hole similar to the entrance. An actual wormhole would be analogous to this, but with the spatial dimensions raised by one. For example, instead of circular holes on a 2D plane, the entry and exit points could be visualized as spheres in 3D space.

Overview

The equations of the theory of general relativity have valid solutions that contain wormholes. The first type of wormhole solution discovered was the Schwarzschild wormhole, which would be present in the Schwarzschild metric describing an eternal black hole, but it was found that it would collapse too quickly for anything to cross from one end to the other. Wormholes that could be crossed in both directions, known as traversable wormholes, would only be possible if exotic matter with negative energy density could be used to stabilize them. Wormholes are also a very powerful mathematical metaphor for teaching general relativity.

The Casimir effect shows that quantum field theory allows the energy density in certain regions of space to be negative relative to the ordinary vacuum energy, and it has been shown theoretically that quantum field theory allows states where energy can be *arbitrarily* negative at a given point. Many physicists, such as Stephen Hawking, Kip Thorne and others, therefore argue that such effects might make it possible to stabilize a traversable wormhole. Physicists have not found any natural process that would be predicted to form a wormhole naturally in the context of general relativity, although the quantum foam hypothesis is sometimes used to suggest that tiny wormholes might appear and disappear spontaneously at the Planck scale, and stable versions of such wormholes have been suggested as dark matter candidates. It has also been proposed that, if a tiny wormhole held open by a negative mass cosmic string had appeared around the time of the Big Bang, it could have been inflated to macroscopic size by cosmic inflation.

The American theoretical physicist John Archibald Wheeler coined the term *wormhole* in 1957; the German mathematician Hermann Weyl, however, had proposed the wormhole theory in 1921, in connection with mass analysis of electromagnetic field energy.

This analysis forces one to consider situations... where there is a net flux of lines of force, through what topologists would call "a handle" of the multiply-connected space, and what physicists might perhaps be excused for more vividly terming a "wormhole".

—John Wheeler in Annals of Physics

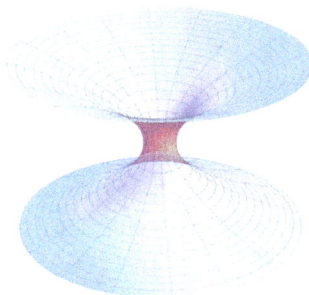

"Embedding diagram" of a Schwarzschild wormhole

Definitions

Topological

An intra-universe wormhole is a compact region of spacetime whose boundary is topologically trivial, but whose interior is not simply connected. Formalizing this idea leads to definitions such as the following, taken from Matt Visser's *Lorentzian Wormholes*.

If a Minkowski spacetime contains a compact region Ω, and if the topology of Ω is of the form $\Omega \sim R \times \Sigma$, where Σ is a three-manifold of the nontrivial topology, whose boundary has topology of the form $\partial\Sigma \sim S^2$, and if, furthermore, the hypersurfaces Σ are all spacelike, then the region Ω contains a quasipermanent intrauniverse wormhole.

Geometric

Wormholes have been defined *geometrically*, as opposed to *topologically*, as regions of spacetime that constrain the incremental deformation of closed surfaces. For example, in Enrico Rodrigo's *The Physics of Stargates,* a wormhole is defined informally as:

a region of spacetime containing a "world tube" (the time evolution of a closed surface) that cannot be continuously deformed (shrunk) to a world line (the time evolution of a point).

Schwarzschild Wormholes

An artist's impression of a wormhole from an observer's perspective, crossing the event horizon of a Schwarzschild wormhole that bridges two different universes. The observer originates from the right, and another universe becomes visible in the center of the wormhole's shadow once the horizon is crossed, the observer seeing light that has fallen into the black hole interior region from the other universe; however, this other universe is unreachable in the case of a Schwarzschild wormhole, as the bridge always collapses before the observer has time to cross it, and everything that has fallen through the event horizon of either universe is inevitably crushed in the singularity.

Lorentzian wormholes known as *Schwarzschild wormholes* or *Einstein–Rosen bridges* are connections between areas of space that can be modeled as vacuum solutions to the Einstein field equations, and that are now understood to be intrinsic parts of the maximally extended version of the Schwarzschild metric describing an eternal black hole with no charge and no rotation. Here, "maximally extended" refers to the idea that the space-time should not have any "edges": it should be possible to continue this path arbitrarily far into the particle's future or past for any possible trajectory of a free-falling particle (following a Geodesic in the spacetime), unless the trajectory hits a gravitational singularity like the one at the center of the black hole's interior.

In order to satisfy this requirement, it turns out that in addition to the black hole interior region that particles enter when they fall through the event horizon from the outside, there must be a separate white hole interior region that allows us to extrapolate the trajectories of particles that an outside observer sees rising up *away* from the event horizon. And just as there are two separate interior regions of the maximally extended spacetime, there are also two separate exterior regions, sometimes called two different "universes", with the second universe allowing us to extrapolate some possible particle trajectories in the two interior regions. This means that the interior black hole region can contain a mix of particles that fell in from either universe (and thus an observer who fell in from one universe might be able to see light that fell in from the other one), and likewise particles from the interior white hole region can escape into either universe. All four regions can be seen in a spacetime diagram that uses Kruskal–Szekeres coordinates.

In this spacetime, it is possible to come up with coordinate systems such that if you pick a hypersurface of constant time (a set of points that all have the same time coordinate, such that every point on the surface has a space-like separation, giving what is called a 'space-like surface') and draw an "embedding diagram" depicting the curvature of space at that time, the embedding diagram will look like a tube connecting the two exterior regions, known as an "Einstein–Rosen bridge". Note that the Schwarzschild metric describes an idealized black hole that exists eternally from the perspective of external observers; a more realistic black hole that forms at some particular time from a collapsing star would require a different metric. When the infalling stellar matter is added to a diagram of a black hole's history, it removes the part of the diagram corresponding to the white hole interior region, along with the part of the diagram corresponding to the other universe.

The Einstein–Rosen bridge was discovered by Ludwig Flamm in 1916, a few months after Schwarzschild published his solution, and was rediscovered (although it is hard to imagine that Einstein had not seen Flamm's paper when it came out) by Albert Einstein and his colleague Nathan Rosen, who published their result in 1935. However, in 1962, John A. Wheeler and Robert W. Fuller published a paper showing that this type of wormhole is unstable if it connects two parts of the same universe, and that it will pinch off too quickly for light (or any particle moving slower than light) that falls in from one exterior region to make it to the other exterior region.

According to general relativity, the gravitational collapse of a sufficiently compact mass forms a singular Schwarzschild black hole. In the Einstein–Cartan–Sciama–Kibble theory of gravity, however, it forms a regular Einstein–Rosen bridge. This theory extends general relativity by removing a constraint of the symmetry of the affine connection and regarding its antisymmetric part, the torsion tensor, as a dynamical variable. Torsion naturally accounts for the quantum-mechanical, intrinsic angular momentum (spin) of matter. The minimal coupling between torsion and Dirac spinors generates a repulsive spin–spin interaction that is significant in fermionic matter at extremely high densities. Such an interaction prevents the formation of a gravitational singularity. Instead, the collapsing matter reaches an enormous but finite density and rebounds, forming the other side of the bridge.

Although Schwarzschild wormholes are not traversable in both directions, their existence inspired Kip Thorne to imagine traversable wormholes created by holding the "throat" of a Schwarzschild wormhole open with exotic matter (material that has negative mass/energy).

Traversable Wormholes

Image of a simulated traversable wormhole that connects the square in front of the physical institutes of University of Tübingen with the sand dunes near Boulogne sur Mer in the north of France. The image is calculated with 4D raytracing in a Morris–Thorne wormhole metric, but the gravitational effects on the wavelength of light have not been simulated.

Lorentzian traversable wormholes would allow travel in both directions from one part of the universe to another part of that same universe very quickly or would allow travel from one universe to another. The possibility of traversable wormholes in general relativity was first demonstrated in a 1973 paper by Homer Ellis and independently in a 1973 paper by K. A. Bronnikov. Ellis thoroughly analyzed the topology and the geodesics of the Ellis drainhole, showing it to be geodesically complete, horizonless, singularity-free, and fully traversable in both directions. The drainhole is a solution manifold of Einstein's field equations for a vacuum space-time, modified by inclusion of a scalar field minimally coupled to the Ricci tensor with antiorthodox polarity (negative instead of positive). (Ellis specifically rejected referring to the scalar field as 'exotic' because of the antiorthodox coupling, finding arguments for doing so unpersuasive.) The solution depends on two parameters: m , which fixes the strength of its gravitational field, and n , which determines the curvature of its spatial cross sections. When m is set equal to 0, the drainhole's gravitational field vanishes. What is left is the Ellis wormhole, a nongravitating, purely geometric, traversable wormhole. Kip Thorne and his graduate student Mike Morris, unaware of the 1973 papers by Ellis and Bronnikov, manufactured, and in 1988 published, a duplicate of the Ellis wormhole for use as a tool for teaching general relativity. For this reason, the type of traversable wormhole they proposed, held open by a spherical shell of exotic matter, was from 1988 to 2015 exclusively referred to in the literature as a *Morris–Thorne wormhole*. Later, other types of traversable wormholes were discovered as allowable solutions to the equations of general relativity, including a variety analyzed in a 1989 paper by Matt Visser, in which a path through the wormhole can be made where the traversing path does not pass through a region of exotic matter. However, in the pure Gauss–Bonnet gravity (a modification to general relativity involving extra spatial dimensions which is sometimes studied in the context of brane cosmology) exotic matter is not needed in order for wormholes to exist—they can exist even with no matter. A type held open by negative mass cosmic strings was put forth by Visser in collaboration with Cramer *et al.*, in which it was proposed that such wormholes could have been naturally created in the early universe.

Wormholes connect two points in spacetime, which means that they would in principle allow travel in time, as well as in space. In 1988, Morris, Thorne and Yurtsever worked out explicitly how to convert a wormhole traversing space into one traversing time. However, according to general relativity, it would not be possible to use a wormhole to travel back to a time earlier than when the wormhole was first converted into a time machine by accelerating one of its two mouths.

Raychaudhuri's Theorem and Exotic Matter

To see why exotic matter is required, consider an incoming light front traveling along geodesics, which then crosses the wormhole and re-expands on the other side. The expansion goes from negative to positive. As the wormhole neck is of finite size, we would not expect caustics to develop, at least within the vicinity of the neck. According to the optical Raychaudhuri's theorem, this requires a violation of the averaged null energy condition. Quantum effects such as the Casimir effect cannot violate the averaged null energy condition in any neighborhood of space with zero curvature, but calculations in semiclassical gravity suggest that quantum effects may be able to violate this condition in curved spacetime. Although it was hoped recently that quantum effects could not violate an achronal version of the averaged null energy condition, violations have nevertheless been found, so it remains an open possibility that quantum effects might be used to support a wormhole.

Modified General Relativity

In some theories where general relativity is modified, it is possible to have a wormhole that does not collapse without having to resort to exotic matter. For example, this is possible with R^2 gravity, a form of f(R) gravity.

Faster-than-light Travel

The impossibility of faster-than-light relative speed only applies locally. Wormholes might allow effective superluminal (faster-than-light) travel by ensuring that the speed of light is not exceeded locally at any time. While traveling through a wormhole, subluminal (slower-than-light) speeds are used. If two points are connected by a wormhole whose length is shorter than the distance between them *outside* the wormhole, the time taken to traverse it could be less than the time it would take a light beam to make the journey if it took a path through the space *outside* the wormhole. However, a light beam traveling through the wormhole would of course beat the traveler.

Time Travel

The theory of general relativity predicts that if traversable wormholes exist, they can also alter the speed of time. They could allow time travel. This would be accomplished by accelerating one end of the wormhole to a high velocity relative to the other, and then sometime later bringing it back; relativistic time dilation would result in the accelerated wormhole mouth aging less than the stationary one as seen by an external observer, similar to what is seen in the twin paradox. However, time connects differently through the wormhole than outside it, so that synchronized clocks at each mouth will remain synchronized to someone traveling through the wormhole itself, no matter how the mouths move around. This means that anything which entered the accelerated wormhole mouth would exit the stationary one at a point in time prior to its entry.

For example, consider two clocks at both mouths both showing the date as 2000. After being taken on a trip at relativistic velocities, the accelerated mouth is brought back to the same region as the stationary mouth with the accelerated mouth's clock reading 2004 while the stationary mouth's clock read 2012. A traveler who entered the accelerated mouth at this moment would exit the stationary mouth when its clock also read 2004, in the same region but now eight years in the past. Such a configuration of wormholes would allow for a particle's world line to form a closed loop in spacetime, known as a closed timelike curve. An object traveling through a wormhole could carry energy or charge from one time to another, but this would not violate conservation of energy or charge in each time, because the energy/charge of the wormhole mouth itself would change to compensate for the object that fell into it or emerged from it.

It is thought that it may not be possible to convert a wormhole into a time machine in this manner; the predictions are made in the context of general relativity, but general relativity does not include quantum effects. Analyses using the semiclassical approach to incorporating quantum effects into general relativity have sometimes indicated that a feedback loop of virtual particles would circulate through the wormhole and pile up on themselves, driving the energy density in the region very high and possibly destroying it before any information could be passed through it, in keeping with the chronology protection conjecture. The debate on this matter is described by Kip S. Thorne in the book *Black Holes and Time Warps*, and a more technical discussion can be found in *The quantum physics of chronology protection* by Matt Visser. There is also the Roman ring, which is a configuration of more than one wormhole. This ring seems to allow a closed time loop with stable wormholes when analyzed using semiclassical gravity, although without a full theory of quantum gravity it is uncertain whether the semiclassical approach is reliable in this case.

Interuniversal Travel

A possible resolution to the paradoxes resulting from wormhole-enabled time travel rests on the many-worlds interpretation of quantum mechanics. In 1991 David Deutsch showed that quantum theory is fully consistent (in the sense that the so-called density matrix can be made free of discontinuities) in spacetimes with closed timelike curves. However, later it was shown that such model of closed timelike curve can have internal inconsistencies as it will lead to strange phenomena like distinguishing non orthogonal quantum states and distinguishing proper and improper mixture. Accordingly, the destructive positive feedback loop of virtual particles circulating through a wormhole time machine, a result indicated by semi-classical calculations, is averted. A particle returning from the future does not return to its universe of origination but to a parallel universe. This suggests that a wormhole time machine with an exceedingly short time jump is a theoretical bridge between contemporaneous parallel universes. Because a wormhole time-machine introduces a type of nonlinearity into quantum theory, this sort of communication between parallel universes is consistent with Joseph Polchinski's discovery of an "Everett phone" in Steven Weinberg's formulation of nonlinear quantum mechanics. Such a possibility is depicted in the science-fiction 2014 movie *Interstellar*.

Metrics

Theories of *wormhole metrics* describe the spacetime geometry of a wormhole and serve as theoretical models for time travel. An example of a (traversable) wormhole metric is the following:

$$ds^2 = -c^2 dt^2 + dl^2 + (k^2 + l^2)(d\theta^2 + \sin^2 \theta d\phi^2),$$

first presented by Ellis as a special case of the Ellis drainhole.

One type of non-traversable wormhole metric is the Schwarzschild solution

$$ds^2 = -c^2 \left(1 - \frac{2GM}{rc^2}\right) dt^2 + \frac{dr^2}{1 - \frac{2GM}{rc^2}} + r^2(d\theta^2 + \sin^2 \theta d\phi^2).$$

The original Einstein-Rosen bridge was described in an article published in July 1935.

For the Schwartzschild spherically symmetric static solution

$$ds^2 = -\frac{1}{1 - \frac{2m}{r}} dr^2 - r^2(d\theta^2 + \sin^2 \theta d\phi^2) + (1 - \frac{2m}{r}) dt^2$$

(ds = proper time, c = 1)

If one replaces r with u according to $u^2 = r - 2m$

$$ds^2 = -4(u^2 + 2m)du^2 - (u^2 + 2m)^2(d\theta^2 + \sin^2 \theta d\phi^2) + \frac{u^2}{u^2 + 2m} dt^2$$

The four-dimensional space is described mathematically by two congruent parts or "sheets", corresponding to u > 0 and u < 0, which are joined by a hyperplane r = 2m or u = 0 in which g vanishes. We call such a connection between the two sheets a "bridge".

—*A.Einstein,N.Rosen - The Particle Problem in the General Theory of Relativity,*
http://adsabs.harvard.edu/abs/1935PhRv...48...73E

For the combined field, gravity and electricity, Einstein and Rosen derived the following Schwarzschild static spherically symmetric solution

$$\phi_1 = \phi_2 = \phi_3 = 0, \phi_4 = \frac{\epsilon}{4},$$

$$ds^2 = -\frac{1}{(1 - \frac{2m}{r} - \frac{\epsilon^2}{2r^2})} dr^2 - r^2(d\theta^2 + \sin^2\theta d\phi^2) + (1 - \frac{2m}{r} - \frac{\epsilon^2}{2r^2})dt^2$$

(ϵ = electrical charge)

The field equations without denominators in the case when m = 0 can be written

$$\phi_{\mu\nu} = \phi_{\mu,\nu} - \phi_{\nu,\mu}$$

$$g^2 \phi_{\mu\nu;\sigma} g^{\nu\sigma} = 0$$

$$g^2(R_{ik} + \phi_{i\alpha}\phi_k^\alpha - \frac{1}{4}g_{ik}\phi_{\alpha\beta}\phi^{ab}) = 0$$

In order to eliminate singularities, if one replaces r by u according to the equation:

$$u^2 = r^2 - \frac{\epsilon^2}{2}$$

and with m = 0 one obtains

$$\phi_1 = \phi_2 = \phi_3 = 0, \phi_4 = \epsilon / (u^2 + \frac{\epsilon^2}{2})^{\frac{1}{2}}$$

$$ds^2 = -du^2 - (u^2 + \frac{\epsilon^2}{2})(d\theta^2 + \sin^2\theta d\phi^2) + (\frac{2u^2}{2u^2 + \epsilon^2})dt^2$$

The solution is free from singularities for all finite points in the space of the two sheets

—*A.Einstein,N.Rosen - The Particle Problem in the*
General Theory of Relativity

In Fiction

Wormholes are a common element in science fiction because they allow interstellar, intergalactic, and sometimes even interuniversal travel within human lifetime scales. They have also served as a method for time travel.

References

- Ellis, G. F. R.; Williams, Ruth M. (2000). Flat and curved space–times (2nd ed.). Oxford University Press. p. 9. ISBN 0-19-850657-0.

- Petkov, Vesselin (2010). Minkowski Spacetime: A Hundred Years Later. Springer. p. 70. ISBN 90-481-3474-9. Retrieved 2016-02-28., Section 3.4, p. 70

- Berry, Michael V. (1989). Principles of Cosmology and Gravitation. CRC Press. p. 58. ISBN 0-85274-037-9. Retrieved 2016-02-28. Extract of page 58, caption of Fig. 25

- I. M. Yaglom (1979) A Simple Non-Euclidean Geometry and its Physical Basis, page 178, Springer, ISBN 0387-90332-1, MR 520230

- McComb, W. D. (1999). Dynamics and relativity. Oxford [etc.]: Oxford University Press. pp. 22–24. ISBN 0-19-850112-9.

- MP Hobson; GP Efstathiou; AN Lasenby (2006). General Relativity: An introduction for physicists (Reprinted with corrections 2007 ed.). Cambridge University Press. p. 187. ISBN 978-0-521-82951-9.

- D.S.A. Freed; K.K.A. Uhlenbeck. Geometry and quantum field theory (2nd ed.). Institute For Advanced Study (Princeton, N.J.): American Mathematical Society. ISBN 0-8218-8683-5.

- R. Penrose (2005). The Road to Reality. vintage books. p. 433. ISBN 978-0-09-944068-0. Penrose includes a factor of 2 in the wedge product, other authors may also.

4

Mathematical Tools of General Relativity

General relativity uses a number of mathematical tools. Some of these tools are Einstein field equations, covariant derivative, parallel transport, geodesics in general relativity and the Schwarzschild metric. Einstein field equations are a set of 10 equations, which help in the understanding of the fundamental interaction of gravitation and parallel transport is the conveying of geometrical data along even curves in a manifold.

Introduction to the Mathematics of General Relativity

The mathematics of general relativity are complex. In Newton's theories of motion, an object's length and the rate at which time passes remain constant while the object accelerates, meaning that many problems in Newtonian mechanics may be solved by algebra alone. In relativity, however, an object's length and the rate at which time passes both change appreciably as the object's speed approaches the speed of light, meaning that more variables and more complicated mathematics are required to calculate the object's motion. As a result, relativity requires the use of concepts such as vectors, tensors, pseudotensors and curvilinear coordinates.

For an introduction based on the example of particles following circular orbits about a large mass, nonrelativistic and relativistic treatments are given in, respectively, Newtonian motivations for general relativity and Theoretical motivation for general relativity.

Vectors and Tensors

Vectors

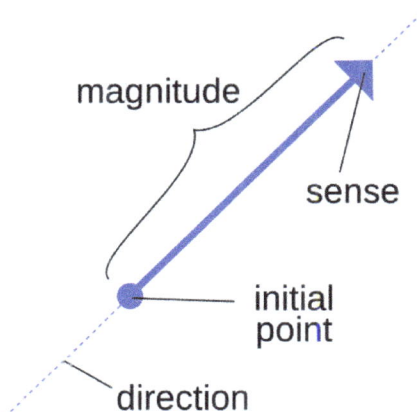

Illustration of a typical vector.

In mathematics, physics, and engineering, a Euclidean vector (sometimes called a geometric or spatial vector, or – as here – simply a vector) is a geometric object that has both a magnitude (or length) and direction. A vector is what is needed to "carry" the point A to the point B; the Latin word *vector* means "one who carries". The magnitude of the vector is the distance between the two points and the direction refers to the direction of displacement from A to B. Many algebraic operations on real numbers such as addition, subtraction, multiplication, and negation have close analogues for vectors, operations which obey the familiar algebraic laws of commutativity, associativity, and distributivity.

Tensors

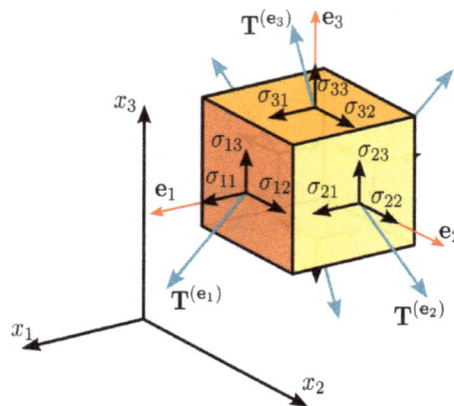

Stress, a second-order tensor. Stress is here shown as a series of vectors on each side of the box

A tensor extends the concept of a vector to additional dimensions. A scalar, that is, a simple number without a direction, would be shown on a graph as a point, a zero-dimensional object. A vector, which has a magnitude and direction, would appear on a graph as a line, which is a one-dimensional object. A tensor extends this concept to additional dimensions. A two-dimensional tensor would be called a second-order tensor. This can be viewed as a set of related vectors, moving in multiple directions on a plane.

Applications

Vectors are fundamental in the physical sciences. They can be used to represent any quantity that has both a magnitude and direction, such as velocity, the magnitude of which is speed. For example, the velocity *5 meters per second upward* could be represented by the vector (0, 5) (in 2 dimensions with the positive y axis as 'up'). Another quantity represented by a vector is force, since it has a magnitude and direction. Vectors also describe many other physical quantities, such as displacement, acceleration, momentum, and angular momentum. Other physical vectors, such as the electric and magnetic field, are represented as a system of vectors at each point of a physical space; that is, a vector field.

Tensors also have extensive applications in physics:

- Electromagnetic tensor (or Faraday's tensor) in electromagnetism

- Finite deformation tensors for describing deformations and strain tensor for strain in continuum mechanics

- Permittivity and electric susceptibility are tensors in anisotropic media

- Stress–energy tensor in general relativity, used to represent momentum fluxes

- Spherical tensor operators are the eigenfunctions of the quantum angular momentum operator in spherical coordinates

- Diffusion tensors, the basis of diffusion tensor imaging, represent rates of diffusion in biologic environments

Dimensions

In general relativity, four-dimensional vectors, or four-vectors, are required. These four dimensions are length, height, width and time. A "point" in this context would be an event, as it has both a location and a time. Similar to vectors, tensors in relativity require four dimensions. One example is the Riemann curvature tensor.

Coordinate Transformation

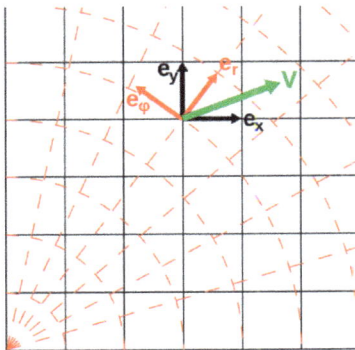

- A vector **v**, is shown with two coordinate grids, e_x and e_r. In space, there is no clear coordinate grid to use. This means that the coordinate system changes based on the location and orientation of the observer. Observer e_x and e_r in this image are facing different directions.

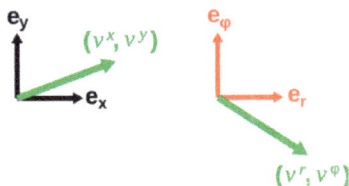

- Here we see that e_x and e_r see the vector differently. The direction of the vector is the same. But to e_x, the vector is moving to its left. To e_r, the vector is moving to its right.

In physics, as well as mathematics, a vector is often identified with a tuple, or list of numbers, which depend on some auxiliary coordinate system or reference frame. When the coordinates are transformed, for example by rotation or stretching, then the components of the vector also transform. The vector itself has not changed, but the reference frame has, so the components of the vector (or measurements taken with respect to the reference frame) must change to compensate.

The vector is called *covariant* or *contravariant* depending on how the transformation of the vector's components is related to the transformation of coordinates.

- Contravariant vectors are "regular vectors" with units of distance (such as a displacement) or distance times some other unit (such as velocity or acceleration). For example, in changing units from meters to millimeters, a displacement of 1 m becomes 1000 mm.

- Covariant vectors, on the other hand, have units of one-over-distance (typically such as gradient). For example, in changing again from meters to millimeters, a gradient of 1 K/m becomes 0.001 K/mm.

In Einstein notation, contravariant vectors and components of tensors are shown with superscripts, e.g. x^i, and covariant vectors and components of tensors with subscripts, e.g. x_i. Indices are "raised" or "lowered" by multiplication by an appropriate matrix, often the identity matrix.

Coordinate transformation is important because relativity states that there is no one correct reference point in the universe. On earth, we use dimensions like north, east, and elevation, which are used throughout the entire planet. There is no such system for space. Without a clear reference grid, it becomes more accurate to describe the four dimensions as towards/away, left/right, up/down and past/future. As an example event, take the signing of the Declaration of Independence. To a modern observer on Mount Rainier looking east, the event is ahead, to the right, below, and in the past. However, to an observer in medieval England looking north, the event is behind, to the left, neither up nor down, and in the future. The event itself has not changed, the location of the observer has.

Oblique Axes

An oblique coordinate system is one in which the axes are not necessarily orthogonal to each other; that is, they meet at angles other than right angles. When using coordinate transformations as described above, the new coordinate system will often appear to have oblique axes compared to the old system.

Nontensors

A nontensor is a tensor-like quantity that behaves like a tensor in the raising and lowering of indices, but that does not transform like a tensor under a coordinate transformation. For example, Christoffel symbols cannot be tensors themselves if the coordinates don't change in a linear way.

In general relativity, one cannot describe the energy and momentum of the gravitational field by an energy–momentum tensor. Instead, one introduces objects that behave as tensors only with respect to restricted coordinate transformations. Strictly speaking, such objects are not tensors at all. A famous example of such a pseudotensor is the Landau–Lifshitz pseudotensor.

Curvilinear Coordinates and Curved Spacetime

Curvilinear coordinates are coordinates in which the angles between axes can change from point to point. This means that rather than having a grid of straight lines, the grid instead has curvature.

High-precision test of general relativity by the Cassini space probe (artist's impression): radio signals sent between the Earth and the probe (green wave) are delayed by the warping of space and time (blue lines) due to the Sun's mass. That is, the Sun's mass causes the regular grid coordinate system (in blue) to distort and have curvature. The radio wave then follows this curvature and moves toward the Sun.

A good example of this is the surface of the Earth. While maps frequently portray north, south, east and west as a simple square grid, that is not in fact the case. Instead, the longitude lines running north and south are curved and meet at the north pole. This is because the Earth is not flat, but instead round.

In general relativity, gravity has curvature effects on the four dimensions of the universe. A common analogy is placing a heavy object on a stretched out rubber sheet, causing the sheet to bend downward. This curves the coordinate system around the object, much like an object in the universe curves the coordinate system it sits in. The mathematics here are conceptually more complex than on Earth, as it results in four dimensions of curved coordinates instead of three as used to describe a curved 2D surface.

Parallel Transport

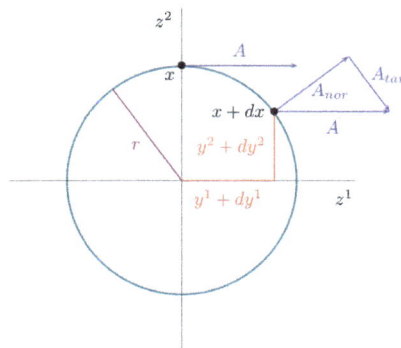

Example: Parallel displacement along a circle of a three-dimensional ball embedded in two dimensions. The circle of radius r is embedded in a two-dimensional space characterized by the coordinates z^1 and z^2. The circle itself is characterized by coordinates y^1 and y^2 in the two-dimensional space. The circle itself is one-dimensional and can be characterized by its arc length x. The coordinate y is related to the coordinate x through the relation $y^1 = r \cos x/r$ and $y^2 = r \sin x/r$. This gives $\partial y^{1}/\partial x = -\sin x/r$ and $\partial y^{2}/\partial x = \cos x/r$ In this case the metric is a scalar and is given by $g = \cos^2 x/r + \sin^2 x/r = 1$. The interval is then $ds^2 = g\, dx^2 = dx^2$. The interval is just equal to the arc length as expected.

The Interval in a High-dimensional Space

In a Euclidean space, the separation between two points is measured by the distance between the two points. The distance is purely spatial, and is always positive. In spacetime, the separation between two events is measured by the *invariant interval* between the two events, which takes into account not only the spatial separation between the events, but also their temporal separation. The interval, s^2, between two events is defined as:

$$s^2 = \Delta r^2 - c^2 \Delta t^2 \quad \text{(spacetime interval)},$$

where c is the speed of light, and Δr and Δt denote differences of the space and time coordinates, respectively, between the events. The choice of signs for s^2 above follows the space-like convention $(-+++)$. A notation like Δr^2 means $(\Delta r)^2$. The reason s^2 is called the interval and not s is that s^2 can be positive, zero or negative.

Spacetime intervals may be classified into three distinct types, based on whether the temporal separation $(c^2\Delta t^2)$ or the spatial separation (Δr^2) of the two events is greater: time-like, light-like or space-like.

Certain types of world lines are called geodesics of the spacetime – straight lines in the case of Minkowski space and their closest equivalent in the curved spacetime of general relativity. In the case of purely time-like paths, geodesics are (locally) the paths of greatest separation (spacetime interval) as measured along the path between two events, whereas in Euclidean space and Riemannian manifolds, geodesics are paths of shortest distance between two points. The concept of geodesics becomes central in general relativity, since geodesic motion may be thought of as "pure motion" (inertial motion) in spacetime, that is, free from any external influences.

The Covariant Derivative

The covariant derivative is a generalization of the directional derivative from vector calculus. As with the directional derivative, the covariant derivative is a rule, which takes as its inputs: (1) a vector, u, (along which the derivative is taken) defined at a point P, and (2) a vector field, v, defined in a neighborhood of P. The output is a vector, also at the point P. The primary difference from the usual directional derivative is that the covariant derivative must, in a certain precise sense, be independent of the manner in which it is expressed in a coordinate system.

Parallel Transport

Given the covariant derivative, one can define the parallel transport of a vector v at a point P along a curve γ starting at P. For each point x of γ, the parallel transport of v at x will be a function of x, and can be written as v(x), where v(0) = v. The function v is determined by the requirement that the covariant derivative of v(x) along γ is 0. This is similar to the fact the a constant function is one whose derivative is constantly 0.

Christoffel Symbols

The equation for the covariant derivative can be written down in terms of Christoffel symbols. The Christoffel symbols find frequent use in Einstein's theory of general relativity, where spacetime is represented by a curved 4-dimensional Lorentz manifold with a Levi-Civita connection. The Einstein field equations – which determine the geometry of spacetime in the presence of matter – contain the Ricci tensor, and so calculating the Christoffel symbols is essential. Once the geometry is determined, the paths of particles and light beams are calculated by solving the geodesic equations in which the Christoffel symbols explicitly appear.

Geodesics

In general relativity, a geodesic generalizes the notion of a "straight line" to curved spacetime. Importantly, the world line of a particle free from all external, non-gravitational force, is a particular type of geodesic. In other words, a freely moving or falling particle always moves along a geodesic.

In general relativity, gravity can be regarded as not a force but a consequence of a curved spacetime geometry where the source of curvature is the stress–energy tensor (representing matter, for instance). Thus, for example, the path of a planet orbiting around a star is the projection of a geodesic of the curved 4-dimensional spacetime geometry around the star onto 3-dimensional space.

A curve is a geodesic if the tangent vector of the curve at any point is equal to the parallel transport of the tangent vector of the base point.

Curvature Tensor

The Riemann tensor tells us, mathematically, how much curvature there is in any given region of space. Contracting the tensor produces 3 different mathematical objects:

1. The Riemann curvature tensor: $R^\rho{}_{\sigma\mu\nu}$, which gives the most information on the curvature of a space and is derived from derivatives of the metric tensor. In flat space this tensor is zero.

2. The Ricci tensor: $R_{\sigma\nu}$, comes from the need in Einstein's theory for a curvature tensor with only 2 indices. It is obtained by averaging certain portions of the Riemann curvature tensor.

3. The scalar curvature: R, the simplest measure of curvature, assigns a single scalar value to each point in a space. It is obtained by averaging the Ricci tensor.

The Riemann curvature tensor can be expressed in terms of the covariant derivative.

The Einstein tensor G is a rank-2 tensor defined over pseudo-Riemannian manifolds. In index-free notation it is defined as

$$\mathbf{G} = \mathbf{R} - \tfrac{1}{2}\mathbf{g}R,$$

where R is the Ricci tensor, g is the metric tensor and R is the scalar curvature. It is used in the Einstein field equations.

Stress–energy Tensor

The stress–energy tensor (sometimes stress–energy–momentum tensor or energy–momentum tensor) is a tensor quantity in physics that describes the density and flux of energy and momentum in spacetime, generalizing the stress tensor of Newtonian physics. It is an attribute of matter, radiation, and non-gravitational force fields. The stress–energy tensor is the source of the gravitational field in the Einstein field equations of general relativity, just as mass density is the source of such a field in Newtonian gravity.

Contravariant components of the stress–energy tensor.

Einstein Equation

The Einstein field equations (EFE) or Einstein's equations are a set of 10 equations in Albert Einstein's general theory of relativity which describe the fundamental interaction of gravitation as a result of spacetime being curved by matter and energy. First published by Einstein in 1915 as a tensor equation, the EFE equate local spacetime curvature (expressed by the Einstein tensor) with the local energy and momentum within that spacetime (expressed by the stress–energy tensor).

The Einstein Field Equations can be written as

$$G_{\mu\nu} = \frac{8\pi G}{c^4} T_{\mu\nu},$$

where $G_{\mu\nu}$ is the Einstein tensor and $T_{\mu\nu}$ is the stress–energy tensor.

This implies that the curvature of space (represented by the Einstein tensor) is directly connected to the presence of matter and energy (represented by the stress–energy tensor).

Schwarzschild Solution and Black Holes

In Einstein's theory of general relativity, the Schwarzschild metric (also Schwarzschild vacuum or Schwarzschild solution), is a solution to the Einstein field equations which describes the gravitational field outside a spherical mass, on the assumption that the electric charge of the mass, angular momentum of the mass, and universal cosmological constant are all zero. The solution is a useful approximation for describing slowly rotating astronomical objects such as many stars and planets, including Earth and the Sun. The solution is named after Karl Schwarzschild, who first published the solution in 1916.

According to Birkhoff's theorem, the Schwarzschild metric is the most general spherically symmetric, vacuum solution of the Einstein field equations. A Schwarzschild black hole or static black hole is a black hole that has no charge or angular momentum. A Schwarzschild black hole is described by the Schwarzschild metric, and cannot be distinguished from any other Schwarzschild black hole except by its mass.

Tensor

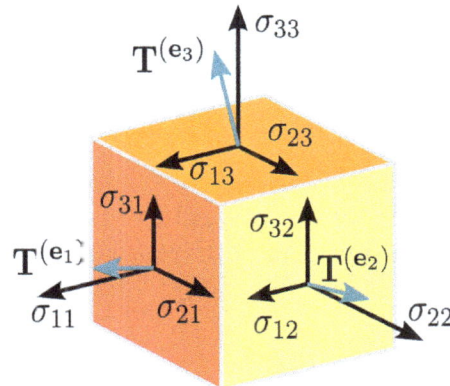

Cauchy stress tensor, a second-order tensor. The tensor's components, in a three-dimensional Cartesian coordinate system, form the matrix

$$\sigma = \left[\mathbf{T}^{(\mathbf{e}_1)} \mathbf{T}^{(\mathbf{e}_2)} \mathbf{T}^{(\mathbf{e}_3)} \right]$$

$$= \begin{bmatrix} \sigma_{11} & \sigma_{12} & \sigma_{13} \\ \sigma_{21} & \sigma_{22} & \sigma_{23} \\ \sigma_{31} & \sigma_{32} & \sigma_{33} \end{bmatrix}$$

whose columns are the stresses (forces per unit area) acting on the $\mathbf{e}_1$, $\mathbf{e}_2$, and $\mathbf{e}_3$ faces of the cube.

Tensors are geometric objects that describe linear relations between geometric vectors, scalars, and other tensors. Elementary examples of such relations include the dot product, the cross product, and linear maps. Geometric vectors, often used in physics and engineering applications, and scalars themselves are also tensors. A more sophisticated example is the Cauchy stress tensor T, which takes a direction v as input and produces the stress $T^{(v)}$ on the surface normal to this vector for output, thus expressing a relationship between these two vectors.

Given a coordinate basis or fixed frame of reference, a tensor can be represented as an organized multidimensional array of numerical values. The order (also *degree* or *rank*) of a tensor is the dimensionality of the array needed to represent it, or equivalently, the number of indices needed to label a component of that array. For example, a linear map is represented by a matrix (a 2-dimensional array) in a basis, and therefore is a 2nd-order tensor. A vector is represented as a 1-dimensional array in a basis, and is a 1st-order tensor. Scalars are single numbers and are thus 0th-order tensors. Because they express a relationship between vectors, tensors themselves must be independent of a particular choice of coordinate system. The coordinate independence of a tensor then takes the form of a covariant and/or contravariant transformation law that relates the array computed in one coordinate system to that computed in another one. The precise form of the transformation law determines the *type* (or *valence*) of the tensor. The tensor type is a pair of natural numbers (n, m), where n is the number of contravariant indices and m is the number of covariant indices. The total order of a tensor is the sum of these two numbers.

Tensors are important in physics because they provide a concise mathematical framework for formulating and solving physics problems in areas such as elasticity, fluid mechanics, and general relativity. Tensors were first conceived by Tullio Levi-Civita and Gregorio Ricci-Curbastro, who continued the earlier work of Bernhard Riemann and Elwin Bruno Christoffel and others, as part of the *absolute differential calculus*. The concept enabled an alternative formulation of the intrinsic differential geometry of a manifold in the form of the Riemann curvature tensor.

Definition

There are several approaches to defining tensors. Although seemingly different, the approaches just describe the same geometric concept using different languages and at different levels of abstraction.

As Multidimensional Arrays

Just as a vector in an n-dimensional space is represented by a one-dimensional array of length n with respect to a given basis, any tensor with respect to a basis is represented by a multidimensional array. For example, a linear operator is represented in a basis as a two-dimensional square $n \times n$ array. The numbers in the multidimensional array are known as the *scalar components* of the tensor or simply its *components*. They are denoted by indices giving their position in the array, as subscripts and superscripts, following the symbolic name of the tensor. For example, the components of an order 2 tensor T could be denoted T_{ij}, where i and j are indices running from 1 to n, or also by T_i^j. Whether an index is displayed as a superscript or subscript depends on the transformation properties of the tensor, described below. The total number of indices required to identify each component uniquely is equal to the dimension of the array, and is called the *order*, *degree* or *rank* of the tensor. However, the term "rank" generally has another meaning in the context of matrices and tensors.

Just as the components of a vector change when we change the basis of the vector space, the components of a tensor also change under such a transformation. Each tensor comes equipped with a *transformation law* that details how the components of the tensor respond to a change of basis. The components of a vector can respond in two distinct ways to a change of basis, where the new basis vectors $\hat{\mathbf{e}}_i$ are expressed in terms of the old basis vectors $\mathbf{e}_j$ as,

$$\hat{\mathbf{e}}_i = \sum_{j=1}^{n} \mathbf{e}_j R_i^j = \mathbf{e}_j R_i^j.$$

Here R_i^j are the entries of the change of basis matrix, and in the rightmost expression the summation sign was suppressed: this is the Einstein summation convention, which will be used throughout this article. The components v^i of a column vector v transform with the inverse of the matrix R,

$$\hat{v}^i = (R^{-1})_j^i v^j,$$

where the hat denotes the components in the new basis. This is called a *contravariant* transformation law, because the vector transforms by the *inverse* of the change of basis. In contrast, the components, w_i, of a covector (or row vector), w transform with the matrix R itself,

$$\hat{w}_i = w_j R_i^j.$$

This is called a *covariant* transformation law, because the covector transforms by the *same matrix* as the change of basis matrix. The components of a more general tensor transform by some combination of covariant and contravariant transformations, with one transformation law for each index. If the transformation matrix of an index is the inverse matrix of the basis transformation, then the index is called *contravariant* and is traditionally denoted with an upper index (superscript). If the transformation matrix of an index is the basis transformation itself, then the index is called *covariant* and is denoted with a lower index (subscript).

The transformation law for an order $p + q$ tensor with p contravariant indices and q covariant indices is thus given as,

$$\hat{T}_{j_1',\dots,j_q'}^{i_1',\dots,i_p'} = (R^{-1})_{i_1}^{i_1'} \cdots (R^{-1})_{i_p}^{i_p'}\ T_{j_1,\dots,j_q}^{i_1,\dots,i_p} R_{j_1'}^{j_1} \cdots R_{j_q'}^{j_q}.$$

Here the primed indices denote components in the new coordinates, and the unprimed indices denote the components in the old coordinates. Such a tensor is said to be of order or *type* (p, q). The terms "order", "type", "rank", "valence", and "degree" are all sometimes used for the same concept. Here, the term "order" or "total order" will be used for the total dimension of the array (or its generalisation in other definitions), $p+q$ in the preceding example, and the term "type" for the pair giving the number of contravariant and covariant indices. A tensor of type (p, q) is also called as a (p, q)-tensor for short.

As an example, the matrix of a linear operator in a basis is a rectangular array T that transforms under a change of basis matrix $R = (R_i^j)$ by $\hat{T} = R^{-1}TR$. For the individual matrix entries, this transformation law has the form $\hat{T}_{j'}^{i'} = (R^{-1})_i^{i'} T_j^i R_{j'}^j$ so the tensor corresponding to the matrix of a

linear operator has one covariant and one contravariant index: it is of type (1,1). A linear operator itself does not actually depend on a basis: it is just a linear map that accepts a vector as an argument and produces another vector. The transformation law for the matrix of a linear operator is consistent with the transformation law for a contravariant vector, so that the action of a linear operator on a contravariant vector is represented in coordinates as the matrix product of their

respective coordinate representations. That is, the components $(Tv)^i$ are given by $(Tv)^i = T_j^i v^j$. These components transform contravariantly, since

$$(\widehat{Tv})^{i'} = \hat{T}_{j'}^{i'}\hat{v}^{j'} = \left[(R^{-1})_i^{i'} T_j^i R_{j'}^j \right]\left[(R^{-1})_j^{j'} v^j \right] = (R^{-1})_i^{i'} (Tv)^i.$$

This discussion motivates the following formal definition:

Definition. A tensor of type (p, q) is an assignment of a multidimensional array

$$T^{i_1 \dots i_p}_{j_1 \dots j_q}[\mathbf{f}]$$

to each basis $\mathbf{f} = (\mathbf{e}_1, \dots, \mathbf{e}_n)$ of a fixed n-dimensional vector space such that, if we apply the change of basis

$$\mathbf{f} \mapsto \mathbf{f} \cdot R = \left(\mathbf{e}_i R^i_1, \dots, \mathbf{e}_i R^i_n \right)$$

then the multidimensional array obeys the transformation law

$$T^{i'_1 \dots i'_p}_{j'_1 \dots j'_q}[\mathbf{f} \cdot R] = (R^{-1})^{i'_1}_{i_1} \cdots (R^{-1})^{i'_p}_{i_p} \, T^{i_1, \dots, i_p}_{j_1, \dots, j_q}[\mathbf{f}] R^{j_1}_{j'_1} \cdots R^{j_q}_{j'_q}.$$

The definition of a tensor as a multidimensional array satisfying a transformation law traces back to the work of Ricci. This definition is still used in some physics and engineering text books.

Tensor Fields

In many applications, especially in differential geometry and physics, it is natural to consider a tensor with components that are functions of the point in a space. This was the setting of Ricci's original work. In modern mathematical terminology such an object is called a tensor field, often referred to simply as a tensor.

In this context, a coordinate basis is often chosen for the tangent vector space. The transformation law may then be expressed in terms of partial derivatives of the coordinate functions,

$$\overline{x}^i(x^1, \dots, x^n),$$

defining a coordinate transformation,

$$\hat{T}^{i'_1 \dots i'_p}_{j'_1 \dots j'_q}(\overline{x}^1, \dots, \overline{x}^n) = \frac{\partial \overline{x}^{i'_1}}{\partial x^{i_1}} \cdots \frac{\partial \overline{x}^{i'_p}}{\partial x^{i_p}} \frac{\partial x^{j_1}}{\partial \overline{x}^{j'_1}} \cdots \frac{\partial x^{j_q}}{\partial \overline{x}^{j'_q}} T^{i_1 \dots i_p}_{j_1 \dots j_q}(x^1, \dots, x^n).$$

As Multilinear Maps

A downside to the definition of a tensor using the multidimensional array approach is that it is not apparent from the definition that the defined object is indeed basis independent, as is expected from an intrinsically geometric object. Although it is possible to show that transformation laws indeed ensure independence from the basis, sometimes a more intrinsic definition is preferred. One approach is to define a tensor as a multilinear map. In that approach a type (p, q) tensor T is defined as a map,

$$T : \underbrace{V^* \times \dots \times V^*}_{p \text{ copies}} \times \underbrace{V \times \dots \times V}_{q \text{ copies}} \to \mathbf{R},$$

where V is a (finite-dimensional) vector space and V^* is the corresponding dual space of covectors, which is linear in each of its arguments.

By applying a multilinear map T of type (p, q) to a basis $\{e_j\}$ for V and a canonical cobasis $\{\varepsilon^i\}$ for V^*,

$$T^{i_1\ldots i_p}_{j_1\ldots j_q} \equiv T(\epsilon^{i_1},\ldots,\epsilon^{i_p},\mathbf{e}_{j_1},\ldots,\mathbf{e}_{j_q}),$$

a $(p+q)$-dimensional array of components can be obtained. A different choice of basis will yield different components. But, because T is linear in all of its arguments, the components satisfy the tensor transformation law used in the multilinear array definition. The multidimensional array of components of T thus form a tensor according to that definition. Moreover, such an array can be realized as the components of some multilinear map T. This motivates viewing multilinear maps as the intrinsic objects underlying tensors.

In viewing a tensor as a multilinear map, it is conventional to identify the vector space V with the space of linear functionals on the dual of V, the double dual V^{**}. There is always a natural linear map from V to its double dual, given by evaluating a linear form in V^* against a vector in V. This linear mapping is an isomorphism in finite dimensions, and it is often then expedient to identify V with its double dual.

Using Tensor Products

For some mathematical applications, a more abstract approach is sometimes useful. This can be achieved by defining tensors in terms of elements of tensor products of vector spaces, which in turn are defined through a universal property. A type (p, q) tensor is defined in this context as an element of the tensor product of vector spaces,

$$T \in \underbrace{V \otimes \ldots \otimes V}_{p \text{ copies}} \otimes \underbrace{V^* \otimes \ldots \otimes V^*}_{q \text{ copies}}.$$

If v_i is a basis of V and w_j is a basis of W, then the tensor product $V \otimes W$ has a natural basis $v_i \otimes w_j$. The components of a tensor T are the coefficients of the tensor with respect to the basis obtained from a basis $\{e_i\}$ for V and its dual $\{\varepsilon^j\}$, i.e.

$$T = T^{i_1\ldots i_p}_{j_1\ldots j_q}\, \mathbf{e}_{i_1} \otimes \cdots \otimes \mathbf{e}_{i_p} \otimes \epsilon^{j_1} \otimes \cdots \otimes \epsilon^{j_q}.$$

Using the properties of the tensor product, it can be shown that these components satisfy the transformation law for a type (p, q) tensor. Moreover, the universal property of the tensor product gives a 1-to-1 correspondence between tensors defined in this way and tensors defined as multilinear maps.

Tensor products can be defined in great generality – for example, involving arbitrary modules over a ring. In principle, one could define a "tensor" simply to be an element of any tensor product. However, the mathematics literature usually reserves the term *tensor* for an element of a tensor product of a single vector space V and its dual, as above.

Tensors in Infinite Dimensions

This discussion of tensors so far assumes finite dimensionality of the spaces involved, where the spaces of tensors obtained by each of these constructions are naturally isomorphic. Constructions of spaces of tensors based on the tensor product and multilinear mappings can be generalized, essentially without modification, to vector bundles or coherent sheaves. For infinite-dimensional vector spaces, inequivalent topologies lead to inequivalent notions of tensor, and these various isomorphisms may or may not hold depending on what exactly is meant by a tensor. In some applications, it is the tensor product of Hilbert spaces that is intended, whose properties are the most similar to the finite-dimensional case. A more modern view is that it is the tensors' structure as a symmetric monoidal category that encodes their most important properties, rather than the specific models of those categories

Examples

This table shows important examples of tensors, including both tensors on vector spaces and tensor fields on manifolds. The tensors are classified according to their type (n, m), where n is the number of contravariant indices, m is the number of covariant indices, and $n + m$ gives the total order of the tensor. For example, a bilinear form is the same thing as a $(0, 2)$-tensor; an inner product is an example of a $(0, 2)$-tensor, but not all $(0, 2)$-tensors are inner products. In the $(0, M)$-entry of the table, M denotes the dimensionality of the underlying vector space or manifold because for each dimension of the space, a separate index is needed to select that dimension to get a maximally covariant antisymmetric tensor.

n, m	m = 0	m = 1	m = 2	m = 3	...	m = M	...
n = 0	scalar, e.g. scalar curvature	covector, linear functional, 1-form, e.g. gradient of a scalar field	bilinear form, e.g. inner product, metric tensor, Ricci curvature, 2-form, symplectic form	e.g. 3-form		e.g. M-form i.e. volume form	
n = 1	vector, e.g. direction vector	linear transformation, Kronecker delta	e.g. cross product in three dimensions	e.g. Riemann curvature tensor			
n = 2	inverse metric tensor, bivector, e.g., Poisson structure		e.g. elasticity tensor				
...							
n	n-vector, a sum of n-blades						
...							

Raising an index on an (n, m)-tensor produces an $(n + 1, m - 1)$-tensor; this can be visualized as moving diagonally down and to the left on the table. Symmetrically, lowering an index can be visualized as moving diagonally up and to the right on the table. Contraction of an upper with a lower index of an (n, m)-tensor produces an $(n-1, m-1)$-tensor; this can be visualized as moving diagonally up and to the left on the table.

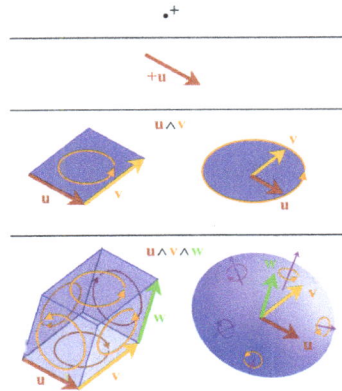

Orientation defined by an ordered set of vectors.

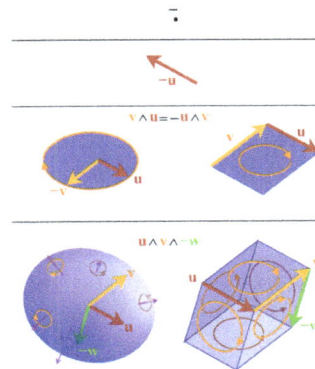

Reversed orientation corresponds to negating the exterior product.

Geometric interpretation of grade n elements in a real exterior algebra for $n = 0$ (signed point), 1 (directed line segment, or vector), 2 (oriented plane element), 3 (oriented volume). The exterior product of n vectors can be visualized as any n-dimensional shape (e.g. n-parallelotope, n-ellipsoid); with magnitude (hypervolume), and orientation defined by that on its $n - 1$-dimensional boundary and on which side the interior is.

Notation

Ricci Calculus

Ricci calculus is the modern formalism and notation for tensor indices: indicating inner and outer products, covariance and contravariance, summations of tensor components, symmetry and antisymmetry, and partial and covariant derivatives.

Einstein Summation Convention

The Einstein summation convention dispenses with writing summation signs, leaving the summation implicit. Any repeated index symbol is summed over: if the index i is used twice in a given term of a tensor expression, it means that the term is to be summed for all i. Several distinct pairs of indices may be summed this way.

Penrose Graphical Notation

Penrose graphical notation is a diagrammatic notation which replaces the symbols for tensors with shapes, and their indices by lines and curves. It is independent of basis elements, and requires no symbols for the indices.

Abstract Index Notation

The abstract index notation is a way to write tensors such that the indices are no longer thought of as numerical, but rather are indeterminates. This notation captures the expressiveness of indices and the basis-independence of index-free notation.

Component-free Notation

A component-free treatment of tensors uses notation that emphasises that tensors do not rely on any basis, and is defined in terms of the tensor product of vector spaces.

Operations

There are a number of basic operations that may be conducted on tensors that again produce a tensor. The linear nature of tensor implies that two tensors of the same type may be added together, and that tensors may be multiplied by a scalar with results analogous to the scaling of a vector. On components, these operations are simply performed component for component. These operations do not change the type of the tensor, however there also exist operations that change the type of the tensors.

Tensor Product

The tensor product takes two tensors, S and T, and produces a new tensor, $S \otimes T$, whose order is the sum of the orders of the original tensors. When described as multilinear maps, the tensor product simply multiplies the two tensors, i.e.

$$(S \otimes T)(v_1, \ldots, v_n, v_{n+1}, \ldots, v_{n+m}) = S(v_1, \ldots, v_n)T(v_{n+1}, \ldots, v_{n+m}),$$

which again produces a map that is linear in all its arguments. On components the effect similarly is to multiply the components of the two input tensors, i.e.

$$(S \otimes T)^{i_1 \ldots i_l i_{l+1} \ldots i_{l+n}}_{j_1 \ldots j_k j_{k+1} \ldots j_{k+m}} = S^{i_1 \ldots i_l}_{j_1 \ldots j_k} T^{i_{l+1} \ldots i_{l+n}}_{j_{k+1} \ldots j_{k+m}},$$

If S is of type (l, k) and T is of type (n, m), then the tensor product $S \otimes T$ has type $(l + n, k + m)$.

Contraction

Tensor contraction is an operation that reduces a type (n, m) tensor to a type $(n - 1, m - 1)$ tensor. It thereby reduces the total order of a tensor by two. The operation is achieved by summing components for which one specified contravariant index is the same as one specified covariant index to produce a new component. Components for which those two indices are different are discarded.

(This is like taking the trace of a matrix.) For example, a (1, 1)-tensor T_i^j T_i^j can be contracted to a scalar through

$$T_i^j.$$

Where the summation is again implied. When the (1, 1)-tensor is interpreted as a linear map, this operation is known as the trace.

The contraction is often used in conjunction with the tensor product to contract an index from each tensor.

The contraction can also be understood using the definition of a tensor as an element of a tensor product of copies of the space V with the space V^* by first decomposing the tensor into a linear combination of simple tensors, and then applying a factor from V^* to a factor from V. For example, a tensor

$$T \in V \otimes V \otimes V^*$$

can be written as a linear combination

$$T = v_1 \otimes w_1 \otimes \alpha_1 + v_2 \otimes w_2 \otimes \alpha_2 + \cdots + v_N \otimes w_N \otimes \alpha_N.$$

The contraction of T on the first and last slots is then the vector

$$\alpha_1(v_1)w_1 + \alpha_2(v_2)w_2 + \cdots + \alpha_N(v_N)w_N.$$

In a vector space with an inner product (also known as a metric) g, the term contraction is used for removing two contravariant or two covariant indices by forming a trace with the metric tensor or its inverse. For example, a (2, 0)-tensor T^{ij} can be contracted to a scalar through

$$T^{ij}g_{ij}$$

(yet again assuming the summation convention).

Raising or Lowering an Index

When a vector space is equipped with a nondegenerate bilinear form (or *metric tensor* as it is often called in this context), operations can be defined that convert a contravariant (upper) index into a covariant (lower) index and vice versa. A metric tensor is a (symmetric) (0, 2)-tensor, it is thus possible to contract an upper index of a tensor with one of lower indices of the metric tensor in the product. This produces a new tensor with the same index structure as the previous, but with lower index in the position of the contracted upper index. This operation is quite graphically known as *lowering an index.*

Conversely, the inverse operation can be defined, and is called *raising an index*. This is equivalent to a similar contraction on the product with a (2, 0)-tensor. This *inverse metric tensor* has components that are the matrix inverse of those if the metric tensor.

Applications

Continuum Mechanics

Important examples are provided by continuum mechanics. The stresses inside a solid body or fluid are described by a tensor. The stress tensor and strain tensor are both second-order tensors, and are related in a general linear elastic material by a fourth-order elasticity tensor. In detail, the tensor quantifying stress in a 3-dimensional solid object has components that can be conveniently represented as a 3×3 array. The three faces of a cube-shaped infinitesimal volume segment of the solid are each subject to some given force. The force's vector components are also three in number. Thus, 3×3, or 9 components are required to describe the stress at this cube-shaped infinitesimal segment. Within the bounds of this solid is a whole mass of varying stress quantities, each requiring 9 quantities to describe. Thus, a second-order tensor is needed.

If a particular surface element inside the material is singled out, the material on one side of the surface will apply a force on the other side. In general, this force will not be orthogonal to the surface, but it will depend on the orientation of the surface in a linear manner. This is described by a tensor of type (2, 0), in linear elasticity, or more precisely by a tensor field of type (2, 0), since the stresses may vary from point to point.

Other Examples from Physics

Common applications include

- Electromagnetic tensor (or Faraday's tensor) in electromagnetism

- Finite deformation tensors for describing deformations and strain tensor for strain in continuum mechanics

- Permittivity and electric susceptibility are tensors in anisotropic media

- Four-tensors in general relativity (e.g. stress–energy tensor), used to represent momentum fluxes

- Spherical tensor operators are the eigenfunctions of the quantum angular momentum operator in spherical coordinates

- Diffusion tensors, the basis of diffusion tensor imaging, represent rates of diffusion in biologic environments

- Quantum mechanics and quantum computing utilise tensor products for combination of quantum states

Applications of Tensors of Order > 2

The concept of a tensor of order two is often conflated with that of a matrix. Tensors of higher order do however capture ideas important in science and engineering, as has been shown successively in numerous areas as they develop. This happens, for instance, in the field of computer vision, with the trifocal tensor generalizing the fundamental matrix.

The field of nonlinear optics studies the changes to material polarization density under extreme electric fields. The polarization waves generated are related to the generating electric fields through the nonlinear susceptibility tensor. If the polarization P is not linearly proportional to the electric field E, the medium is termed *nonlinear*. To a good approximation (for sufficiently weak fields, assuming no permanent dipole moments are present), P is given by a Taylor series in E whose coefficients are the nonlinear susceptibilities:

$$\frac{P_i}{\varepsilon_0} = \sum_j \chi_{ij}^{(1)} E_j + \sum_{jk} \chi_{ijk}^{(2)} E_j E_k + \sum_{jk\ell} \chi_{ijk\ell}^{(3)} E_j E_k E_\ell + \cdots.$$

Here $\chi^{(1)}$ is the linear susceptibility, $\chi^{(2)}$ gives the Pockels effect and second harmonic generation, and $\chi^{(3)}$ gives the Kerr effect. This expansion shows the way higher-order tensors arise naturally in the subject matter.

Generalizations

Tensor Products of Vector Spaces

The vector spaces of a tensor product need not be the same, and sometimes the elements of such a more general tensor product are called "tensors". For example, an element of the tensor product space $V \otimes W$ is a second-order "tensor" in this more general sense, and an order-d tensor may likewise be defined as an element of a tensor product of d different vector spaces. A type (n, m) tensor, in the sense defined previously, is also a tensor of order $n + m$ in this more general sense.

Tensors in Infinite Dimensions

The notion of a tensor can be generalized in a variety of ways to infinite dimensions. One, for instance, is via the tensor product of Hilbert spaces. Another way of generalizing the idea of tensor, common in nonlinear analysis, is via the multilinear maps definition where instead of using finite-dimensional vector spaces and their algebraic duals, one uses infinite-dimensional Banach spaces and their continuous dual. Tensors thus live naturally on Banach manifolds.

Tensor Densities

The concept of a tensor field can be generalized by considering objects that transform differently. An object that transforms as an ordinary tensor field under coordinate transformations, except that it is also multiplied by the determinant of the Jacobian of the inverse coordinate transformation to the wth power, is called a tensor density with weight w. Invariantly, in the language of multilinear algebra, one can think of tensor densities as multilinear maps taking their values in a density bundle such as the (1-dimensional) space of n-forms (where n is the dimension of the space), as opposed to taking their values in just R. Higher "weights" then just correspond to taking additional tensor products with this space in the range.

A special case are the scalar densities. Scalar 1-densities are especially important because it makes sense to define their integral over a manifold. They appear, for instance, in the Einstein−Hilbert action in general relativity. The most common example of a scalar 1-density is the volume element, which in the presence of a metric tensor g is the square root of its determinant in coordinates, denoted $\sqrt{\det g}$. The metric tensor is a covariant tensor of order 2, and so its determinant scales by the square of the coordinate transition:

$$\det(g') = \left(\det \frac{\partial x}{\partial x'} \right)^2 \det(g)$$

which is the transformation law for a scalar density of weight +2.

More generally, any tensor density is the product of an ordinary tensor with a scalar density of the appropriate weight. In the language of vector bundles, the determinant bundle of the tangent bundle is a line bundle that can be used to 'twist' other bundles w times. While locally the more general transformation law can indeed be used to recognise these tensors, there is a global question that arises, reflecting that in the transformation law one may write either the Jacobian determinant, or its absolute value. Non-integral powers of the (positive) transition functions of the bundle of densities make sense, so that the weight of a density, in that sense, is not restricted to integer values. Restricting to changes of coordinates with positive Jacobian determinant is possible on orientable manifolds, because there is a consistent global way to eliminate the minus signs; but otherwise the line bundle of densities and the line bundle of n-forms are distinct.

Spinors

When changing from one orthonormal basis (called a *frame*) to another by a rotation, the components of a tensor transform by that same rotation. This transformation does not depend on the path taken through the space of frames. However, the space of frames is not simply connected: there are continuous paths in the space of frames with the same beginning and ending configurations that are not deformable one into the other. It is possible to attach an additional discrete invariant to each frame called the "spin" that incorporates this path dependence, and which turns out to have values of ±1. A spinor is an object that trans-forms like a tensor under rotations in the frame, apart from a possible sign that is determined by the spin.

History

The concepts of later tensor analysis arose from the work of Carl Friedrich Gauss in differential geometry, and the formulation was much influenced by the theory of algebraic forms and invariants developed during the middle of the nineteenth century. The word "tensor" itself was introduced in 1846 by William Rowan Hamilton to describe something different from what is now meant by a tensor. The contemporary usage was introduced by Woldemar Voigt in 1898.

Tensor calculus was developed around 1890 by Gregorio Ricci-Curbastro under the title *absolute differential calculus*, and originally presented by Ricci in 1892. It was made accessible to many mathematicians by the publication of Ricci and Tullio Levi-Civita's 1900 classic text *Méthodes de calcul différentiel absolu et leurs applications* (Methods of absolute differential calculus and their applications).

In the 20th century, the subject came to be known as *tensor analysis,* and achieved broader acceptance with the introduction of Einstein's theory of general relativity, around 1915. General relativity is formulated completely in the language of tensors. Einstein had learned about them, with great difficulty, from the geometer Marcel Grossmann. Levi-Civita then initiated a correspondence with Einstein to correct mistakes Einstein had made in his use of tensor analysis. The correspondence lasted 1915–17, and was characterized by mutual respect:

I admire the elegance of your method of computation; it must be nice to ride through these fields upon the horse of true mathematics while the like of us have to make our way laboriously on foot.

> —*Albert Einstein, The Italian Mathematicians of Relativity*

Tensors were also found to be useful in other fields such as continuum mechanics. Some well-known examples of tensors in differential geometry are quadratic forms such as metric tensors, and the Riemann curvature tensor. The exterior algebra of Hermann Grassmann, from the middle of the nineteenth century, is itself a tensor theory, and highly geometric, but it was some time before it was seen, with the theory of differential forms, as naturally unified with tensor calculus. The work of Élie Cartan made differential forms one of the basic kinds of tensors used in mathematics.

From about the 1920s onwards, it was realised that tensors play a basic role in algebraic topology (for example in the Künneth theorem). Correspondingly there are types of tensors at work in many branches of abstract algebra, particularly in homological algebra and representation theory. Multilinear algebra can be developed in greater generality than for scalars coming from a field. For example, scalars can come from a ring. But the theory is then less geometric and computations more technical and less algorithmic. Tensors are generalized within category theory by means of the concept of monoidal category, from the 1960s.

Riemann Curvature Tensor

In the mathematical field of differential geometry, the Riemann curvature tensor or Riemann–Christoffel tensor (after Bernhard Riemann and Elwin Bruno Christoffel) is the most common method used to express the curvature of Riemannian manifolds. It associates a tensor to each point of a Riemannian manifold (i.e., it is a tensor field), that measures the extent to which the metric tensor is not locally isometric to that of Euclidean space. The curvature tensor can also be defined for any pseudo-Riemannian manifold, or indeed any manifold equipped with an affine connection.

It is a central mathematical tool in the theory of general relativity, the modern theory of gravity, and the curvature of spacetime is in principle observable via the geodesic deviation equation. The curvature tensor represents the tidal force experienced by a rigid body moving along a geodesic in a sense made precise by the Jacobi equation.

The curvature tensor is given in terms of the Levi-Civita connection ∇ by the following formula:

$$R(u, v)w = \nabla_u \nabla_v w - \nabla_v \nabla_u w - \nabla_{[u,v]} w$$

where $[u,v]$ is the Lie bracket of vector fields. For each pair of tangent vectors u, v, $R(u,v)$ is a linear transformation of the tangent space of the manifold. It is linear in u and v, and so defines a tensor. Occasionally, the curvature tensor is defined with the opposite sign.

If $u = \partial / \partial x^i$ and $v = \partial / \partial x^j$ are coordinate vector fields then $[u,v] = 0$ and therefore the formula simplifies to

$$R(u,v)w = \nabla_u \nabla_v w - \nabla_v \nabla_u w.$$

The curvature tensor measures *noncommutativity of the covariant derivative*, and as such is the integrability obstruction for the existence of an isometry with Euclidean space (called, in this context, *flat* space). The linear transformation is also called the curvature transformation or endomorphism.

The curvature formula can also be expressed in terms of the second covariant derivative defined as:

$$\nabla^2_{u,v} w = \nabla_u \nabla_v w - \nabla_{\nabla_u v} w$$

which is linear in u and v. Then:

$$R(u,v) = \nabla^2_{u,v} - \nabla^2_{v,u}$$

Thus in the general case of non-coordinate vectors u and v, the curvature tensor measures the noncommutativity of the second covariant derivative.

Geometric Meaning

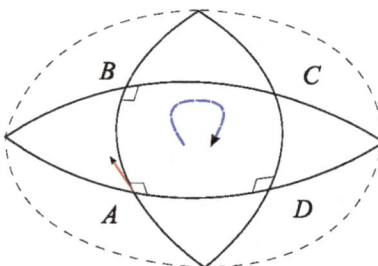

An illustration of the motivation of Riemann curvature on a sphere-like manifold. The fact that this transport may define two different vectors at the start point gives rise to Riemann curvature tensor. The right angle symbol denotes that the inner product (given by the metric tensor) between transported vectors (or tangent vectors of the curves) is 0.

Informally

Imagine walking around the bounding white line of a tennis court with a stick held out in front of you. When you reach the first corner of the court, you turn to follow the white line, but you keep the stick held out in the same direction, which means you are now holding the stick out to your side. You do the same when you reach each corner of the court. When you get back to where you started, you are holding the stick out in exactly the same direction as you were when you started (no surprise there).

Now imagine you are standing on the equator of the earth, facing north with the stick held out in front of you. You walk north up along a line of longitude until you get to the north pole. At that point you turn right, ninety degrees, but you keep the stick held out in the same direction, which means you are now holding the stick out to your left. You keep walking until you get to the equator. There, you turn right again (and so now you have to hold the stick pointing out behind you) and walk along the equator until you get back to where you started from. But here is the thing: the stick is pointing back along the equator from where you just came, not north up to the pole how it was when you started!

The reason for the difference is that the surface of the earth is curved, whereas the surface of a tennis court is flat, but it is not quite that simple. Imagine that the tennis court is slightly humped along its centre-line so that it is like part of the surface of a cylinder. If you walk around the court again, the stick still points in the same direction as it did when you started. This is a consequence of that the tennis court still has zero Gaussian curvature (such as for the surface of a sheet of paper that is bent but not stretched) and the Gauss–Bonnet theorem.

The Riemann curvature tensor is a way to capture a measure of the intrinsic curvature. When you write it down in terms of its components (like writing down the components of a vector), it consists of a multi-dimensional array of sums and products of partial derivatives (some of those partial derivatives can be thought of as akin to capturing the curvature imposed upon someone walking in straight lines on a curved surface).

Formally

When a vector in a Euclidean space is parallel transported around a loop, it will again point in the initial direction after returning to its original position. However, this property does not hold in the general case. The Riemann curvature tensor directly measures the failure of this in a general Riemannian manifold. This failure is known as the non-holonomy of the manifold.

Let x_t be a curve in a Riemannian manifold M. Denote by $\tau_{xt} : T_{xo}M \to T_{xt}M$ the parallel transport map along x_t. The parallel transport maps are related to the covariant derivative by

$$\nabla_{\dot{x}_0} Y = \lim_{h \to 0} \frac{1}{h}\left(Y_{x_0} - \tau_{x_h}^{-1}(Y_{x_h})\right) = \frac{d}{dt}(\tau_{x_t} Y)\bigg|_{t=0}$$

for each vector field Y defined along the curve.

Suppose that X and Y are a pair of commuting vector fields. Each of these fields generates a one-parameter group of diffeomorphisms in a neighborhood of x_0. Denote by τ_{tX} and τ_{tY}, respectively, the parallel transports along the flows of X and Y for time t. Parallel transport of a vector $Z \in T_{xo}M$ around the quadrilateral with sides tY, sX, $-tY$, $-sX$ is given by

$$\tau_{sX}^{-1}\tau_{tY}^{-1}\tau_{sX}\tau_{tY}Z.$$

This measures the failure of parallel transport to return Z to its original position in the tangent space $T_{xo}M$. Shrinking the loop by sending s, $t \to 0$ gives the infinitesimal description of this deviation:

$$\frac{d}{ds}\frac{d}{dt}\tau_{sX}^{-1}\tau_{tY}^{-1}\tau_{sX}\tau_{tY}Z\bigg|_{s=t=0} = (\nabla_X\nabla_Y - \nabla_Y\nabla_X - \nabla_{[X,Y]})Z = R(X,Y)Z$$

where R is the Riemann curvature tensor.

Coordinate Expression

Converting to the tensor index notation, the Riemann curvature tensor is given by

$$R^\rho{}_{\sigma\mu\nu} = dx^\rho(R(\partial_\mu,\partial_\nu)\partial_\sigma)$$

where $\partial_\mu = \partial/\partial x^\mu$ are the coordinate vector fields. The above expression can be written using Christoffel symbols:

$$R^\rho{}_{\sigma\mu\nu} = \partial_\mu\Gamma^\rho{}_{\nu\sigma} - \partial_\nu\Gamma^\rho{}_{\mu\sigma} + \Gamma^\rho{}_{\mu\lambda}\Gamma^\lambda{}_{\nu\sigma} - \Gamma^\rho{}_{\nu\lambda}\Gamma^\lambda{}_{\mu\sigma}$$

The Riemann curvature tensor is also the commutator of the covariant derivative of an arbitrary covector A_ν with itself:

$$A_{\nu;\rho\sigma} - A_{\nu;\sigma\rho} = A_\beta R^\beta{}_{\nu\rho\sigma},$$

since the connection $\tilde{A}^\alpha{}_{\beta\mu}$ is torsionless, which means that the torsion tensor $\Gamma^\lambda{}_{\mu\nu} - \Gamma^\lambda{}_{\nu\mu}$ vanishes.

This formula is often called the *Ricci identity*. This is the classical method used by Ricci and Levi-Civita to obtain an expression for the Riemann curvature tensor. In this way, the tensor character of the set of quantities $R^\beta{}_{\nu\rho\sigma}$ is proved.

This identity can be generalized to get the commutators for two covariant derivatives of arbitrary tensors as follows

$$\nabla_\delta\nabla_\gamma T^{\alpha_1\cdots\alpha_r}{}_{\beta_1\cdots\beta_s} - \nabla_\gamma\nabla_\delta T^{\alpha_1\cdots\alpha_r}{}_{\beta_1\cdots\beta_s} = -R^{\alpha_1}{}_{\rho\gamma\delta}T^{\rho\alpha_2\cdots\alpha_r}{}_{\beta_1\cdots\beta_s} - \cdots - R^{\alpha_r}{}_{\rho\gamma\delta}T^{\alpha_1\cdots\alpha_{r-1}\rho}{}_{\beta_1\cdots\beta_s}$$
$$+ R^\sigma{}_{\beta_1\gamma\delta}T^{\alpha_1\cdots\alpha_r}{}_{\sigma\beta_2\cdots\beta_s} + \cdots + R^\sigma{}_{\beta_s\gamma\delta}T^{\alpha_1\cdots\alpha_r}{}_{\beta_1\cdots\beta_{s-1}\sigma}.$$

This formula also applies to tensor densities without alteration, because for the Levi-Civita (*not generic*) connection one gets:

$$\nabla_\mu(\sqrt{g}) \equiv (\sqrt{g})_{;\mu} = 0, \quad \text{where} \quad g = |\det(g_{\mu\nu})|.$$

It is sometimes convenient to also define the purely covariant version by

$$R_{\rho\sigma\mu\nu} = g_{\rho\zeta}R^\zeta{}_{\sigma\mu\nu}.$$

Symmetries and Identities

The Riemann curvature tensor has the following symmetries:

$$R(u,v) = -R(v,u)$$

$$\langle R(u,v)w, z \rangle = -\langle R(u,v)z, w \rangle$$

$$R(u,v)w + R(v,w)u + R(w,u)v = 0 .$$

Here the bracket $\langle , \rangle$ refers to the inner product on the tangent space induced by the metric tensor. The last identity was discovered by Ricci, but is often called the first Bianchi identity or algebraic Bianchi identity, because it looks similar to the Bianchi identity below. (Also, if there is nonzero torsion, the first Bianchi identity becomes a differential identity of the torsion tensor.) These three identities form a complete list of symmetries of the curvature tensor, i.e. given any tensor which satisfies the identities above, one can find a Riemannian manifold with such a curvature tensor at some point. Simple calculations show that such a tensor has $n^2(n^2-1)/12$ independent components.

Yet another useful identity follows from these three:

$$\langle R(u,v)w, z \rangle = \langle R(w,z)u, v \rangle.$$

On a Riemannian manifold one has the covariant derivative $\nabla_u R$ and the Bianchi identity (often called the second Bianchi identity or differential Bianchi identity) takes the form:

$$(\nabla_u R)(v, w) + (\nabla_v R)(w, u) + (\nabla_w R)(u, v) = 0.$$

Given any coordinate chart about some point on the manifold, the above identities may be written in terms of the components of the Riemann tensor at this point as:

Skew symmetry

$$R_{abcd} = -R_{bacd} = -R_{abdc}$$

Interchange symmetry

$$R_{abcd} = R_{cdab}$$

First Bianchi identity

$$R_{abcd} + R_{acdb} + R_{adbc} = 0$$

This is often written

$$R_{a[bcd]} = 0,$$

where the brackets denote the antisymmetric part on the indicated indices. This is equivalent to the previous version of the identity because the Riemann tensor is already skew on its last two indices.

Second Bianchi identity

$$R_{abcd;e} + R_{abde;c} + R_{abec;d} = 0$$

The semi-colon denotes a covariant derivative. Equivalently,

$$R_{ab[cd;e]} = 0$$

again using the antisymmetry on the last two indices of R.

The algebraic symmetries are also equivalent to saying that R belongs to the image of the Young symmetrizer corresponding to the partition 2+2.

Ricci Curvature

The Ricci curvature tensor is the contraction of the first and third indices of the Riemann tensor.

$$Ric_{ab} \equiv Riem^c{}_{acb} = g^{cd} Riem_{cadb}$$

Special Cases

Surfaces

For a two-dimensional surface, the Bianchi identities imply that the Riemann tensor has only one independent component which means the Ricci scalar completely determines the Riemann tensor. There is only one valid expression for the Riemann tensor which fits the required symmetries:

$$R_{abcd} = f(R)(g_{ac}g_{db} - g_{ad}g_{cb})$$

and by contracting with the metric twice we find the explicit form:

$$R_{abcd} = K(g_{ac}g_{db} - g_{ad}g_{cb})$$

where g_{ab} is the metric tensor and $K = R/2$ is a function called the Gaussian curvature and a, b, c and d take values either 1 or 2. The Riemann tensor has only one functionally independent component. The Gaussian curvature coincides with the sectional curvature of the surface. It is also exactly half the scalar curvature of the 2-manifold, while the Ricci curvature tensor of the surface is simply given by

$$Ric_{ab} = Kg_{ab}.$$

Space forms

A Riemannian manifold is a space form if its sectional curvature is equal to a constant K. The Riemann tensor of a space form is given by

$$R_{abcd} = K(g_{ac}g_{db} - g_{ad}g_{cb}).$$

Conversely, except in dimension 2, if the curvature of a Riemannian manifold has this form for some function K, then the Bianchi identities imply that K is constant and thus that the manifold is (locally) a space form.

Stress–energy Tensor

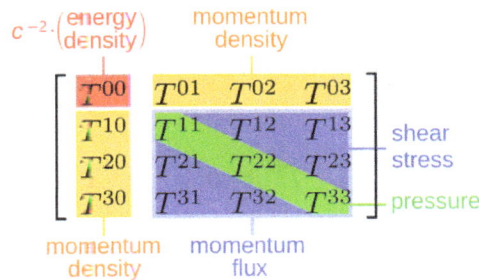

Contravariant components of the stress-energy tensor.

The stress–energy tensor (sometimes stress–energy–momentum tensor or energy–momentum tensor) is a tensor quantity in physics that describes the density and flux of energy and momentum in spacetime, generalizing the stress tensor of Newtonian physics. It is an attribute of matter, radiation, and non-gravitational force fields. The stress–energy tensor is the source of the gravitational field in the Einstein field equations of general relativity, just as mass density is the source of such a field in Newtonian gravity.

Definition

The stress–energy tensor involves the use of superscripted variables. If Cartesian coordinates in SI units are used, then the components of the position four-vector are given by: $x^0 = t$, $x^1 = x$, $x^2 = y$, and $x^3 = z$, where t is time in seconds, and x, y, and z are distances in meters.

The stress–energy tensor is defined as the tensor $T^{\alpha\beta}$ of order two that gives the flux of the αth component of the momentum vector across a surface with constant x^β coordinate. In the theory of relativity, this momentum vector is taken as the four-momentum. In general relativity, the stress–energy tensor is symmetric,

$$T^{\alpha\beta} = T^{\beta\alpha}.$$

In some alternative theories like Einstein–Cartan theory, the stress–energy tensor may not be perfectly symmetric because of a nonzero spin tensor, which geometrically corresponds to a nonzero torsion tensor.

Identifying the Components of the Tensor

Because the stress–energy tensor is of order two, its components can be displayed in 4×4 matrix form:

$$(T^{\mu\nu})_{\mu,\nu=0,1,2,3} = \begin{pmatrix} T^{00} & T^{01} & T^{02} & T^{03} \\ T^{10} & T^{11} & T^{12} & T^{13} \\ T^{20} & T^{21} & T^{22} & T^{23} \\ T^{30} & T^{31} & T^{32} & T^{33} \end{pmatrix}.$$

In the following, i and k range from 1 through 3.

The time–time component is the density of relativistic mass, i.e. the energy density divided by the speed of light squared. It is of special interest because it has a simple physical interpretation. In the case of a perfect fluid this component is

$$T^{00} = \rho,$$

and for an electromagnetic field in otherwise empty space this component is

$$T^{00} = \frac{1}{c^2}\left(\frac{1}{2}\epsilon_0 E^2 + \frac{1}{2\mu_0}B^2\right),$$

where E and B are the electric and magnetic fields, respectively.

The flux of relativistic mass across the x^i surface is equivalent to the density of the ith component of linear momentum,

$$T^{0i} = T^{i0}.$$

The components

$$T^{ik}$$

represent flux of ith component of linear momentum across the x^k surface. In particular,

$$T^{ii}$$

(not summed) represents normal stress, which is called pressure when it is independent of direction. The remaining components

$$T^{ik} \quad i \neq k$$

represent shear stress (compare with the stress tensor).

In solid state physics and fluid mechanics, the stress tensor is defined to be the spatial components of the stress–energy tensor in the proper frame of reference. In other words, the stress energy tensor in engineering *differs* from the stress–energy tensor here by a momentum convective term.

Covariant and Mixed Forms

In most of this article we work with the contravariant form, $T^{\mu\nu}$ of the stress–energy tensor. However, it is often necessary to work with the covariant form,

$$T_{\mu\nu} = T^{\alpha\beta} g_{\alpha\mu} g_{\beta\nu},$$

or the mixed form,

$$T^{\mu}{}_{\nu} = T^{\mu\alpha} g_{\alpha\nu},$$

or as a mixed tensor density

$$\mathfrak{T}^{\mu}{}_{\nu} = T^{\mu}{}_{\nu} \sqrt{-g}.$$

In this article we use the spacelike sign convention $(-+++)$ for the metric signature.

Conservation Law

In Special Relativity

The stress–energy tensor is the conserved Noether current associated with spacetime translations.

The divergence of the non-gravitational stress–energy is zero. In other words, non-gravitational energy and momentum are conserved,

$$0 = T^{\mu\nu}{}_{;\nu} = \nabla_{\nu} T^{\mu\nu}.$$

When gravity is negligible and using a Cartesian coordinate system for spacetime, this may be expressed in terms of partial derivatives as

$$0 = T^{\mu\nu}{}_{,\nu} = \partial_{\nu} T^{\mu\nu}.$$

The integral form of this is

$$0 = \int_{\partial N} T^{\mu\nu} \mathrm{d}^3 s_{\nu}$$

where N is any compact four-dimensional region of spacetime; ∂N is its boundary, a three-dimensional hypersurface; and $\mathrm{d}^3 s_{\nu}$ is an element of the boundary regarded as the outward pointing normal.

In flat spacetime and using Cartesian coordinates, if one combines this with the symmetry of the stress–energy tensor, one can show that angular momentum is also conserved:

$$0 = (x^{\alpha} T^{\mu\nu} - x^{\mu} T^{\alpha\nu})_{,\nu}.$$

In General Relativity

When gravity is non-negligible or when using arbitrary coordinate systems, the divergence of the stress–energy still vanishes. But in this case, a coordinate free definition of the divergence is used which incorporates the covariant derivative

$$0 = \mathrm{div}\, T = T^{\mu\nu}{}_{;\nu} = \nabla_\nu T^{\mu\nu} = T^{\mu\nu}{}_{,\nu} + \Gamma^\mu{}_{\sigma\nu} T^{\sigma\nu} + \Gamma^\nu{}_{\sigma\nu} T^{\mu\sigma}$$

where $\Gamma^\mu{}_{\sigma\nu}$ is the Christoffel symbol which is the gravitational force field.

Consequently, if ξ^μ is any Killing vector field, then the conservation law associated with the symmetry generated by the Killing vector field may be expressed as

$$0 = \nabla_\nu (\xi^\mu T^\nu_\mu) = \frac{1}{\sqrt{-g}} \partial_\nu (\sqrt{-g}\, \xi^\mu T^\nu_\mu)$$

The integral form of this is

$$0 = \int_{\partial N} \sqrt{-g}\, \xi^\mu T^\nu_\mu\, \mathrm{d}^3 s_\nu = \int_{\partial N} \xi^\mu \mathfrak{T}^\nu_\mu\, \mathrm{d}^3 s_\nu$$

In General Relativity

In general relativity, the symmetric stress–energy tensor acts as the source of spacetime curvature, and is the current density associated with gauge transformations of gravity which are general curvilinear coordinate transformations. (If there is torsion, then the tensor is no longer symmetric. This corresponds to the case with a nonzero spin tensor in Einstein–Cartan gravity theory.)

In general relativity, the partial derivatives used in special relativity are replaced by covariant derivatives. What this means is that the continuity equation no longer implies that the non-gravitational energy and momentum expressed by the tensor are absolutely conserved, i.e. the gravitational field can do work on matter and vice versa. In the classical limit of Newtonian gravity, this has a simple interpretation: energy is being exchanged with gravitational potential energy, which is not included in the tensor, and momentum is being transferred through the field to other bodies. In general relativity the Landau–Lifshitz pseudotensor is a unique way to define the *gravitational* field energy and momentum densities. Any such stress–energy pseudotensor can be made to vanish locally by a coordinate transformation.

In curved spacetime, the spacelike integral now depends on the spacelike slice, in general. There is in fact no way to define a global energy–momentum vector in a general curved spacetime.

The Einstein Field Equations

In general relativity, the stress tensor is studied in the context of the Einstein field equations which are often written as

$$R_{\mu\nu} - \tfrac{1}{2} R g_{\mu\nu} = \frac{8\pi G}{c^4} T_{\mu\nu},$$

where $R_{\mu\nu}$ is the Ricci tensor, R is the Ricci scalar (the tensor contraction of the Ricci tensor), $g_{\mu\nu}$ the metric tensor, and G is the universal gravitational constant.

Stress–energy in Special Situations

Isolated Particle

In special relativity, the stress–energy of a non-interacting particle with mass m and trajectory $\mathbf{x}_p(t)$ is:

$$T^{\alpha\beta}(\mathbf{x},t) = \frac{mv^\alpha(t)v^\beta(t)}{\sqrt{1-(v/c)^2}}\,\delta(\mathbf{x}-\mathbf{x}_p(t)) = E\frac{v^\alpha(t)v^\beta(t)}{c^2}\,\delta(\mathbf{x}-\mathbf{x}_p(t))$$

where $(v^\alpha)_{\alpha=0,1,2,3}$ is the velocity vector (which should not be confused with four-velocity)

$$(v^\alpha)_{\alpha=0,1,2,3} = \left(1, \frac{d\mathbf{x}_p}{dt}(t)\right),$$

δ is the Dirac delta function and $E = \sqrt{p^2c^2 + m^2c^4}$ is the energy of the particle.

Stress–energy of a Fluid in Equilibrium

For a perfect fluid in thermodynamic equilibrium, the stress–energy tensor takes on a particularly simple form

$$T^{\alpha\beta} = \left(\rho + \frac{p}{c^2}\right)u^\alpha u^\beta + pg^{\alpha\beta}$$

where ρ is the mass–energy density (kilograms per cubic meter), p is the hydrostatic pressure (pascals), u^α is the fluid's four velocity, and $g^{\alpha\beta}$ is the reciprocal of the metric tensor.

$$T = 3p - \rho c^2.$$

The four velocity satisfies

$$u^\alpha u^\beta g_{\alpha\beta} = -c^2.$$

In an inertial frame of reference comoving with the fluid, better known as the fluid's proper frame of reference, the four velocity is

$$(u^\alpha)_{\alpha=0,1,2,3} = (1,0,0,0),$$

the reciprocal of the metric tensor is simply

$$(g^{\alpha\beta})_{\alpha,\beta=0,1,2,3} = \begin{pmatrix} -c^{-2} & 0 & 0 & 0 \\ 0 & 1 & 0 & 0 \\ 0 & 0 & 1 & 0 \\ 0 & 0 & 0 & 1 \end{pmatrix}$$

and the stress–energy tensor is a diagonal matrix

$$(T^{\alpha\beta})_{\alpha,\beta=0,1,2,3} = \begin{pmatrix} \rho & 0 & 0 & 0 \\ 0 & p & 0 & 0 \\ 0 & 0 & p & 0 \\ 0 & 0 & 0 & p \end{pmatrix}.$$

Electromagnetic Stress–energy Tensor

The Hilbert stress–energy tensor of a source-free electromagnetic field is

$$T^{\mu\nu} = \frac{1}{\mu_0}\left(F^{\mu\alpha} g_{\alpha\beta} F^{\nu\beta} - \frac{1}{4} g^{\mu\nu} F_{\delta\gamma} F^{\delta\gamma} \right)$$

where $F_{\mu\nu}$ is the electromagnetic field tensor.

Variant Definitions of Stress–energy

There are a number of inequivalent definitions of non-gravitational stress–energy:

Hilbert Stress–energy Tensor

It is defined as a functional derivative

$$T_{\mu\nu} = \frac{-2}{\sqrt{-g}} \frac{\delta(\mathcal{L}_{\text{matter}}\sqrt{-g})}{\delta g^{\mu\nu}} = -2\frac{\delta\mathcal{L}_{\text{matter}}}{\delta g^{\mu\nu}} + g_{\mu\nu}\mathcal{L}_{\text{matter}}.$$

where $\mathcal{L}_{\text{matter}}$ is the nongravitational part of the Lagrangian density of the action. This is symmetric and gauge-invariant.

Canonical Stress–energy Tensor

Noether's theorem implies that there is a conserved current associated with translations through space and time. This is called the canonical stress–energy tensor. Generally, this is not symmetric and if we have some gauge theory, it may not be gauge invariant because space-dependent gauge transformations do not commute with spatial translations.

In general relativity, the translations are with respect to the coordinate system and as such, do not transform covariantly.

Belinfante–Rosenfeld Stress–energy Tensor

In the presence of spin or other intrinsic angular momentum, the canonical Noether stress energy tensor fails to be symmetric. The Belinfante–Rosenfeld stress energy tensor is constructed from the canonical stress–energy tensor and the spin current in such a way as to be symmetric and still conserved. In general relativity, this modified tensor agrees with the Hilbert stress–energy tensor.

Gravitational Stress–energy

By the equivalence principle gravitational stress–energy will always vanish locally at any chosen point in some chosen frame, therefore gravitational stress–energy cannot be expressed as a non-zero tensor; instead we have to use a pseudotensor.

In general relativity, there are many possible distinct definitions of the gravitational stress–energy–momentum pseudotensor. These include the Einstein pseudotensor and the Landau–Lifshitz pseudotensor. The Landau–Lifshitz pseudotensor can be reduced to zero at any event in spacetime by choosing an appropriate coordinate system.

Covariant Derivative

In mathematics, the covariant derivative is a way of specifying a derivative along tangent vectors of a manifold. Alternatively, the covariant derivative is a way of introducing and working with a connection on a manifold by means of a differential operator, to be contrasted with the approach given by a principal connection on the frame bundle. In the special case of a manifold isometrically embedded into a higher-dimensional Euclidean space, the covariant derivative can be viewed as the orthogonal projection of the Euclidean derivative along a tangent vector onto the manifold's tangent space. In this case the Euclidean derivative is broken into two parts, the extrinsic normal component and the intrinsic covariant derivative component.

This article presents an introduction to the covariant derivative of a vector field with respect to a vector field, both in a coordinate free language and using a local coordinate system and the traditional index notation. The covariant derivative of a tensor field is presented as an extension of the same concept. The covariant derivative generalizes straightforwardly to a notion of differentiation associated to a connection on a vector bundle, also known as a Koszul connection.

Introduction and History

Historically, at the turn of the 20th century, the covariant derivative was introduced by Gregorio Ricci-Curbastro and Tullio Levi-Civita in the theory of Riemannian and pseudo-Riemannian geometry. Ricci and Levi-Civita (following ideas of Elwin Bruno Christoffel) observed that the Christoffel symbols used to define the curvature could also provide a notion of differentiation which generalized the classical directional derivative of vector fields on a manifold. This new derivative – the Levi-Civita connection – was *covariant* in the sense that it satisfied Riemann's requirement that objects in geometry should be independent of their description in a particular coordinate system.

It was soon noted by other mathematicians, prominent among these being Hermann Weyl, Jan Arnoldus Schouten, and Élie Cartan, that a covariant derivative could be defined abstractly without the presence of a metric. The crucial feature was not a particular dependence on the metric, but that the Christoffel symbols satisfied a certain precise second order transformation law. This transformation law could serve as a starting point for defining the derivative in a covariant manner. Thus the theory of covariant differentiation forked off from the strictly Riemannian context to include a wider range of possible geometries.

In the 1940s, practitioners of differential geometry began introducing other notions of covariant differentiation in general vector bundles which were, in contrast to the classical bundles of interest to geometers, not part of the tensor analysis of the manifold. By and large, these generalized covariant derivatives had to be specified *ad hoc* by some version of the connection concept. In 1950, Jean-Louis Koszul unified these new ideas of covariant differentiation in a vector bundle by means of what is known today as a Koszul connection or a connection on a vector bundle. Using ideas from Lie algebra cohomology, Koszul successfully converted many of the analytic features of covariant differentiation into algebraic ones. In particular, Koszul connections eliminated the need for awkward manipulations of Christoffel symbols (and other analogous non-tensorial) objects in differential geometry. Thus they quickly supplanted the classical notion of covariant derivative in many post-1950 treatments of the subject.

Motivation

The covariant derivative is a generalization of the directional derivative from vector calculus. As with the directional derivative, the covariant derivative is a rule, $\nabla_u \mathbf{v}$, which takes as its inputs: (1) a vector, u, defined at a point P, and (2) a vector field, v, defined in a neighborhood of P. The output is the vector $\nabla_u \mathbf{v}(P)$, also at the point P. The primary difference from the usual directional derivative is that $\nabla_u \mathbf{v}$ must, in a certain precise sense, be *independent* of the manner in which it is expressed in a coordinate system.

A vector may be *described* as a list of numbers in terms of a basis, but as a geometrical object a vector retains its own identity regardless of how one chooses to describe it in a basis. This persistence of identity is reflected in the fact that when a vector is written in one basis, and then the basis is changed, the components of the vector transform according to a change of basis formula. Such a transformation law is known as a covariant transformation. The covariant derivative is required to transform, under a change in coordinates, in the same way as a basis does: the covariant derivative must change by a covariant transformation (hence the name).

In the case of Euclidean space, one tends to define the derivative of a vector field in terms of the difference between two vectors at two nearby points. In such a system one translates one of the vectors to the origin of the other, keeping it parallel. With a Cartesian (fixed orthonormal) coordinate system "keeping it parallel" amounts to keeping the components constant. Thus is obtained the simplest example: a covariant derivative which is obtained by taking the ordinary directional derivative of the components in the direction of the displacement vector between the two nearby points.

In the general case, however, one must take into account the change of the coordinate system. For example, if the same covariant derivative is written in polar coordinates in a two dimensional Euclidean plane, then it contains extra terms that describe how the coordinate grid itself "rotates". In other cases the extra terms describe how the coordinate grid expands, contracts, twists, interweaves, etc. In this case "keeping it parallel" does *not* amount to keeping components constant under translation.

Consider the example of moving along a curve $\gamma(t)$ in the Euclidean plane. In polar coordinates, γ may be written in terms of its radial and angular coordinates by $\gamma(t) = (r(t), \theta(t))$. A vector at a particular time t (for instance, the acceleration of the curve) is expressed in terms of $(\mathbf{e}_r, \mathbf{e}_\theta)$, where $\mathbf{e}_r$ and $\mathbf{e}_\theta$ are unit tangent vectors for the polar coordinates, serving as a basis to decompose a vector in terms of radial and tangential components. At a slightly later time, the new basis in polar coordinates appears slightly rotated with respect to the first set. The covariant derivative of the basis vectors (the Christoffel symbols) serve to express this change.

In a curved space, such as the surface of the Earth (regarded as a sphere), the translation is not well defined and its analog, parallel transport, depends on the path along which the vector is translated.

A vector e on a globe on the equator at point Q is directed to the north. Suppose we parallel transport the vector first along the equator until at point P and then (keeping it parallel to itself) drag it along a meridian to the pole N and (keeping the direction there) subsequently transport it along another meridian back to Q. Then we notice that the parallel-transported vector along a closed circuit does not return as the same vector; instead, it has another orientation. This would not happen in Euclidean space and is caused by the *curvature* of the surface of the globe. The same effect can be noticed if we drag the vector along an infinitesimally small closed surface subsequently along two directions and then back. The infinitesimal change of the vector is a measure of the curvature.

Remarks

- The definition of the covariant derivative does not use the metric in space. However, for each metric there is a unique torsion-free covariant derivative called the Levi-Civita connection such that the covariant derivative of the metric is zero.

- The properties of a derivative imply that $\nabla_v \mathbf{u}$ depends on an arbitrarily small neighborhood of a point p in the same way as e.g. the derivative of a scalar function along a curve at a given point p depends on an arbitrarily small neighborhood of p.

- The information on the neighborhood of a point p in the covariant derivative can be used to define parallel transport of a vector. Also the curvature, torsion, and geodesics may be defined only in terms of the covariant derivative or other related variation on the idea of a linear connection.

Informal Definition using an Embedding into Euclidean Space

Suppose a (pseudo) Riemann manifold M, is embedded into Euclidean space $(\mathbb{R}^n, \langle \cdot; \cdot \rangle)$ via a (twice continuously) differentiable mapping $\vec{\Psi} : \mathbb{R}^d \supset U \to \mathbb{R}^n$ such that the tangent space at $\vec{\Psi}(p) \in M$ is spanned by the vectors

$$\left\{ \left. \frac{\partial \vec{\Psi}}{\partial x^i} \right|_p : i \in \{1, \ldots, d\} \right\}$$

and the scalar product on $\mathbb{R}^n$ is compatible with the metric on M: $g_{ij} = \left\langle \frac{\partial \vec{\Psi}}{\partial x^i} ; \frac{\partial \vec{\Psi}}{\partial x^j} \right\rangle$. (Since the manifold metric is always assumed to be regular, the compatibility condition implies linear independence of the partial derivative tangent vectors.)

For a tangent vector field

$$\vec{V} = v^j \frac{\partial \vec{\Psi}}{\partial x^j} \, ,$$

one has $\quad \dfrac{\partial \vec{V}}{\partial x^i} = \dfrac{\partial v^j}{\partial x^i} \dfrac{\partial \vec{\Psi}}{\partial x^j} + v^j \dfrac{\partial^2 \vec{\Psi}}{\partial x^i \partial x^j}$. The last term is not tangential to M, but can be expressed as a linear combination of the tangent space base vectors using the Christoffel symbols as linear factors plus a vector normal to the tangent space:

$$\frac{\partial^2 \vec{\Psi}}{\partial x^i \partial x^j} = \Gamma^k{}_{ij} \frac{\partial \vec{\Psi}}{\partial x^k} + \vec{n}.$$

The covariant derivative $\nabla_{\mathbf{e}_i} \vec{V}$, also written $\nabla_i \vec{V}$, is defined as just a tangential portion of the usual derivative:

$$\nabla_{\mathbf{e}_i} \vec{V} := \frac{\partial \vec{V}}{\partial x^i} - \vec{n} = \left(\frac{\partial v^k}{\partial x^i} + v^j \Gamma^k{}_{ij} \right) \frac{\partial \vec{\Psi}}{\partial x^k}.$$

In the case of the Levi-Civita connection is required to be orthogonal to tangent space, so

$$\left\langle \frac{\partial^2 \vec{\Psi}}{\partial x^i \partial x^j} ; \frac{\partial \vec{\Psi}}{\partial x^l} \right\rangle = \Gamma^k{}_{ij} \left\langle \frac{\partial \vec{\Psi}}{\partial x^k} ; \frac{\partial \vec{\Psi}}{\partial x^l} \right\rangle = \Gamma^k{}_{ij} g_{kl}..$$

On the other hand,

$$\frac{\partial g_{ab}}{\partial x^c} = \left\langle \frac{\partial^2 \vec{\Psi}}{\partial x^c \partial x^a} ; \frac{\partial \vec{\Psi}}{\partial x^b} \right\rangle + \left\langle \frac{\partial \vec{\Psi}}{\partial x^a} ; \frac{\partial^2 \vec{\Psi}}{\partial x^c \partial x^b} \right\rangle$$

implies (using the symmetry of the scalar product and swapping the order of partial differentiations)

$$\frac{\partial g_{jk}}{\partial x^i} + \frac{\partial g_{ki}}{\partial x^j} - \frac{\partial g_{ij}}{\partial x^k} = 2 \left\langle \frac{\partial^2 \vec{\Psi}}{\partial x^i \partial x^j} ; \frac{\partial \vec{\Psi}}{\partial x^k} \right\rangle$$

and yields the Christoffel symbols for the Levi-Civita connection in terms of the metric:

$$g_{kl} \Gamma^k{}_{ij} = \frac{1}{2} \left(\frac{\partial g_{jl}}{\partial x^i} + \frac{\partial g_{li}}{\partial x^j} - \frac{\partial g_{ij}}{\partial x^l} \right).$$

For a very simple example that captures the essence of the description above, draw a circle on a flat sheet of paper. Travel around the circle at a constant speed. The derivative of your velocity, your acceleration vector, always points radially inward. Roll this sheet of paper into a cylinder. Now the (Euclidean) derivative of your velocity has a component that sometimes points inward toward the axis of the cylinder depending on whether you're near a solstice or an equinox. (At the point of the circle when you are moving parallel to the axis, there is no inward acceleration. At the point, the velocity is along the cylinder's bend, the inward acceleration is maximum.) This is the (Euclidean) normal component. The covariant derivative component is the component parallel to the cylinder's surface, and is the same as that before you rolled the sheet into a cylinder.

Formal Definition

A covariant derivative is a (Koszul) connection on the tangent bundle and other tensor bundles. Thus it has a certain behavior on vector fields that extends that of the usual differential on functions. It also extends in a unique way to the duals of vector fields (i.e., covector fields), and to arbitrary tensor fields, that ensures compatibility with the tensor product and trace operations (tensor contraction).

Functions

Given a point p of the manifold, a real function f on the manifold, and a tangent vector v at p, the covariant derivative of f at p along v is the scalar at p, denoted $(\nabla_v f)_p$, that represents the principal part of the change in the value of f when the argument of f is changed by the infinitesimal displacement vector v. (This is the differential of f evaluated against the vector v.) Formally, there is a differentiable curve $\phi : [-1,1] \to M$ such that $\phi(0) = p$ and $\phi'(0) = v$, and the covariant derivative of f at p is defined by

$$(\nabla_v f)_p = (f \circ \phi)'(0) = \lim_{t \to 0} t^{-1}(f(\phi(t)) - f(p)).$$

When v is a vector field, the covariant derivative $\nabla_v f$ is the function that associates with each point p in the common domain of f and v the scalar $(\nabla_v f)_p$. This coincides with the usual Lie derivative of f along the vector field v.

Vector Fields

A covariant derivative ∇ at a point p in a smooth manifold assigns a tangent vector $(\nabla_v u)_p$ to each pair (u, v), consisting of a tangent vector v at p and vector field u defined in a neighborhood of p, such that the following properties hold (for any vectors v, x and y at p, vector fields u and w defined in a neighborhood of p, scalar values g and h at p, and scalar function f defined in a neighborhood of p):

1. $(\nabla_v u)_p$ is linear in v so $(\nabla_{gx+hy} u)_p = (\nabla_x u)_p g + (\nabla_y u)_p h$

2. $(\nabla_v u)_p$ is additive in u so $(\nabla_v (u + w))_p = (\nabla_v u)_p + (\nabla_v w)_p$

3. $(\nabla_v u)_p$ obeys the product rule, i.e., $(\nabla_v (fu))_p = f(p)(\nabla_v u)_p + (\nabla_v f)_p u_p$, where $\nabla_v f$ is defined above.

If u and v are both vector fields defined over a common domain, then $\nabla_v u$ denotes the vector field whose value at each point p of the domain is the tangent vector $(\nabla_v u)_p$. Note that $(\nabla_v u)_p$ depends not only on the value of v at p but also on values of u in an infinitesimal neighbourhood of p because of the last property, the product rule.

Covector Fields

Given a field of covectors (or one-form) α defined in a neighborhood of p, its covariant derivative $(\nabla_v \alpha)_p$ is defined in a way to make the resulting operation compatible with tensor contraction and the product rule. That is, $(\nabla_v \alpha)_p$ is defined as the unique one-form at p such that the following identity is satisfied for all vector fields u in a neighborhood of p

$$(\nabla_v \alpha)_p (\mathbf{u}_p) = \nabla_v (\alpha(\mathbf{u}))_p - \alpha_p ((\nabla_v \mathbf{u})_p).$$

The covariant derivative of a covector field along a vector field v is again a covector field.

Tensor Fields

Once the covariant derivative is defined for fields of vectors and covectors it can be defined for arbitrary tensor fields by imposing the following identities for every pair of tensor fields φ and ψ in a neighborhood of the point p:

$$\nabla_v (\varphi \otimes \psi)_p = (\nabla_v \varphi)_p \otimes \psi(p) + \varphi(p) \otimes (\nabla_v \psi)_p,$$

and for φ and ψ of the same valence

$$\nabla_v (\varphi + \psi)_p = (\nabla_v \varphi)_p + (\nabla_v \psi)_p.$$

The covariant derivative of a tensor field along a vector field v is again a tensor field of the same type.

Explicitly, let T be a tensor field of type (p, q). Consider T to be a differentiable multilinear map of smooth sections $\alpha^1, \alpha^2, ..., \alpha^q$ of the cotangent bundle T^*M and of sections $X_1, X_2, ... X_p$ of the tangent bundle TM, written $T(\alpha^1, \alpha^2, ..., X_1, X_2, ...)$ into R. The covariant derivative of T along Y is given by the formula

$$(\nabla_Y T)(\alpha_1, \alpha_2, ..., X_1, X_2, ...) = Y(T(\alpha_1, \alpha_2, ..., X_1, X_2, ...))$$

$$-T(\nabla_Y \alpha_1, \alpha_2, ..., X_1, X_2, ...) - T(\alpha_1, \nabla_Y \alpha_2, ..., X_1, X_2, ...) - ...$$

$$-T(\alpha_1, \alpha_2, ..., \nabla_Y X_1, X_2, ...) - T(\alpha_1, \alpha_2, ..., X_1, \nabla_Y X_2, ...) - ...$$

Coordinate Description

Given coordinate functions

$$x^i, i = 0, 1, 2, ..., ,$$

any tangent vector can be described by its components in the basis

$$\mathbf{e}_i = \frac{\partial}{\partial x^i}.$$

The covariant derivative of a basis vector along a basis vector is again a vector and so can be expressed as a linear combination $\tilde{A}^k \mathbf{e}_k$. To specify the covariant derivative it is enough to specify the covariant derivative of each basis vector field $\mathbf{e}_j$ along $\mathbf{e}_i$.

$$\nabla_{\mathbf{e}_i} \mathbf{e}_j = \Gamma^k_{ij} \mathbf{e}_k,$$

the coefficients Γ^k_{ij} are called Christoffel symbols of the second kind. Then using the rules in the definition, we find that for general vector fields $\mathbf{v} = v^i \mathbf{e}_i$ and $\mathbf{u} = u^j \mathbf{e}_j$ we get

$$\nabla_{\mathbf{v}} \mathbf{u} = \nabla_{v^i \mathbf{e}_i} u^j \mathbf{e}_j = v^i \nabla_{\mathbf{e}_i} u^j \mathbf{e}_j = v^i u^j \nabla_{\mathbf{e}_i} \mathbf{e}_j + v^i \mathbf{e}_j \nabla_{\mathbf{e}_i} u^j = v^i u^j \Gamma^k_{ij} \mathbf{e}_k + v^i \frac{\partial u^j}{\partial x^i} \mathbf{e}_j$$

so

$$\nabla_{\mathbf{v}} \mathbf{u} = \left(v^i u^j \Gamma^k_{ij} + v^i \frac{\partial u^k}{\partial x^i} \right) \mathbf{e}_k$$

The first term in this formula is responsible for "twisting" the coordinate system with respect to the covariant derivative and the second for changes of components of the vector field u. In particular

$$\nabla_{\mathbf{e}_j} \mathbf{u} = \nabla_j \mathbf{u} = \left(\frac{\partial u^i}{\partial x^j} + u^k \Gamma^i_{jk} \right) \mathbf{e}_i$$

In words: the covariant derivative is the usual derivative along the coordinates with correction terms which tell how the coordinates change.

For covectors similarly we have

$$\nabla_{\mathbf{e}_j} \theta = \left(\frac{\partial \theta_i}{\partial x^j} - \theta_k \Gamma^k_{ij} \right) \mathbf{e}^{*i}$$

where $\mathbf{e}^{*i}(\mathbf{e}_j) = \delta^i_j$..

The covariant derivative of a type (r, s) tensor field along e_c is given by the expression:

$$(\nabla_{e_c} T)^{a_1 \ldots a_r}{}_{b_1 \ldots b_s} = \frac{\partial}{\partial x^c} T^{a_1 \ldots a_r}{}_{b_1 \ldots b_s} + \Gamma^{a_1}_{dc} T^{d a_2 \ldots a_r}{}_{b_1 \ldots b_s} + \cdots + \Gamma^{a_r}_{dc} T^{a_1 \ldots a_{r-1} d}{}_{b_1 \ldots b_s}$$

$$- \Gamma^d_{b_1 c} T^{a_1 \ldots a_r}{}_{d b_2 \ldots b_s} - \cdots - \Gamma^d_{b_s c} T^{a_1 \ldots a_r}{}_{b_1 \ldots b_{s-1} d}.$$

Or, in words: take the partial derivative of the tensor and add: a $+\Gamma^{a_i}_{dc}$ for every upper index a_i, and a $-\Gamma^d_{b_i c}$ for every lower index b_i.

If instead of a tensor, one is trying to differentiate a *tensor density* (of weight +1), then you also add a term

$$-\Gamma^d_{\ dc} T^{a_1...a_r}_{\quad b_1...b_s}.$$

If it is a tensor density of weight W, then multiply that term by W. For example, $\sqrt{-g}$ is a scalar density (of weight +1), so we get:

$$(\sqrt{-g})_{;c} = (\sqrt{-g})_{,c} - \sqrt{-g}\,\Gamma^d_{\ dc}$$

where semicolon ";" indicates covariant differentiation and comma "," indicates partial differentiation. Incidentally, this particular expression is equal to zero, because the covariant derivative of a function solely of the metric is always zero.

Examples

For a scalar field ϕ, covariant differentiation is simply partial differentiation:

$$\phi_{;a} \equiv \partial_a \phi$$

For a contravariant vector field λ^a, we have:

$$\lambda^a_{\ ;b} \equiv \partial_b \lambda^a + \Gamma^a_{\ bc} \lambda^c$$

For a covariant vector field λ_a, we have:

$$\lambda_{a;c} \equiv \partial_c \lambda_a - \Gamma^b_{\ ca} \lambda_b$$

For a type (2,0) tensor field τ^{ab}, we have:

$$\tau^{ab}_{\ \ ;c} \equiv \partial_c \tau^{ab} + \Gamma^a_{\ cd} \tau^{db} + \Gamma^b_{\ cd} \tau^{ad}$$

For a type (0,2) tensor field τ_{ab}, we have:

$$\tau_{ab;c} \equiv \partial_c \tau_{ab} - \Gamma^d_{\ ca} \tau_{db} - \Gamma^d_{\ cb} \tau_{ad}$$

For a type (1,1) tensor field $\tau^a_{\ b}$, we have:

$$\tau^a_{\ b;c} \equiv \partial_c \tau^a_{\ b} + \Gamma^a_{\ cd} \tau^d_{\ b} - \Gamma^d_{\ cb} \tau^a_{\ d}$$

The notation above is meant in the sense

$$\tau^{ab}_{\ \ ;c} \equiv (\nabla_{\mathbf{e}_c} \tau)^{ab}$$

One must always remember that covariant derivatives do not commute, i.e. $\lambda_{a;bc} \neq \lambda_{a;cb}.$. It is actually easy to show that:

$$\lambda_{a;bc} - \lambda_{a;cb} = R^d{}_{abc}\lambda_d$$

where $R^d{}_{abc}$ is the Riemann tensor. Similarly,

$$\lambda^a{}_{;bc} - \lambda^a{}_{;cb} = -R^a{}_{dbc}\lambda^d$$

and

$$\tau^{ab}{}_{;cd} - \tau^{ab}{}_{;dc} = -R^a{}_{ecd}\tau^{eb} - R^b{}_{ecd}\tau^{ae}$$

The latter can be shown by taking (without loss of generality) that $\tau^{ab} = \lambda^a\mu^b$.

Notation

In textbooks on physics, the covariant derivative is sometimes simply stated in terms of its components in this equation.

Often a notation is used in which the covariant derivative is given with a semicolon, while a normal partial derivative is indicated by a comma. In this notation we write the same as:

$$\nabla_{e_j}\mathbf{v} \stackrel{def}{=} v^s{}_{;j}e_s \qquad v^i{}_{;j} = v^i{}_{,j} + v^k\Gamma^i{}_{kj}$$

Once again this shows that the covariant derivative of a vector field is not just simply obtained by differentiating to the coordinates $v^i{}_{,j}$, but also depends on the vector v itself through $v^k\Gamma^i{}_{kj}$.

In some older texts (notably Adler, Bazin & Schiffer, *Introduction to General Relativity*), the covariant derivative is denoted by a double pipe:

$$\nabla_{e_j}\mathbf{v} \stackrel{def}{=} v^i{}_{\|j}$$

Derivative Along Curve

Since the covariant derivative $\nabla_X T$ of a tensor field T at a point p depends only on value of the vector field X at p one can define the covariant derivative along a smooth curve $\gamma(t)$ in a manifold:

$$D_t T = \nabla_{\dot\gamma(t)} T.$$

Note that the tensor field T only needs to be defined on the curve $\gamma(t)$ for this definition to make sense.

In particular, $\dot\gamma(t)$ is a vector field along the curve γ itself. If $\nabla_{\dot\gamma(t)}\dot\gamma(t)$ vanishes then the curve is called a geodesic of the covariant derivative. If the covariant derivative is the Levi-Civita connection of a certain metric then the geodesics for the connection are precisely the geodesics of the metric that are parametrised by arc length.

The derivative along a curve is also used to define the parallel transport along the curve.

Sometimes the covariant derivative along a curve is called absolute or intrinsic derivative.

Relation to Lie Derivative

A covariant derivative introduces an extra geometric structure on a manifold which allows vectors in neighboring tangent spaces to be compared. This extra structure is necessary because there is no canonical way to compare vectors from different vector spaces, as is necessary for this generalization of the directional derivative. There is however another generalization of directional derivatives which *is* canonical: the Lie derivative. The Lie derivative evaluates the change of one vector field along the flow of another vector field. Thus, one must know both vector fields in an open neighborhood. The covariant derivative on the other hand introduces its own change for vectors in a given direction, and it only depends on the vector direction at a single point, rather than a vector field in an open neighborhood of a point. In other words, the covariant derivative is linear (over $C^\infty(M)$) in the direction argument, while the Lie derivative is linear in neither argument.

Note that the antisymmetrized covariant derivative $\nabla_u v - \nabla_v u$, and the Lie derivative $L_u v$ differ by the torsion of the connection, so that if a connection is torsion free, then its antisymmetrization *is* the Lie derivative.

Parallel Transport

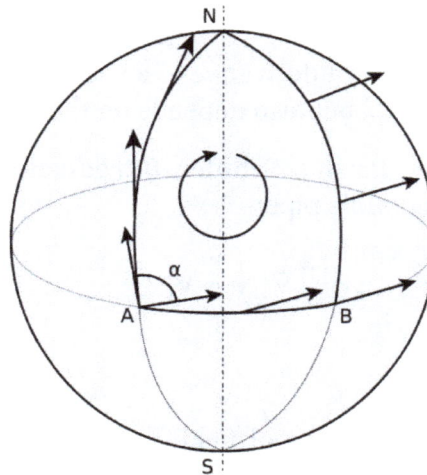

Parallel transport of a vector around a closed loop (from A to N to B and back to A) on the sphere. The angle by which it twists, α, is proportional to the area inside the loop.

In geometry, parallel transport is a way of transporting geometrical data along smooth curves in a manifold. If the manifold is equipped with an affine connection (a covariant derivative or connection on the tangent bundle), then this connection allows one to transport vectors of the manifold along curves so that they stay *parallel* with respect to the connection.

The parallel transport for a connection thus supplies a way of, in some sense, moving the local geometry of a manifold along a curve: that is, of *connecting* the geometries of nearby points. There may be many notions of parallel transport available, but a specification of one — one way of connecting up the geometries of points on a curve — is tantamount to providing a *connection*. In fact, the usual notion of connection is the infinitesimal analog of parallel transport. Or, *vice versa*, parallel transport is the local realization of a connection.

As parallel transport supplies a local realization of the connection, it also supplies a local realization of the curvature known as holonomy. The Ambrose-Singer theorem makes explicit this relationship between curvature and holonomy.

Other notions of connection come equipped with their own parallel transportation systems as well. For instance, a Koszul connection in a vector bundle also allows for the parallel transport of vectors in much the same way as with a covariant derivative. An Ehresmann or Cartan connection supplies a *lifting of curves* from the manifold to the total space of a principal bundle. Such curve lifting may sometimes be thought of as the parallel transport of reference frames.

Parallel Transport on a Vector Bundle

Let M be a smooth manifold. Let $E{\to}M$ be a vector bundle with covariant derivative ∇ and $\gamma: I{\to}M$ a smooth curve parameterized by an open interval I. A section X of E along γ is called parallel if

$$\nabla_{\dot\gamma(t)}X = 0 \text{ for } t \in I.$$

Suppose we are given an element $e_0 \in E_P$ at $P = \gamma(0) \in M$, rather than a section. The parallel transport of e_0 along γ is the extension of e_0 to a parallel *section* X on γ. More precisely, X is the unique section of E along γ such that

$$\nabla_{\dot\gamma}X = 0$$

$$X_{\gamma(0)} = e_0.$$

Note that in any given coordinate patch, (1) defines an ordinary differential equation, with the initial condition given by (2). Thus the Picard–Lindelöf theorem guarantees the existence and uniqueness of the solution.

Thus the connection ∇ defines a way of moving elements of the fibers along a curve, and this provides linear isomorphisms between the fibers at points along the curve:

$$\Gamma(\gamma)_s^t : E_{\gamma(s)} \to E_{\gamma(t)}$$

from the vector space lying over $\gamma(s)$ to that over $\gamma(t)$. This isomorphism is known as the parallel transport map associated to the curve. The isomorphisms between fibers obtained in this way will in general depend on the choice of the curve: if they do not, then parallel transport along every curve can be used to define parallel sections of E over all of M. This is only possible if the curvature of ∇ is zero.

In particular, parallel transport around a closed curve starting at a point x defines an automorphism of the tangent space at x which is not necessarily trivial. The parallel transport automorphisms defined by all closed curves based at x form a transformation group called the holonomy group of ∇ at x. There is a close relation between this group and the value of the curvature of ∇ at x; this is the content of the Ambrose-Singer holonomy theorem.

Recovering the Connection from the Parallel Transport

Given a covariant derivative ∇, the parallel transport along a curve γ is obtained by integrating the condition $\nabla_{\dot{\gamma}} = 0$. Conversely, if a suitable notion of parallel transport is available, then a corresponding connection can be obtained by differentiation.

Consider an assignment to each curve γ in the manifold a collection of mappings

$$\Gamma(\gamma)_s^t : E_{\gamma(s)} \to E_{\gamma(t)}$$

such that

1. $\Gamma(\gamma)_s^s = Id$, the identity transformation of $E_{\gamma(s)}$.

2. $\Gamma(\gamma)_u^t \circ \Gamma(\gamma)_s^u = \Gamma(\gamma)_s^t$.

3. The dependence of Γ on γ, s, and t is "smooth."

The notion of smoothness in condition 3. is somewhat difficult to pin down. In particular, modern authors such as Kobayashi and Nomizu generally view the parallel transport of the connection as coming from a connection in some other sense, where smoothness is more easily expressed.

Nevertheless, given such a rule for parallel transport, it is possible to recover the associated infinitesimal connection in E as follows. Let γ be a differentiable curve in M with initial point $\gamma(0)$ and initial tangent vector $X = \gamma'(0)$. If V is a section of E over γ, then let

$$\nabla_X V = \lim_{h \to 0} \frac{\Gamma(\gamma)_h^0 V_{\gamma(h)} - V_{\gamma(0)}}{h} = \frac{d}{dt} \Gamma(\gamma)_t^0 V_{\gamma(t)} \bigg|_{t=0} .$$

This defines the associated infinitesimal connection ∇ on E. One recovers the same parallel transport Γ from this infinitesimal connection.

Special Case: The Tangent Bundle

Let M be a smooth manifold. Then a connection on the tangent bundle of M, called an affine connection, distinguishes a class of curves called (affine) geodesics (Kobayashi & Nomizu, Volume 1, Chapter III). A smooth curve $\gamma: I \to M$ is an affine geodesic if $\dot{\gamma}$ is parallel transported along γ, that is

$$\Gamma(\gamma)_s^t \dot{\gamma}(s) = \dot{\gamma}(t).$$

Taking the derivative with respect to time, this takes the more familiar form

$$\nabla_{\dot{\gamma}(t)} \dot{\gamma} = 0.$$

Parallel Transport in Riemannian Geometry

In (pseudo) Riemannian geometry, a metric connection is any connection whose parallel transport mappings preserve the metric tensor. Thus a metric connection is any connection Γ such that, for any two vectors $X, Y \in T_{\gamma(s)}$

$$\langle \Gamma(\gamma)_s^t X, \Gamma(\gamma)_s^t Y \rangle_{\gamma(t)} = \langle X, Y \rangle_{\gamma(s)}.$$

Taking the derivative at $t=0$, the associated differential operator ∇ must satisfy a product rule with respect to the metric:

$$Z\langle X, Y \rangle = \langle \nabla_Z X, Y \rangle + \langle X, \nabla_Z Y \rangle.$$

Geodesics

If ∇ is a metric connection, then the affine geodesics are the usual geodesics of Riemannian geometry and are the locally distance minimizing curves. More precisely, first note that if $\gamma: I \to M$, where I is an open interval, is a geodesic, then the norm of $\dot{\gamma}$ is constant on I. Indeed,

$$\frac{d}{dt}\langle \dot{\gamma}(t), \dot{\gamma}(t) \rangle = 2\langle \nabla_{\dot{\gamma}(t)} \dot{\gamma}(t), \dot{\gamma}(t) \rangle = 0.$$

It follows from an application of Gauss's Lemma that if A is the norm of $\dot{\gamma}(t)$ then the distance, induced by the metric, between two *close enough* points on the curve γ, say $\gamma(t_1)$ and $\gamma(t_2)$, is given by

$$\mathrm{dist}\big(\gamma(t_1), \gamma(t_2)\big) = A \,|\, t_1 - t_2 \,|.$$

The formula above might not be true for points which are not close enough since the geodesic might for example wrap around the manifold (e.g. on a sphere).

Generalizations

The parallel transport can be defined in greater generality for other types of connections, not just those defined in a vector bundle. One generalization is for principal connections (Kobayashi & Nomizu 1996, Volume 1, Chapter II). Let $P \to M$ be a principal bundle over a manifold M with structure Lie group G and a principal connection ω. As in the case of vector bundles, a principal connection ω on P defines, for each curve γ in M, a mapping

$$\Gamma(\gamma)_s^t : P_{\gamma(s)} \to P_{\gamma(t)}$$

from the fibre over $\gamma(s)$ to that over $\gamma(t)$, which is an isomorphism of homogeneous spaces: i.e. $\Gamma_{\gamma(s)} gu = g\Gamma_{\gamma(s)}$ for each $g \in G$.

Further generalizations of parallel transport are also possible. In the context of Ehresmann connections, where the connection depends on a special notion of "horizontal lifting" of tangent spaces, one can define parallel transport via horizontal lifts. Cartan connections are Ehresmann connections with additional structure which allows the parallel transport to be though of as a map "rolling" a certain model space along a curve in the manifold. This rolling is called development.

Approximation: Schild's Ladder

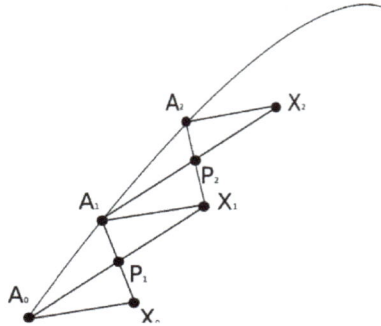

Two rungs of Schild's ladder. The segments A_1X_1 and A_2X_2 are an approximation to first order of the parallel transport of A_0X_0 along the curve.

Parallel transport can be discretely approximated by Schild's ladder, which takes finite steps along a curve, and approximates Levi-Civita parallelogramoids by approximate parallelograms.

Geodesics in General Relativity

In general relativity, a geodesic generalizes the notion of a "straight line" to curved spacetime. Importantly, the world line of a particle free from all external, non-gravitational force, is a particular type of geodesic. In other words, a freely moving or falling particle always moves along a geodesic.

In general relativity, gravity can be regarded as not a force but a consequence of a curved space-time geometry where the source of curvature is the stress–energy tensor (representing matter, for instance). Thus, for example, the path of a planet orbiting around a star is the projection of a geodesic of the curved 4-D spacetime geometry around the star onto 3-D space.

Mathematical Expression

The full geodesic equation is this:

$$\frac{d^2 x^\mu}{ds^2} = -\Gamma^\mu{}_{\alpha\beta} \frac{dx^\alpha}{ds} \frac{dx^\beta}{ds}.$$

where s is a scalar parameter of motion (e.g. the proper time), and $\Gamma^\mu{}_{\alpha\beta}$ are Christoffel symbols (sometimes called the affine connection or Levi-Civita connection) which is symmetric in the two lower indices. Greek indices take the values [0,1,2,3]. The quantity on the left-hand-side of this equation is the acceleration of a particle, and so this equation is analogous to Newton's laws of motion which likewise provide formulae for the acceleration of a particle. This equation of motion employs the Einstein notation, meaning that repeated indices are summed (i.e. from zero to three). The Christoffel symbols are functions of the four space-time coordinates, and so are independent of the velocity or acceleration or other characteristics of a test particle whose motion is described by the geodesic equation.

Equivalent Mathematical Expression using Coordinate Time as Parameter

So far the geodesic equation of motion has been written in terms of a scalar parameter s. It can alternatively be written in terms of the time coordinate, $t \equiv x^0$ (here we have used the triple bar to signify a definition). The geodesic equation of motion then becomes:

$$\frac{d^2 x^\mu}{dt^2} = -\Gamma^\mu_{\ \alpha\beta} \frac{dx^\alpha}{dt} \frac{dx^\beta}{dt} + \Gamma^0_{\ \alpha\beta} \frac{dx^\alpha}{dt} \frac{dx^\beta}{dt} \frac{dx^\mu}{dt}.$$

This formulation of the geodesic equation of motion can be useful for computer calculations and to compare General Relativity with Newtonian Gravity. It is straightforward to derive this form of the geodesic equation of motion from the form which uses proper time as a parameter, using the chain rule. Notice that both sides of this last equation vanish when the mu index is set to zero. If the particle's velocity is small enough, then the geodesic equation reduces to this:

$$\frac{d^2 x^n}{dt^2} = -\Gamma^n_{\ 00}.$$

Here the Latin index n takes the values [1,2,3]. This equation simply means that all test particles at a particular place and time will have the same acceleration, which is a well-known feature of Newtonian gravity. For example, everything floating around in the international space station will undergo roughly the same acceleration due to gravity.

Derivation Directly from the Equivalence Principle

Physicist Steven Weinberg has presented a derivation of the geodesic equation of motion directly from the equivalence principle. The first step in such a derivation is to suppose that no particles are accelerating in the neighborhood of a point-event with respect to a freely falling coordinate system (X^μ). Setting $T \equiv X^0$, we have the following equation that is locally applicable in free fall:

$$\frac{d^2 X^\mu}{dT^2} = 0.$$

The next step is to employ the chain rule. We have:

$$\frac{dX^\mu}{dT} = \frac{dx^\nu}{dT} \frac{\partial X^\mu}{\partial x^\nu}$$

Differentiating once more with respect to the time, we have:

$$\frac{d^2 X^\mu}{dT^2} = \frac{d^2 x^\nu}{dT^2} \frac{\partial X^\mu}{\partial x^\nu} + \frac{dx^\nu}{dT} \frac{dx^\alpha}{dT} \frac{\partial^2 X^\mu}{\partial x^\nu \partial x^\alpha}$$

Therefore:

$$\frac{d^2 x^\nu}{dT^2} \frac{\partial X^\mu}{\partial x^\nu} = -\frac{dx^\nu}{dT} \frac{dx^\alpha}{dT} \frac{\partial^2 X^\mu}{\partial x^\nu \partial x^\alpha}$$

Multiply both sides of this last equation by the following quantity:

$$\frac{\partial x^{\lambda}}{\partial X^{\mu}}$$

Consequently, we have this:

$$\frac{d^2 x^{\lambda}}{dT^2} = -\frac{dx^{\nu}}{dT}\frac{dx^{\alpha}}{dT}\left[\frac{\partial^2 X^{\mu}}{\partial x^{\nu}\partial x^{\alpha}}\frac{\partial x^{\lambda}}{\partial X^{\mu}}\right]$$

As before, we can set $t \equiv x^0$. Using the chain rule, the parameter T can be eliminated in favor of the parameter t like so:

$$\frac{d^2 x^{\lambda}}{dt^2} = -\frac{dx^{\nu}}{dt}\frac{dx^{\alpha}}{dt}\left[\frac{\partial^2 X^{\mu}}{\partial x^{\nu}\partial x^{\alpha}}\frac{\partial x^{\lambda}}{\partial X^{\mu}}\right] + \frac{dx^{\nu}}{dt}\frac{dx^{\alpha}}{dt}\frac{dx^{\lambda}}{dt}\left[\frac{\partial^2 X^{\mu}}{\partial x^{\nu}\partial x^{\alpha}}\frac{\partial x^0}{\partial X^{\mu}}\right]$$

The geodesic equation of motion (using the coordinate time as parameter) follows immediately from this last equation, because the bracketed terms (which involve the relationship between local coordinates X and general coordinates x) are functions of the general coordinates. The geodesic equation of motion can alternatively be derived using the concept of parallel transport.

Deriving the Geodesic Equation Via an Action

We can (and this is the most common technique) derive the geodesic equation via the action principle. Consider the case of trying to find a geodesic between two timelike-separated points.

Let the action be

$$S = \int ds$$

where $ds = \sqrt{-g_{\mu\nu}(x)dx^{\mu}dx^{\nu}}$ is the line element. There is a negative sign inside the square root because the curve must be timelike. To get the geodesic equation we must vary this action. To do this let's parameterize this action with respect a parameter λ. Doing this we get:

$$S = \int \sqrt{-g_{\mu\nu}\frac{dx^{\mu}}{d\lambda}\frac{dx^{\nu}}{d\lambda}}d\lambda$$

We can now go ahead and vary this action with respect to the curve x^{μ}. By the principle of least action we get:

$$0 = \delta S = \int \delta\left(\sqrt{-g_{\mu\nu}\frac{dx^{\mu}}{d\lambda}\frac{dx^{\nu}}{d\lambda}}\right)d\lambda = \int \frac{\delta\left(-g_{\mu\nu}\frac{dx^{\mu}}{d\lambda}\frac{dx^{\nu}}{d\lambda}\right)}{2\sqrt{-g_{\mu\nu}\frac{dx^{\mu}}{d\lambda}\frac{dx^{\nu}}{d\lambda}}}d\lambda$$

For concreteness let's parameterize this action w.r.t. the proper time, τ. Since the four-velocity is normalized to -1 (for time-like paths) we can say that the above is equivalent to the action:

$$0 = \int \delta \left(g_{\mu\nu} \frac{dx^\mu}{d\tau} \frac{dx^\nu}{d\tau} \right) d\tau$$

Using the product rule we get:

$$0 = \int \left(\frac{dx^\mu}{d\tau} \frac{dx^\nu}{d\tau} \delta g_{\mu\nu} + g_{\mu\nu} \frac{d\delta x^\mu}{d\tau} \frac{dx^\nu}{d\tau} + g_{\mu\nu} \frac{dx^\mu}{d\tau} \frac{d\delta x^\nu}{d\tau} \right) d\tau = \int \left(\frac{dx^\mu}{d\tau} \frac{dx^\nu}{d\tau} \partial_\alpha g_{\mu\nu} \delta x^\alpha + 2 g_{\mu\nu} \frac{d\delta x^\mu}{d\tau} \frac{dx^\nu}{d\tau} \right) d\tau$$

Integrating by-parts the last term and dropping the total derivative (which equals to zero at the boundaries) we get that:

$$0 = \int d\tau \left(\frac{dx^\mu}{d\tau} \frac{dx^\nu}{d\tau} \partial_\alpha g_{\mu\nu} \delta x^\alpha - 2 \delta x^\mu \frac{d}{d\tau} \left(g_{\mu\nu} \frac{dx^\nu}{d\tau} \right) \right) = \int d\tau \left(\frac{dx^\mu}{d\tau} \frac{dx^\nu}{d\tau} \partial_\alpha g_{\mu\nu} \delta x^\alpha - 2 \delta x^\mu \partial_\alpha g_{\mu\nu} \frac{dx^\alpha}{d\tau} \frac{dx^\nu}{d\tau} - 2 \delta x^\mu g_{\mu\nu} \frac{d^2 x^\nu}{d\tau^2} \right)$$

Simplifying a bit we see that:

$$0 = \int d\tau \delta x^\mu \left(-2 g_{\mu\nu} \frac{d^2 x^\nu}{d\tau^2} + \frac{dx^\alpha}{d\tau} \frac{dx^\nu}{d\tau} \partial_\mu g_{\alpha\nu} - 2 \frac{dx^\alpha}{d\tau} \frac{dx^\nu}{d\tau} \partial_\alpha g_{\mu\nu} \right)$$

so,

$$0 = \int d\tau \delta x^\mu \left(-2 g_{\mu\nu} \frac{d^2 x^\nu}{d\tau^2} + \frac{dx^\alpha}{d\tau} \frac{dx^\nu}{d\tau} \partial_\mu g_{\alpha\nu} - \frac{dx^\alpha}{d\tau} \frac{dx^\nu}{d\tau} \partial_\alpha g_{\mu\nu} - \frac{dx^\nu}{d\tau} \frac{dx^\alpha}{d\tau} \partial_\nu g_{\mu\alpha} \right)$$

multiplying this equation by $-\dfrac{1}{2}$ we get:

$$0 = \int d\tau \delta x^\mu \left(g_{\mu\nu} \frac{d^2 x^\nu}{d\tau^2} + \frac{1}{2} \frac{dx^\alpha}{d\tau} \frac{dx^\nu}{d\tau} \left(\partial_\alpha g_{\mu\nu} + \partial_\nu g_{\mu\alpha} - \partial_\mu g_{\alpha\nu} \right) \right)$$

So by Hamilton's principle we find that the Euler–Lagrange equation is

$$g_{\mu\nu} \frac{d^2 x^\nu}{d\tau^2} + \frac{1}{2} \frac{dx^\alpha}{d\tau} \frac{dx^\nu}{d\tau} \left(\partial_\alpha g_{\mu\nu} + \partial_\nu g_{\mu\alpha} - \partial_\mu g_{\alpha\nu} \right) = 0$$

Multiplying by the inverse metric tensor $g^{\mu\beta}$ we get that

$$\frac{d^2 x^\beta}{d\tau^2} + \frac{1}{2} g^{\mu\beta} \left(\partial_\alpha g_{\mu\nu} + \partial_\nu g_{\mu\alpha} - \partial_\mu g_{\alpha\nu} \right) \frac{dx^\alpha}{d\tau} \frac{dx^\nu}{d\tau} = 0$$

Thus we get the geodesic equation:

$$\frac{d^2 x^\beta}{d\tau^2} + \Gamma^\beta{}_{\alpha v}\frac{dx^\alpha}{d\tau}\frac{dx^v}{d\tau} = 0$$

with the Christoffel symbol defined in terms of the metric tensor as

$$\Gamma^\beta{}_{\alpha v} = \frac{1}{2}g^{\mu\beta}\left(\partial_\alpha g_{\mu v} + \partial_v g_{\mu\alpha} - \partial_\mu g_{\alpha v}\right)$$

Equation of Motion May Follow from the Field Equations for Empty Space

Albert Einstein believed that the geodesic equation of motion can be derived from the field equations for empty space, i.e. from the fact that the Ricci curvature vanishes. He wrote:

It has been shown that this law of motion — generalized to the case of arbitrarily large gravitating masses — can be derived from the field equations of empty space alone. According to this derivation the law of motion is implied by the condition that the field be singular nowhere outside its generating mass points.

Both physicists and philosophers have often repeated the assertion that the geodesic equation can be obtained from the field equations to describe the motion of a gravitational singularity, but this claim remains disputed. Less controversial is the notion that the field equations determine the motion of a fluid or dust, as distinguished from the motion of a point-singularity.

Extension to the Case of a Charged Particle

In deriving the geodesic equation from the equivalence principle, it was assumed that particles in a local inertial coordinate system are not accelerating. However, in real life, the particles may be charged, and therefore may be accelerating locally in accordance with the Lorentz force. That is:

$$\frac{d^2 X^\mu}{ds^2} = \frac{q}{m}F^{\mu\beta}\frac{dX^\alpha}{ds}\eta_{\alpha\beta}.$$

with

$$\eta_{\alpha\beta}\frac{dX^\alpha}{ds}\frac{dX^\beta}{ds} = -1.$$

The Minkowski tensor $\eta_{\alpha\beta}$ is given by:

$$\eta_{\alpha\beta} = \begin{pmatrix} -1 & 0 & 0 & 0 \\ 0 & 1 & 0 & 0 \\ 0 & 0 & 1 & 0 \\ 0 & 0 & 0 & 1 \end{pmatrix}$$

These last three equations can be used as the starting point for the derivation of an equation of motion in General Relativity, instead of assuming that acceleration is zero in free fall. Because the Minkowski tensor is involved here, it becomes necessary to introduce something called the *metric tensor* in General Relativity. The metric tensor g is symmetric, and locally reduces to the Minkowski tensor in free fall. The resulting equation of motion is as follows:

$$\frac{d^2 x^\mu}{ds^2} = -\Gamma^\mu_{\alpha\beta} \frac{dx^\alpha}{ds} \frac{dx^\beta}{ds} + \frac{q}{m} F^{\mu\beta} \frac{dx^\alpha}{ds} g_{\alpha\beta}.$$

with

$$g_{\alpha\beta} \frac{dx^\alpha}{ds} \frac{dx^\beta}{ds} = -1.$$

This last equation signifies that the particle is moving along a timelike geodesic; massless particles like the photon instead follow null geodesics (replace –1 with zero on the right-hand side of the last equation). It is important that the last two equations are consistent with each other, when the latter is differentiated with respect to proper time, and the following formula for the Christoffel symbols ensures that consistency:

$$\Gamma^\lambda_{\alpha\varsigma} = \frac{1}{2} g^{\lambda\tau} \left(\frac{\partial g_{\tau\alpha}}{\partial x^\beta} + \frac{\partial g_{\tau\beta}}{\partial x^\alpha} - \frac{\partial g_{\alpha\beta}}{\partial x^\tau} \right)$$

This last equation does not involve the electromagnetic fields, and it is applicable even in the limit as the electromagnetic fields vanish. The letter g with superscripts refers to the inverse of the metric tensor. In General Relativity, indices of tensors are lowered and raised by contraction with the metric tensor or its inverse, respectively.

Geodesics as Curves of Stationary Interval

A geodesic between two events can also be described as the curve joining those two events which has a stationary interval (4-dimensional "length"). *Stationary* here is used in the sense in which that term is used in the calculus of variations, namely, that the interval along the curve varies minimally among curves that are nearby to the geodesic.

In Minkowski space there is only one time-like geodesic that connects any given pair of time-like separated events, and that geodesic is the curve with the longest proper time between the two events. But in curved spacetime, it's possible for a pair of widely separated events to have more than one time-like geodesic that connects them. In such instances, the proper times along the various geodesics will not in general be the same. And for some geodesics in such instances, it's possible for a curve that connects the two events and is nearby to the geodesic to have either a longer or a shorter proper time than the geodesic.

For a space-like geodesic through two events, there are always nearby curves which go through the two events that have either a longer or a shorter proper length than the geodesic, even in Minkowski space. In Minkowski space, in an inertial frame of reference in which the two events are simultaneous, the geodesic will be the straight line between the two events at the time at which the events occur. Any curve that differs from the geodesic purely spatially (*i.e.* does not change the time coordinate) in that frame of reference will have a longer proper length than the geodesic, but a curve that differs from the geodesic purely temporally (*i.e.* does not change the space coordinate) in that frame of reference will have a shorter proper length.

The interval of a curve in spacetime is

$$l = \int \sqrt{\left| g_{\mu\nu} \dot{x}^\mu \dot{x}^\nu \right|} \, ds \, .$$

Then, the Euler–Lagrange equation,

$$\frac{d}{ds} \frac{\partial}{\partial \dot{x}^\alpha} \sqrt{\left| g_{\mu\nu} \dot{x}^\mu \dot{x}^\nu \right|} = \frac{\partial}{\partial x^\alpha} \sqrt{\left| g_{\mu\nu} \dot{x}^\mu \dot{x}^\nu \right|} \, ,$$

becomes, after some calculation,

$$2(\Gamma^\lambda{}_{\mu\nu} \dot{x}^\mu \dot{x}^\nu + \ddot{x}^\lambda) = U^\lambda \frac{d}{ds} \ln \left| U_\nu U^\nu \right| \, ,$$

where $U^\mu = \dot{x}^\mu$.

Einstein Field Equations

The Einstein field equations (EFE; also known as "Einstein's equations") are the set of 10 equations in Albert Einstein's general theory of relativity that describes the fundamental interaction of gravitation as a result of spacetime being curved by matter and energy. First published by Einstein in 1915 as a tensor equation, the EFE equate local spacetime curvature (expressed by the Einstein tensor) with the local energy and momentum within that spacetime (expressed by the stress–energy tensor).

Similar to the way that electromagnetic fields are determined using charges and currents via Maxwell's equations, the EFE are used to determine the spacetime geometry resulting from the presence of mass–energy and linear momentum, that is, they determine the metric tensor of spacetime for a given arrangement of stress–energy in the spacetime. The relationship between the metric tensor and the Einstein tensor allows the EFE to be written as a set of non-linear partial differential equations when used in this way. The solutions of the EFE are the components of the metric tensor. The inertial trajectories of particles and radiation (geodesics) in the resulting geometry are then calculated using the geodesic equation.

As well as obeying local energy–momentum conservation, the EFE reduce to Newton's law of gravitation where the gravitational field is weak and velocities are much less than the speed of light.

Exact solutions for the EFE can only be found under simplifying assumptions such as symmetry. Special classes of exact solutions are most often studied as they model many gravitational phenomena, such as rotating black holes and the expanding universe. Further simplification is achieved in approximating the actual spacetime as flat spacetime with a small deviation, leading to the linearised EFE. These equations are used to study phenomena such as gravitational waves.

Mathematical Form

The Einstein field equations (EFE) may be written in the form:

$$R_{\mu\nu} - \tfrac{1}{2} R g_{\mu\nu} + \Lambda g_{\mu\nu} = \frac{8\pi G}{c^4} T_{\mu\nu}$$

EFE on a wall in Leiden

where $R_{\mu\nu}$ is the Ricci curvature tensor, R is the scalar curvature, $g_{\mu\nu}$ is the metric tensor, Λ is the cosmological constant, G is Newton's gravitational constant, c is the speed of light in vacuum, and $T_{\mu\nu}$ is the stress–energy tensor.

The EFE is a tensor equation relating a set of symmetric 4×4 tensors. Each tensor has 10 independent components. The four Bianchi identities reduce the number of independent equations from 10 to 6, leaving the metric with four gauge fixing degrees of freedom, which correspond to the freedom to choose a coordinate system.

Although the Einstein field equations were initially formulated in the context of a four-dimensional theory, some theorists have explored their consequences in n dimensions. The equations in contexts outside of general relativity are still referred to as the Einstein field equations. The vacuum field equations (obtained when T is identically zero) define Einstein manifolds.

Despite the simple appearance of the equations they are actually quite complicated. Given a specified distribution of matter and energy in the form of a stress–energy tensor, the EFE are understood to be equations for the metric tensor $g_{\mu\nu}$, as both the Ricci tensor and scalar curvature depend on the metric in a complicated nonlinear manner. In fact, when fully written out, the EFE are a system of 10 coupled, nonlinear, hyperbolic-elliptic partial differential equations.

One can write the EFE in a more compact form by defining the Einstein tensor

$$G_{\mu\nu} = R_{\mu\nu} - \tfrac{1}{2} R g_{\mu\nu},$$

which is a symmetric second-rank tensor that is a function of the metric. The EFE can then be written as

$$G_{\mu\nu} + \Lambda g_{\mu\nu} = \frac{8\pi G}{c^4}T_{\mu\nu}.$$

Using geometrized units where $G = c = 1$, this can be rewritten as

$$G_{\mu\nu} + \Lambda g_{\mu\nu} = 8\pi T_{\mu\nu}.$$

The expression on the left represents the curvature of spacetime as determined by the metric; the expression on the right represents the matter/energy content of spacetime. The EFE can then be interpreted as a set of equations dictating how matter/energy determines the curvature of space-time.

These equations, together with the geodesic equation, which dictates how freely-falling matter moves through space-time, form the core of the mathematical formulation of general relativity.

Sign Convention

The above form of the EFE is the standard established by Misner, Thorne, and Wheeler. The authors analyzed all conventions that exist and classified according to the following three signs (S1, S2, S3):

$$g_{\mu\nu} = [S1] \times \text{diag}(-1, +1, +1, +1)$$

$$R^{\mu}{}_{\alpha\beta\gamma} = [S2] \times (\Gamma^{\mu}_{\alpha\gamma,\beta} - \Gamma^{\mu}_{\alpha\beta,\gamma} + \Gamma^{\mu}_{\sigma\beta}\Gamma^{\sigma}_{\gamma\alpha} - \Gamma^{\mu}_{\sigma\gamma}\Gamma^{\sigma}_{\beta\alpha})$$

$$G_{\mu\nu} = [S3] \times \frac{8\pi G}{c^4}T_{\mu\nu}$$

The third sign above is related to the choice of convention for the Ricci tensor:

$$R_{\mu\nu} = [S2] \times [S3] \times R^{\alpha}{}_{\mu\alpha\nu}$$

With these definitions Misner, Thorne, and Wheeler classify themselves as (+ + +), whereas Weinberg (1972) is (+ − −), Peebles (1980) and Efstathiou (1990) are (− + +), while Peacock (1994), Rindler (1977), Atwater (1974), Collins Martin & Squires (1989) are (− + −).

Authors including Einstein have used a different sign in their definition for the Ricci tensor which results in the sign of the constant on the right side being negative

$$R_{\mu\nu} - \frac{1}{2}R g_{\mu\nu} - \Lambda g_{\mu\nu} = -\frac{8\pi G}{c^4}T_{\mu\nu}.$$

The sign of the (very small) cosmological term would change in both these versions, if the (+ − − −) metric sign convention is used rather than the MTW (− + + +) metric sign convention adopted here.

Equivalent Formulations

Taking the trace with respect to the metric of both sides of the EFE one gets

$$R - \frac{D}{2}R + D\Lambda = \frac{8\pi G}{c^4}T$$

where D is the spacetime dimension. This expression can be rewritten as

$$-R + \frac{D\Lambda}{(\frac{D}{2}-1)} = \frac{8\pi G}{c^4}\frac{T}{\frac{D}{2}-1}.$$

If one adds $-1/2g_{\mu\nu}$ times this to the EFE, one gets the following equivalent "trace-reversed" form

$$R_{\mu\nu} - \frac{\Lambda g_{\mu\nu}}{\frac{D}{2}-1} = \frac{8\pi G}{c^4}\left(T_{\mu\nu} - \frac{1}{D-2}Tg_{\mu\nu}\right).$$

For example, in $D = 4$ dimensions this reduces to

$$R_{\mu\nu} - \Lambda g_{\mu\nu} = \frac{8\pi G}{c^4}\left(T_{\mu\nu} - \tfrac{1}{2}Tg_{\mu\nu}\right).$$

Reversing the trace again would restore the original EFE. The trace-reversed form may be more convenient in some cases (for example, when one is interested in weak-field limit and can replace $g_{\mu\nu}$ in the expression on the right with the Minkowski metric without significant loss of accuracy).

The Cosmological Constant

Einstein modified his original field equations to include a cosmological constant term Λ proportional to the metric

$$R_{\mu\nu} - \frac{1}{2}Rg_{\mu\nu} + \Lambda g_{\mu\nu} = \frac{8\pi G}{c^4}T_{\mu\nu}.$$

Since Λ is constant, the energy conservation law is unaffected.

The cosmological constant term was originally introduced by Einstein to allow for a universe that is not expanding or contracting. This effort was unsuccessful because:

- the universe described by this theory was unstable, and

- observations by Edwin Hubble confirmed that our universe is expanding.

So, Einstein abandoned Λ, calling it the "biggest blunder [he] ever made".

Despite Einstein's motivation for introducing the cosmological constant term, there is nothing inconsistent with the presence of such a term in the equations. For many years the cosmological constant was almost universally considered to be 0. However, recent improved astronomical techniques have found that a positive value of Λ is needed to explain the accelerating universe.

Einstein thought of the cosmological constant as an independent parameter, but its term in the field equation can also be moved algebraically to the other side, written as part of the stress–energy tensor:

$$T_{\mu\nu}^{(\text{vac})} = -\frac{\Lambda c^4}{8\pi G} g_{\mu\nu}.$$

The resulting vacuum energy is constant and given by

$$\rho_{\text{vac}} = \frac{\Lambda c^2}{8\pi G}$$

The existence of a cosmological constant is thus equivalent to the existence of a non-zero vacuum energy. Thus, the terms "cosmological constant" and "vacuum energy" are now used interchangeably in general relativity.

Features

Conservation of Energy and Momentum

General relativity is consistent with the local conservation of energy and momentum expressed as

$$\nabla_\beta T^{\alpha\beta} = T^{\alpha\beta}{}_{;\beta} = 0.$$

which expresses the local conservation of stress–energy. This conservation law is a physical requirement. With his field equations Einstein ensured that general relativity is consistent with this conservation condition.

Nonlinearity

The nonlinearity of the EFE distinguishes general relativity from many other fundamental physical theories. For example, Maxwell's equations of electromagnetism are linear in the electric and magnetic fields, and charge and current distributions (i.e. the sum of two solutions is also a solution); another example is Schrödinger's equation of quantum mechanics which is linear in the wavefunction.

Vacuum Field Equations

A Swiss commemorative coin from 1979, showing the vacuum field equations with zero cosmological constant (top).

If the energy-momentum tensor $T_{\mu\nu}$ is zero in the region under consideration, then the field equations are also referred to as the vacuum field equations. By setting $T_{\mu\nu} = 0$ in the trace-reversed field equations, the vacuum equations can be written as

$$R_{\mu\nu} = 0$$

In the case of nonzero cosmological constant, the equations are

$$R_{\mu\nu} = \frac{\Lambda}{\frac{D}{2} - 1} g_{\mu\nu}.$$

The solutions to the vacuum field equations are called vacuum solutions. Flat Minkowski space is the simplest example of a vacuum solution. Nontrivial examples include the Schwarzschild solution and the Kerr solution.

Manifolds with a vanishing Ricci tensor, $R_{\mu\nu} = 0$, are referred to as Ricci-flat manifolds and manifolds with a Ricci tensor proportional to the metric as Einstein manifolds.

Einstein–Maxwell Equations

If the energy-momentum tensor $T_{\mu\nu}$ is that of an electromagnetic field in free space, i.e. if the electromagnetic stress–energy tensor

$$T^{\alpha\beta} = -\frac{1}{\mu_0}\left(F^{\alpha\psi} F_{\psi}^{\ \beta} + \frac{1}{4} g^{\alpha\beta} F_{\psi\tau} F^{\psi\tau} \right)$$

is used, then the Einstein field equations are called the *Einstein–Maxwell equations* (with cosmological constant Λ, taken to be zero in conventional relativity theory):

$$R^{\alpha\beta} - \frac{1}{2} R g^{\alpha\beta} + \Lambda g^{\alpha\beta} = \frac{8\pi G}{c^4 \mu_0}\left(F^{\alpha\psi} F_{\psi}^{\ \beta} + \frac{1}{4} g^{\alpha\beta} F_{\psi\tau} F^{\psi\tau} \right).$$

Additionally, the covariant Maxwell Equations are also applicable in free space:

$$F^{\alpha\beta}_{\ \ ;\beta} = 0$$

$$F_{[\alpha\beta;\gamma]} = \frac{1}{3}\left(F_{\alpha\beta;\gamma} + F_{\beta\gamma;\alpha} + F_{\gamma\alpha;\beta} \right) = \frac{1}{3}\left(F_{\alpha\beta,\gamma} + F_{\beta\gamma,\alpha} + F_{\gamma\alpha,\beta} \right) = 0.$$

where the semicolon represents a covariant derivative, and the brackets denote anti-symmetrization. The first equation asserts that the 4-divergence of the two-form F is zero, and the second that its exterior derivative is zero. From the latter, it follows by the Poincaré lemma that in a coordinate chart it is possible to introduce an electromagnetic field potential A_α such that

$$F_{\alpha\beta} = A_{\alpha;\beta} - A_{\beta;\alpha} = A_{\alpha,\beta} - A_{\beta,\alpha}$$

in which the comma denotes a partial derivative. This is often taken as equivalent to the covariant Maxwell equation from which it is derived. However, there are global solutions of the equation which may lack a globally defined potential.

Solutions

The solutions of the Einstein field equations are metrics of spacetime. These metrics describe the structure of the spacetime including the inertial motion of objects in the spacetime. As the field equations are non-linear, they cannot always be completely solved (i.e. without making approximations). For example, there is no known complete solution for a spacetime with two massive bodies in it (which is a theoretical model of a binary star system, for example). However, approximations are usually made in these cases. These are commonly referred to as post-Newtonian approximations. Even so, there are numerous cases where the field equations have been solved completely, and those are called exact solutions.

The study of exact solutions of Einstein's field equations is one of the activities of cosmology. It leads to the prediction of black holes and to different models of evolution of the universe.

One can also discover new solutions of the Einstein field equations via the method of orthonormal frames as pioneered by Ellis and MacCallum. In this approach, the Einstein field equations are reduced to a set of coupled, nonlinear, ordinary differential equations. As discussed by Hsu and Wainwright, self-similar solutions to the Einstein field equations are fixed points of the resulting dynamical system. New solutions have been discovered using these methods by LeBlanc and Kohli and Haslam.

The Linearised EFE

The nonlinearity of the EFE makes finding exact solutions difficult. One way of solving the field equations is to make an approximation, namely, that far from the source(s) of gravitating matter, the gravitational field is very weak and the spacetime approximates that of Minkowski space. The metric is then written as the sum of the Minkowski metric and a term representing the deviation of the true metric from the Minkowski metric, with terms that are quadratic in or higher powers of the deviation being ignored. This linearisation procedure can be used to investigate the phenomena of gravitational radiation.

Polynomial Form

One might think that EFE are non-polynomial since they contain the inverse of the metric tensor. However, the equations can be arranged so that they contain only the metric tensor and not its inverse. First, the determinant of the metric in 4 dimensions can be written:

$$\det(g) = \frac{1}{24} \varepsilon^{\alpha\beta\gamma\delta} \varepsilon^{\kappa\lambda\mu\nu} g_{\alpha\kappa} g_{\beta\lambda} g_{\gamma\mu} g_{\delta\nu}$$

using the Levi-Civita symbol; and the inverse of the metric in 4 dimensions can be written as:

$$g^{\alpha\kappa} = \frac{1}{6} \varepsilon^{\alpha\beta\gamma\delta} \varepsilon^{\kappa\lambda\mu\nu} g_{\beta\lambda} g_{\gamma\mu} g_{\delta\nu} / \det(g).$$

Substituting this definition of the inverse of the metric into the equations then multiplying both sides by det(g) until there are none left in the denominator results in polynomial equations in the metric tensor and its first and second derivatives. The action from which the equations are derived can also be written in polynomial form by suitable redefinitions of the fields.

Solutions of the Einstein Field Equations

Solutions of the Einstein field equations are spacetimes that result from solving the Einstein field equations (EFE) of general relativity. Solving the field equations actually gives a Lorentz manifold. Solutions are broadly classed as *exact* or *non-exact*.

The Einstein field equations are

$$G_{ab} = \kappa T_{ab}$$

or more generally, if one allows a nonzero cosmological constant,

$$G_{ab} + \Lambda g_{ab} = \kappa T_{ab}$$

where κ is a constant, and the Einstein tensor on the left side of the equation is equated to the stress–energy tensor representing the energy and momentum present in the spacetime. The Einstein tensor is built up from the metric tensor and its partial derivatives; thus, the EFE are a system of ten partial differential equations to be solved for the metric.

Solving the Equations

It is important to realize that the Einstein field equations alone are not enough to determine the evolution of a gravitational system in many cases. They depend on the stress–energy tensor, which depends on the dynamics of matter and energy (such as trajectories of moving particles), which in turn depends on the gravitational field. If one is only interested in the weak field limit of the theory, the dynamics of matter can be computed using special relativity methods and/or Newtonian laws of gravity and then the resulting stress–energy tensor can be plugged into the Einstein field equations. But if the exact solution is required or a solution describing strong fields, the evolution of the metric and the stress–energy tensor must be solved for together.

To obtain solutions, the relevant equations are the above quoted EFE (in either form) plus the continuity equation (to determine evolution of the stress–energy tensor):

$$T^{ab}_{\;;b} = 0.$$

This is clearly not enough, as there are only 14 equations (10 from the field equations and 4 from the continuity equation) for 20 unknowns (10 metric components and 10 stress–energy tensor components). Equations of state are missing. In the most general case, it's easy to see that at least 6 more equations are required, possibly more if there are internal degrees of freedom (such as temperature) which may vary throughout space-time.

In practice, it is usually possible to simplify the problem by replacing the full set of equations of state with a simple approximation. Some common approximations are:

- Vacuum:

  $$T_{ab} = 0$$

- Perfect fluid:

  $$T_{ab} = (\rho + p)u_a u_b + p g_{ab} \text{ where } u^a u_a = -1$$

Here ρ is the mass-energy density measured in a momentary co-moving frame, u_a is the fluid's 4-velocity vector field, and p is the pressure.

- Non-interacting dust (a special case of perfect fluid):

  $$T_{ab} = \rho u_a u_b$$

For a perfect fluid, another equation of state relating density ρ and pressure p must be added. This equation will often depend on temperature, so a heat transfer equation is required or the postulate that heat transfer can be neglected.

Next, notice that only 10 of the original 14 equations are independent, because the continuity equation $T^{ab}{}_{;b} = 0$ is a consequence of Einstein's equations. This reflects the fact that the system is gauge invariant (in general, absent some symmetry, any choice of a curvilinear coordinate net on the same system would correspond to a numerically different solution.) A "gauge fixing" is needed, i.e. we need to impose 4 (arbitrary) constraints on the coordinate system in order to obtain unequivocal results. These constraints are known as coordinate conditions.

A popular choice of gauge is the so-called "De Donder gauge", also known as the harmonic condition or harmonic gauge

$$g^{\mu\nu}\Gamma^{\sigma}_{\mu\nu} = 0.$$

In numerical relativity, the preferred gauge is the so-called "3+1 decomposition", based on the ADM formalism. In this decomposition, metric is written in the form

$$ds^2 = (-N + N^i N^j \gamma_{ij})dt^2 + 2N^i \gamma_{ij} dt dx^j + \gamma_{ij} dx^i dx^j \text{, where } i, j = 1...3.$$

N and N^i are functions of spacetime coordinates and can be chosen arbitrarily in each point. The remaining physical degrees of freedom are contained in γ_{ij}, which represents the Riemannian metric on 3-hypersurfaces $t = const$. For example, a naive choice of $N = 1$, $N_i = 0$, would correspond to a so-called synchronous coordinate system: one where t-coordinate coincides with proper time for any comoving observer (particle that moves along a fixed x^i trajectory.)

Once equations of state are chosen and the gauge is fixed, the complete set of equations can be solved for. Unfortunately, even in the simplest case of gravitational field in the vacuum (vanishing stress–energy tensor), the problem turns out too complex to be exactly solvable. To get physical results, we can either turn to numerical methods; try to find exact solutions by imposing symmetries; or try middle-ground approaches such as perturbation methods or linear approximations of the Einstein tensor.

Exact Solutions

Exact solutions are Lorentz metrics that are conformable to a physically realistic stress–energy tensor and which are obtained by solving the EFE exactly in closed form.

Non-exact Solutions

Those solutions that are not exact are called *non-exact solutions*. Such solutions mainly arise due to the difficulty of solving the EFE in closed form and often take the form of approximations to ideal systems. Many non-exact solutions may be devoid of physical content, but serve as useful counterexamples to theoretical conjectures.

Applications

There are practical as well as theoretical reasons for studying solutions of the Einstein field equations.

From a purely mathematical viewpoint, it is interesting to know the set of solutions of the Einstein field equations. Some of these solutions are parametrised by one or more parameters.

Schwarzschild Metric

In Einstein's theory of general relativity, the Schwarzschild metric (also known as the Schwarzschild vacuum or Schwarzschild solution) is the solution to the Einstein field equations that describes the gravitational field outside a spherical mass, on the assumption that the electric charge of the mass, angular momentum of the mass, and universal cosmological constant are all zero. The solution is a useful approximation for describing slowly rotating astronomical objects such as many stars and planets, including Earth and the Sun. The solution is named after Karl Schwarzschild, who first published the solution in 1916.

According to Birkhoff's theorem, the Schwarzschild metric is the most general spherically symmetric, vacuum solution of the Einstein field equations. A Schwarzschild black hole or static black hole is a black hole that has no charge or angular momentum. A Schwarzschild black hole is described by the Schwarzschild metric, and cannot be distinguished from any other Schwarzschild black hole except by its mass.

The Schwarzschild black hole is characterized by a surrounding spherical boundary, called the event horizon, which is situated at the Schwarzschild radius, often called the radius of a black hole. The boundary is not a physical surface, and if a person fell through the event horizon (before being torn apart by tidal forces), they would not notice any physical surface at that position; it is a mathematical surface which is significant in determining the black hole's properties. Any non-rotating and non-charged mass that is smaller than its Schwarzschild radius forms a black hole.

The solution of the Einstein field equations is valid for any mass M, so in principle (according to general relativity theory) a Schwarzschild black hole of any mass could exist if conditions became sufficiently favorable to allow for its formation.

The Schwarzschild Metric

In Schwarzschild coordinates, with signature (1, –1, –1, –1), the line element for the Schwarzschild metric has the form

$$c^2 d\tau^2 = \left(1-\frac{r_s}{r}\right)c^2 dt^2 - \left(1-\frac{r_s}{r}\right)^{-1} dr^2 - r^2\left(d\theta^2 + \sin^2\theta d\varphi^2\right),$$

where

- when $d\tau^2$ is positive, τ is the proper time (time measured by a clock moving along the same world line with the test particle),

- c is the speed of light,

- t is the time coordinate (measured by a stationary clock located infinitely far from the massive body),

- r is the radial coordinate (measured as the circumference, divided by 2π, of a sphere centered around the massive body),

- θ is the colatitude (angle from north, in units of radians),

- φ is the longitude (also in radians), and

- r_s is the Schwarzschild radius of the massive body, a scale factor which is related to its mass M by $r_s = 2GM/c^2$, where G is the gravitational constant.

The analogue of this solution in classical Newtonian theory of gravity corresponds to the gravitational field around a point particle.

In practice, the ratio $r_{s/r}$ is almost always extremely small. For example, the Schwarzschild radius r_s of the Earth is roughly 8.9 mm, while the Sun, which is 3.3×10^5 times as massive has a Schwarzschild radius of approximately 3.0 km. Even at the surface of the Earth, the corrections to Newtonian gravity are only one part in a billion. The ratio only becomes large close to black holes and other ultra-dense objects such as neutron stars.

The Schwarzschild metric is a solution of Einstein's field equations in empty space, meaning that it is valid only *outside* the gravitating body. That is, for a spherical body of radius R the solution is valid for $r > R$. To describe the gravitational field both inside and outside the gravitating body the Schwarzschild solution must be matched with some suitable interior solution at $r = R$, such as the interior Schwarzschild solution.

History

The Schwarzschild solution is named in honor of Karl Schwarzschild, who found the exact solution in 1915 and published it in 1916, a little more than a month after the publication of Einstein's theory of general relativity. It was the first exact solution of the Einstein field equations other than the trivial flat space solution. Schwarzschild died shortly after his paper was published, as a result of a disease he contracted while serving in the German army during World War I.

Johannes Droste in 1916 independently produced the same solution as Schwarzschild, using a simpler, more direct derivation.

In the early years of general relativity there was a lot of confusion about the nature of the singularities found in the Schwarzschild and other solutions of the Einstein field equations. In Schwarzschild's original paper, he put what we now call the event horizon at the origin of his coordinate system. In this paper he also introduced what is now known as the Schwarzschild radial coordinate (r in the equations above), as an auxiliary variable. In his equations, Schwarzschild was using a different radial coordinate that was zero at the Schwarzschild radius.

A more complete analysis of the singularity structure was given by David Hilbert in the following year, identifying the singularities both at $r = 0$ and $r = r_s$. Although there was general consensus that the singularity at $r = 0$ was a 'genuine' physical singularity, the nature of the singularity at $r = r_s$ remained unclear.

In 1921 Paul Painlevé and in 1922 Allvar Gullstrand independently produced a metric, a spherically symmetric solution of Einstein's equations, which we now know is coordinate transformation of the Schwarzschild metric, Gullstrand–Painlevé coordinates, in which there was no singularity at $r = r_s$. They, however, did not recognize that their solutions were just coordinate transforms, and in fact used their solution to argue that Einstein's theory was wrong. In 1924 Arthur Eddington produced the first coordinate transformation (Eddington–Finkelstein coordinates) that showed that the singularity at $r = r_s$ was a coordinate artifact, although he also seems to have been unaware of the significance of this discovery. Later, in 1932, Georges Lemaître gave a different coordinate transformation (Lemaître coordinates) to the same effect and was the first to recognize that this implied that the singularity at $r = r_s$ was not physical. In 1939 Howard Robertson showed that a free falling observer descending in the Schwarzschild metric would cross the $r = r_s$ singularity in a finite amount of proper time even though this would take an infinite amount of time in terms of coordinate time t.

In 1950, John Synge produced a paper that showed the maximal analytic extension of the Schwarzschild metric, again showing that the singularity at $r = r_s$ was a coordinate artifact and that it represented two horizons. A similar result was later rediscovered by George Szekeres, and independently Martin Kruskal. The new coordinates nowadays known as Kruskal-Szekeres coordinates were much simpler than Synge's but both provided a single set of coordinates that covered the entire spacetime. However, perhaps due to the obscurity of the journals in which the papers of Lemaître and Synge were published their conclusions went unnoticed, with many of the major players in the field including Einstein believing that singularity at the Schwarzschild radius was physical.

Progress was only made in the 1960s when the more exact tools of differential geometry entered the field of general relativity, allowing more exact definitions of what it means for a Lorentzian manifold to be singular. This led to definitive identification of the $r = r_s$ singularity in the Schwarzschild metric as an event horizon (a hypersurface in spacetime that can only be crossed in one direction).

Singularities and Black Holes

The Schwarzschild solution appears to have singularities at $r = 0$ and $r = r_s$; some of the metric components "blow up" at these radii. Since the Schwarzschild metric is only expected to be valid for radii larger than the radius R of the gravitating body, there is no problem as long as $R > r_s$. For ordinary stars and planets this is always the case. For example, the radius of the Sun is approximately 700000 km, while its Schwarzschild radius is only 3 km.

The singularity at $r = r_s$ divides the Schwarzschild coordinates in two disconnected patches. The *exterior Schwarzschild solution* with $r > r_s$ is the one that is related to the gravitational fields of stars and planets. The *interior Schwarzschild solution* with $0 \leq r < r_s$, which contains the singularity at $r = 0$, is completely separated from the outer patch by the singularity at $r = r_s$. The Schwarzschild coordinates therefore give no physical connection between the two patches, which may be viewed as separate solutions. The singularity at $r = r_s$ is an illusion however; it is an instance of what is called a *coordinate singularity*. As the name implies, the singularity arises from a bad choice of coordinates or coordinate conditions. When changing to a different coordinate system (for example Lemaitre coordinates, Eddington–Finkelstein coordinates, Kruskal–Szekeres coordinates, Novikov coordinates, or Gullstrand–Painlevé coordinates) the metric becomes regular at $r = r_s$ and can extend the external patch to values of r smaller than r_s. Using a different coordinate transformation one can then relate the extended external patch to the inner patch.

The case $r = 0$ is different, however. If one asks that the solution be valid for all r one runs into a true physical singularity, or *gravitational singularity*, at the origin. To see that this is a true singularity one must look at quantities that are independent of the choice of coordinates. One such important quantity is the Kretschmann invariant, which is given by

$$R^{\alpha\beta\gamma\delta} R_{\alpha\beta\gamma\delta} = \frac{12r_s^2}{r^6} = \frac{48G^2M^2}{c^4 r^6}.$$

At $r = 0$ the curvature becomes infinite, indicating the presence of a singularity. At this point the metric, and spacetime itself, is no longer well-defined. For a long time it was thought that such a solution was non-physical. However, a greater understanding of general relativity led to the realization that such singularities were a generic feature of the theory and not just an exotic special case.

The Schwarzschild solution, taken to be valid for all $r > 0$, is called a Schwarzschild black hole. It is a perfectly valid solution of the Einstein field equations, although it has some rather bizarre properties. For $r < r_s$ the Schwarzschild radial coordinate r becomes timelike and the time coordinate t becomes spacelike. A curve at constant r is no longer a possible worldline of a particle or observer, not even if a force is exerted to try to keep it there; this occurs because spacetime has been curved so much that the direction of cause and effect (the particle's future light cone) points into the singularity. The surface $r = r_s$ demarcates what is called the *event horizon* of the black hole. It represents the point past which light can no longer escape the gravitational field. Any physical object whose radius R becomes less than or equal to the Schwarzschild radius will undergo gravitational collapse and become a black hole.

Alternative Coordinates

The Schwarzschild solution can be expressed in a range of different choices of coordinates besides the Schwarzschild coordinates used above. Different choices tend to highlight different features of the solution. The table below shows some popular choices.

Alternative coordinates			
Coordinates	Line element	Notes	Features
Eddington–Finkelstein coordinates (ingoing)	$\left(1-\dfrac{r_s}{r}\right)dv^2 - 2dvdr - r^2 d\Omega^2$		regular at horizon extends across future horizon
Eddington–Finkelstein coordinates (outgoing)	$\left(1-\dfrac{r_s}{r}\right)du^2 + 2du\,dr - r^2 d\Omega^2$		regular at horizon extends across past horizon
Gullstrand–Painlevé coordinates	$\left(1-\dfrac{r_s}{r}\right)dT^2 - 2\sqrt{\dfrac{r_s}{r}}\,dT\,dr - dr^2 - r^2 d\Omega^2$		regular at horizon
Isotropic coordinates	$\dfrac{\left(1-\dfrac{r_s}{4R}\right)^2}{\left(1+\dfrac{r_s}{4R}\right)^2}dt^2 - \left(1+\dfrac{r_s}{4R}\right)^4\left(dx^2 + dy^2 + dz^2\right)$	$R = \sqrt{x^2 + y^2 + z^2}$	isotropic lightcones on constant time slices
Kruskal–Szekeres coordinates	$\dfrac{4r_s^3}{r}e^{-\frac{r}{r_s}}\left(dT^2 - dR^2\right) - r^2 d\Omega^2$	$T^2 - R^2 = \left(1-\dfrac{r}{r_s}\right)e^{\frac{r}{r_s}}$	regular at horizon Maximally extends to full spacetime
Lemaître coordinates	$dT^2 - \dfrac{r_s}{r}dR^2 - r^2 d\Omega^2$	$r = \left(\tfrac{3}{2}(R-T)\right)^{\frac{2}{3}} r_s^{\frac{1}{3}}$	regular at horizon

In table above, some shorthand has been introduced for brevity. The speed of light c has been set to one. The notation

$$d\Omega^2 = d\theta^2 + \sin(\theta)^2 d\varphi^2$$

is used for the metric of a two dimensional sphere. Moreover, in each entry R and T denote alternative choices of radial and time coordinate for the particular coordinates. Note, the R and/or T may vary from entry to entry.

Flamm's Paraboloid

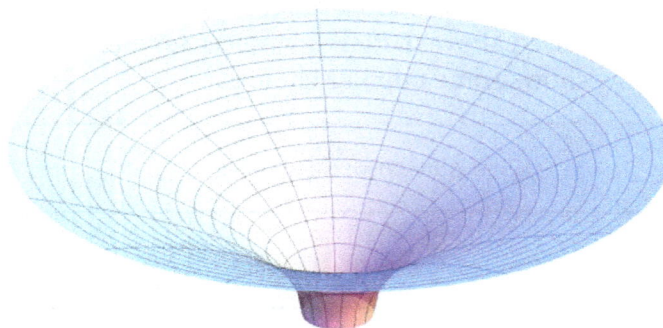

A plot of Flamm's paraboloid. It should not be confused with the unrelated concept of a gravity well.

The spatial curvature of the Schwarzschild solution for $r > r_s$ can be visualized as the graphic shows. Consider a constant time equatorial slice through the Schwarzschild solution ($\theta = \pi/2$, t = constant) and let the position of a particle moving in this plane be described with the remaining Schwarzschild coordinates (r, φ). Imagine now that there is an additional Euclidean dimension w, which has no physical reality (it is not part of spacetime). Then replace the (r, φ) plane with a surface dimpled in the w direction according to the equation (*Flamm's paraboloid*)

$$w = 2\sqrt{r_s \left(r - r_s\right)}.$$

This surface has the property that distances measured within it match distances in the Schwarzschild metric, because with the definition of w above,

$$dw^2 + dr^2 + r^2 d\varphi^2 = -c^2 d\tau^2 = \frac{dr^2}{1 - \dfrac{r_s}{r}} + r^2 d\varphi^2$$

Thus, Flamm's paraboloid is useful for visualizing the spatial curvature of the Schwarzschild metric. It should not, however, be confused with a gravity well. No ordinary (massive or massless) particle can have a worldline lying on the paraboloid, since all distances on it are spacelike (this is a cross-section at one moment of time, so any particle moving on it would have an infinite velocity). Even a tachyon would not move along the path that one might naively expect from a "rubber sheet" analogy: in particular, if the dimple is drawn pointing upward rather than downward, the tachyon's path still curves toward the central mass, not away.

Flamm's paraboloid may be derived as follows. The Euclidean metric in the cylindrical coordinates (r, φ, w) is written

$$ds^2 = dw^2 + dr^2 + r^2 d\varphi^2.$$

Letting the surface be described by the function $w = w(r)$, the Euclidean metric can be written as

$$ds^2 = \left(1 + \left(\frac{dw}{dr}\right)^2\right) dr^2 + r^2 d\varphi^2,$$

Comparing this with the Schwarzschild metric in the equatorial plane ($\theta = \pi/2$) at a fixed time (t = constant, $dt = 0$)

$$ds^2 = \left(1 - \frac{r_s}{r}\right)^{-1} dr^2 + r^2 d\varphi^2,$$

yields an integral expression for $w(r)$:

$$w(r) = \int \frac{dr}{\sqrt{\frac{r}{r_s} - 1}} = 2r_s \sqrt{\frac{r}{r_s} - 1} + \text{constant}$$

whose solution is Flamm's paraboloid.

Orbital Motion

A particle orbiting in the Schwarzschild metric can have a stable circular orbit with $r > 3r_s$. Circular orbits with r between $3/2r_s$ and $3r_s$ are unstable, and no circular orbits exist for $r < 3/2r_s$. The circular orbit of minimum radius $3/2r_s$ corresponds to an orbital velocity approaching the speed of light. It is possible for a particle to have a constant value of r between r_s and $3/2r_s$, but only if some force acts to keep it there.

Noncircular orbits, such as Mercury's, dwell longer at small radii than would be expected classically. This can be seen as a less extreme version of the more dramatic case in which a particle passes through the event horizon and dwells inside it forever. Intermediate between the case of Mercury and the case of an object falling past the event horizon, there are exotic possibilities such as knife-edge orbits, in which the satellite can be made to execute an arbitrarily large number of nearly circular orbits, after which it flies back outward.

Symmetries

The group of isometries of the Schwarzschild metric is the subgroup of the ten-dimensional Poincaré group which takes the time axis (trajectory of the star) to itself. It omits the spatial translations (three dimensions) and boosts (three dimensions). It retains the time translations (one dimension) and rotations (three dimensions). Thus it has four dimensions. Like the Poincaré group, it has four connected components: the component of the identity; the time reversed component; the spatial inversion component; and the component which is both time reversed and spatially inverted.

Quotes

"Es ist immer angenehm, über strenge Lösungen einfacher Form zu verfügen." (It is always pleasant to have exact solutions in simple form at your disposal.) – Karl Schwarzschild, 1916.

Curvatures

The Ricci curvature scalar and the Ricci curvature tensor are both zero. The non-zero components of the Riemann tensor are

$$R^t_{\;rrt} = 2R^{\theta}_{\;r\theta r} = 2R^{\phi}_{\;r\phi r} = \frac{r_s}{r^2(r_s - r)},$$

$$2R^t_{\;\theta\theta t} = 2R^r_{\;\theta\theta r} = R^{\phi}_{\;\theta\phi\theta} = \frac{r_s}{r},$$

$$2R^t_{\;\phi\phi t} = 2R^r_{\;\phi\phi r} = -R^{\theta}_{\;\phi\phi\theta} = \frac{r_s \sin^2(\theta)}{r},$$

$$R^r_{\;trt} = -2R^{\theta}_{\;t\theta t} = -2R^{\phi}_{\;t\phi t} = c^2\frac{r_s(r_s - r)}{r^4}$$

References

- Misner, Charles; Thorne, Kip S.; Wheeler, John Archibald (1973). Gravitation. San Francisco: W. H. Freeman. ISBN 0-7167-0344-0.

- Landau, L. D.; Lifshitz, E. M. (1975). Classical Theory of Fields (Fourth Revised English Edition). Oxford: Pergamon. ISBN 0-08-018176-7.

- R. P. Feynman; F. B. Moringo; W. G. Wagner (1995). Feynman Lectures on Gravitation. Addison-Wesley. ISBN 0-201-62734-5.

- Danielson, Donald A. (2003). Vectors and Tensors in Engineering and Physics (2/e ed.). Westview (Perseus). ISBN 978-0-8133-4080-7.

- Dimitrienko, Yuriy (2002). Tensor Analysis and Nonlinear Tensor Functions. Kluwer Academic Publishers (Springer). ISBN 1-4020-1015-X.

- Jeevanjee, Nadir (2011). An Introduction to Tensors and Group Theory for Physicists. Birkhauser. ISBN 978-0-8176-4714-8.

- Lawden, D. F. (2003). Introduction to Tensor Calculus, Relativity and Cosmology (3/e ed.). Dover. ISBN 978-0-486-42540-5.

- Lovelock, David; Hanno Rund (1989) [1975]. Tensors, Differential Forms, and Variational Principles. Dover. ISBN 978-0-486-65840-7.

- Kline, Morris (1972). Mathematical thought from ancient to modern times, Vol. 3. Oxford University Press. pp. 1122–1127. ISBN 0195061373.

Special Relativity: An Overview

The universally accepted relational theory between space and time is known as special relativity. The term was coined by Albert Einstein. Some of the aspects of special relativity explained in the following chapter are Lorentz transformation, ladder paradox, time dilation, length contraction, mass-energy equivalence and classical electromagnetism and special relativity.

Special Relativity

In physics, special relativity (SR, also known as the special theory of relativity or STR) is the generally accepted and experimentally well-confirmed physical theory regarding the relationship between space and time. In Albert Einstein's original pedagogical treatment, it is based on two postulates:

1. The laws of physics are invariant (i.e. identical) in all inertial systems (non-accelerating frames of reference).

2. The speed of light in a vacuum is the same for all observers, regardless of the motion of the light source.

It was originally proposed in 1905 by Albert Einstein in the paper "On the Electrodynamics of Moving Bodies". The inconsistency of Newtonian mechanics with Maxwell's equations of electromagnetism and the lack of experimental confirmation for a hypothesized luminiferous aether led to the development of special relativity, which corrects mechanics to handle situations involving motions nearing the speed of light. As of today, special relativity is the most accurate model of motion at any speed. Even so, the Newtonian mechanics model is still useful (due to its simplicity and high accuracy) as an approximation at small velocities relative to the speed of light.

Special relativity implies a wide range of consequences, which have been experimentally verified, including length contraction, time dilation, relativistic mass, mass–energy equivalence, a universal speed limit and relativity of simultaneity. It has replaced the conventional notion of an absolute universal time with the notion of a time that is dependent on reference frame and spatial position. Rather than an invariant time interval between two events, there is an invariant spacetime interval. Combined with other laws of physics, the two postulates of special relativity predict the equivalence of mass and energy, as expressed in the mass–energy equivalence formula $E = mc^2$, where c is the speed of light in a vacuum.

A defining feature of special relativity is the replacement of the Galilean transformations of Newtonian mechanics with the Lorentz transformations. Time and space cannot be defined separately from each other. Rather space and time are interwoven into a single continuum known as space-time. Events that occur at the same time for one observer can occur at different times for another.

The theory is "special" in that it only applies in the special case where the curvature of spacetime due to gravity is negligible. In order to include gravity, Einstein formulated general relativity in 1915. Special relativity, contrary to some outdated descriptions, is capable of handling accelerated frames of reference.

As Galilean relativity is now considered an approximation of special relativity that is valid for low speeds, special relativity is considered an approximation of general relativity that is valid for weak gravitational fields, i.e. at a sufficiently small scale and in conditions of free fall. Whereas general relativity incorporates noneuclidean geometry in order to represent gravitational effects as the geometric curvature of spacetime, special relativity is restricted to the flat spacetime known as Minkowski space. A locally Lorentz-invariant frame that abides by special relativity can be defined at sufficiently small scales, even in curved spacetime.

Galileo Galilei had already postulated that there is no absolute and well-defined state of rest (no privileged reference frames), a principle now called Galileo's principle of relativity. Einstein extended this principle so that it accounted for the constant speed of light, a phenomenon that had been recently observed in the Michelson–Morley experiment. He also postulated that it holds for all the laws of physics, including both the laws of mechanics and of electrodynamics.

Albert Einstein around 1905, the year his "*Annus Mirabilis* papers" – which included *Zur Elektrodynamik bewegter Körper*, the paper founding special relativity – were published.

Postulates

66 Reflections of this type made it clear to me as long ago as shortly after 1900, i.e., shortly after Planck's trailblazing work, that neither mechanics nor electrodynamics could (except in limiting cases) claim exact validity. Gradually I despaired of the possibility of discovering the true laws by means of constructive efforts based on known facts. The longer and the more desperately I tried, the more I came to the conviction that only the discovery of a universal formal principle could lead us to assured results... How, then, could such a universal principle be found? 99

— Albert Einstein

Einstein discerned two fundamental propositions that seemed to be the most assured, regardless of the exact validity of the (then) known laws of either mechanics or electrodynamics. These propositions were the constancy of the speed of light and the independence of physical laws (especially the constancy of the speed of light) from the choice of inertial system. In his initial presentation of special relativity in 1905 he expressed these postulates as:

The Principle of Relativity – The laws by which the states of physical systems undergo change are not affected, whether these changes of state be referred to the one or the other of two systems in uniform translatory motion relative to each other.

The Principle of Invariant Light Speed – "... light is always propagated in empty space with a definite velocity [speed] c which is independent of the state of motion of the emitting body" (from the preface). That is, light in vacuum propagates with the speed c (a fixed constant, independent of direction) in at least one system of inertial coordinates (the "stationary system"), regardless of the state of motion of the light source.

The derivation of special relativity depends not only on these two explicit postulates, but also on several tacit assumptions (made in almost all theories of physics), including the isotropy and homogeneity of space and the independence of measuring rods and clocks from their past history.

Following Einstein's original presentation of special relativity in 1905, many different sets of postulates have been proposed in various alternative derivations. However, the most common set of postulates remains those employed by Einstein in his original paper. A more mathematical statement of the Principle of Relativity made later by Einstein, which introduces the concept of simplicity not mentioned above is:

Special principle of relativity: If a system of coordinates K is chosen so that, in relation to it, physical laws hold good in their simplest form, the *same* laws hold good in relation to any other system of coordinates K' moving in uniform translation relatively to K.

Henri Poincaré provided the mathematical framework for relativity theory by proving that Lorentz transformations are a subset of his Poincaré group of symmetry transformations. Einstein later derived these transformations from his axioms.

Many of Einstein's papers present derivations of the Lorentz transformation based upon these two principles.

Einstein consistently based the derivation of Lorentz invariance (the essential core of special relativity) on just the two basic principles of relativity and light-speed invariance. He wrote:

The insight fundamental for the special theory of relativity is this: The assumptions relativity and light speed invariance are compatible if relations of a new type ("Lorentz transformation") are postulated for the conversion of coordinates and times of events... The universal principle of the special theory of relativity is contained in the postulate: The laws of physics are invariant with respect to Lorentz transformations (for the transition from one inertial system to any other arbitrarily chosen inertial system). This is a restricting principle for natural laws...

Thus many modern treatments of special relativity base it on the single postulate of universal Lorentz covariance, or, equivalently, on the single postulate of Minkowski spacetime.

From the principle of relativity alone without assuming the constancy of the speed of light (i.e. using the isotropy of space and the symmetry implied by the principle of special relativity) one can show that the spacetime transformations between inertial frames are either Euclidean, Galilean, or Lorentzian. In the Lorentzian case, one can then obtain relativistic interval conservation and a certain finite limiting speed. Experiments suggest that this speed is the speed of light in vacuum.

The constancy of the speed of light was motivated by Maxwell's theory of electromagnetism and the lack of evidence for the luminiferous ether. There is conflicting evidence on the extent to which Einstein was influenced by the null result of the Michelson–Morley experiment. In any case, the null result of the Michelson–Morley experiment helped the notion of the constancy of the speed of light gain widespread and rapid acceptance.

Lack of an Absolute Reference Frame

The principle of relativity, which states that there is no preferred inertial reference frame, dates back to Galileo, and was incorporated into Newtonian physics. However, in the late 19th century, the existence of electromagnetic waves led physicists to suggest that the universe was filled with a substance that they called "aether", which would act as the medium through which these waves, or vibrations travelled. The aether was thought to constitute an absolute reference frame against which speeds could be measured, and could be considered fixed and motionless. Aether supposedly possessed some wonderful properties: it was sufficiently elastic to support electromagnetic waves, and those waves could interact with matter, yet it offered no resistance to bodies passing through it. The results of various experiments, including the Michelson–Morley experiment, led to the theory of special relativity, by showing that there was no aether. Einstein's solution was to discard the notion of an aether and the absolute state of rest. In relativity, any reference frame moving with uniform motion will observe the same laws of physics. In particular, the speed of light in vacuum is always measured to be c, even when measured by multiple systems that are moving at different (but constant) velocities.

Reference Frames, Coordinates, and the Lorentz Transformation

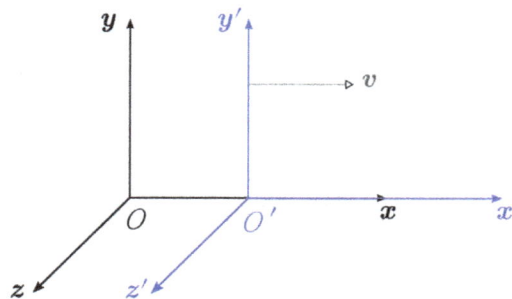

The primed system is in motion relative to the unprimed system with constant velocity v only along the x-axis, from the perspective of an observer stationary in the unprimed system. By the principle of relativity, an observer stationary in the primed system will view a likewise construction except that the velocity they record will be –v. The changing of the speed of propagation of interaction from infinite in non-relativistic mechanics to a finite value will require a modification of the transformation equations mapping events in one frame to another.

Reference frames play a crucial role in relativity theory. The term reference frame as used here is an observational perspective in space which is not undergoing any change in motion (acceleration), from which a position can be measured along 3 spatial axes. In addition, a reference frame has the ability to determine measurements of the time of events using a 'clock' (any reference device with uniform periodicity).

An event is an occurrence that can be assigned a single unique time and location in space relative to a reference frame: it is a "point" in spacetime. Since the speed of light is constant in relativity in each and every reference frame, pulses of light can be used to unambiguously measure distances and refer back the times that events occurred to the clock, even though light takes time to reach the clock after the event has transpired.

For example, the explosion of a firecracker may be considered to be an "event". We can completely specify an event by its four spacetime coordinates: The time of occurrence and its 3-dimensional spatial location define a reference point. Let's call this reference frame S.

In relativity theory we often want to calculate the position of a point from a different reference point.

Suppose we have a second reference frame S', whose spatial axes and clock exactly coincide with that of S at time zero, but it is moving at a constant velocity v with respect to S along the x-axis.

Since there is no absolute reference frame in relativity theory, a concept of 'moving' doesn't strictly exist, as everything is always moving with respect to some other reference frame. Instead, any two frames that move at the same speed in the same direction are said to be *comoving*. Therefore, S and S' are not *comoving*.

Define the event to have spacetime coordinates (t,x,y,z) in system S and (t',x',y',z') in S'. Then the Lorentz transformation specifies that these coordinates are related in the following way:

$$
\begin{aligned}
t' &= \gamma\,(t - vx/c^2)\\
x' &= \gamma\,(x - vt)\\
y' &= y\\
z' &= z,
\end{aligned}
$$

where

$$
\gamma = \frac{1}{\sqrt{1 - \dfrac{v^2}{c^2}}}
$$

is the Lorentz factor and c is the speed of light in vacuum, and the velocity v of S' is parallel to the x-axis. The y and z coordinates are unaffected; only the x and t coordinates are transformed. These Lorentz transformations form a one-parameter group of linear mappings, that parameter being called rapidity.

There is nothing special about the x-axis, the transformation can apply to the y or z axes, or indeed in any direction, which can be done by directions parallel to the motion (which are warped by the γ factor) and perpendicular.

A quantity invariant under Lorentz transformations is known as a Lorentz scalar.

Writing the Lorentz transformation and its inverse in terms of coordinate differences, where for instance one event has coordinates (x_1, t_1) and (x'_1, t'_1), another event has coordinates (x_2, t_2) and (x'_2, t'_2), and the differences are defined as

$$\Delta x' = x'_2 - x'_1, \quad \Delta x = x_2 - x_1,$$
$$\Delta t' = t'_2 - t'_1, \quad \Delta t = t_2 - t_1,$$

we get

$$\Delta x' = \gamma \left(\Delta x - v\Delta t \right), \quad \Delta x = \gamma \left(\Delta x' + v\Delta t' \right),$$
$$\Delta t' = \gamma \left(\Delta t - \frac{v\Delta x}{c^2} \right), \quad \Delta t = \gamma \left(\Delta t' + \frac{v\Delta x'}{c^2} \right).$$

These effects are not merely appearances; they are explicitly related to our way of measuring *time intervals* between events which occur at the same place in a given coordinate system (called "co-local" events). These time intervals will be *different* in another coordinate system moving with respect to the first, unless the events are also simultaneous. Similarly, these effects also relate to our measured distances between separated but simultaneous events in a given coordinate system of choice. If these events are not co-local, but are separated by distance (space), they will *not* occur at the same *spatial distance* from each other when seen from another moving coordinate system. However, the spacetime interval will be the same for all observers.

Consequences Derived from the Lorentz Transformation

The consequences of special relativity can be derived from the Lorentz transformation equations. These transformations, and hence special relativity, lead to different physical predictions than those of Newtonian mechanics when relative velocities become comparable to the speed of light. The speed of light is so much larger than anything humans encounter that some of the effects predicted by relativity are initially counterintuitive.

Relativity of Simultaneity

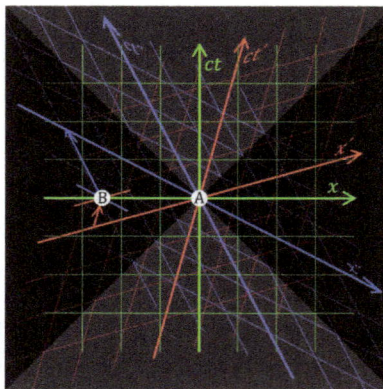

Event B is simultaneous with A in the green reference frame, but it occurs before A in the blue frame, and occurs after A in the red frame.

Two events happening in two different locations that occur simultaneously in the reference frame of one inertial observer, may occur non-simultaneously in the reference frame of another inertial observer (lack of absolute simultaneity).

From the first equation of the Lorentz transformation in terms of coordinate differences

$$\Delta t' = \gamma \left(\Delta t - \frac{v \Delta x}{c^2} \right)$$

it is clear that two events that are simultaneous in frame S (satisfying $\Delta t = 0$), are not necessarily simultaneous in another inertial frame S' (satisfying $\Delta t' = 0$). Only if these events are additionally co-local in frame S (satisfying $\Delta x = 0$), will they be simultaneous in another frame S'.

Time Dilation

The time lapse between two events is not invariant from one observer to another, but is dependent on the relative speeds of the observers' reference frames (e.g., the twin paradox which concerns a twin who flies off in a spaceship traveling near the speed of light and returns to discover that his or her twin sibling has aged much more).

Suppose a clock is at rest in the unprimed system S. The location of the clock on two different ticks is then characterized by $\Delta x = 0$. To find the relation between the times between these ticks as measured in both systems, the first equation can be used to find:

$$\Delta t' = \gamma \Delta t \quad \text{for events satisfying} \quad \Delta x = 0.$$

This shows that the time ($\Delta t'$) between the two ticks as seen in the frame in which the clock is moving (S'), is *longer* than the time (Δt) between these ticks as measured in the rest frame of the clock (S). Time dilation explains a number of physical phenomena; for example, the lifetime of muons produced by cosmic rays impinging on the Earth's atmosphere is measured to be greater than the lifetimes of muons measured in the laboratory.

Length Contraction

The dimensions (e.g., length) of an object as measured by one observer may be smaller than the results of measurements of the same object made by another observer (e.g., the ladder paradox involves a long ladder traveling near the speed of light and being contained within a smaller garage).

Similarly, suppose a measuring rod is at rest and aligned along the x-axis in the unprimed system S. In this system, the length of this rod is written as Δx. To measure the length of this rod in the system S', in which the rod is moving, the distances x' to the end points of the rod must be measured simultaneously in that system S'. In other words, the measurement is characterized by $\Delta t' = 0$, which can be combined with the fourth equation to find the relation between the lengths Δx and $\Delta x'$:

$$\Delta x' = \frac{\Delta x}{\gamma} \quad \text{for events satisfying} \quad \Delta t' = 0.$$

This shows that the length ($\Delta x'$) of the rod as measured in the frame in which it is moving (S'), is *shorter* than its length (Δx) in its own rest frame (S).

Composition of Velocities

Velocities (speeds) do not simply add. If the observer in S measures an object moving along the x axis at velocity u, then the observer in the S' system, a frame of reference moving at velocity v in the x direction with respect to S, will measure the object moving with velocity u' where (from the Lorentz transformations above):

$$u' = \frac{dx'}{dt'} = \frac{\gamma(dx - vdt)}{\gamma(dt - vdx/c^2)} = \frac{(dx/dt) - v}{1 - (v/c^2)(dx/dt)} = \frac{u - v}{1 - uv/c^2}.$$

The other frame S will measure:

$$u = \frac{dx}{dt} = \frac{\gamma(dx' + vdt')}{\gamma(dt' + vdx'/c^2)} = \frac{(dx'/dt') + v}{1 + (v/c^2)(dx'/dt')} = \frac{u' + v}{1 + u'v/c^2}.$$

Notice that if the object were moving at the speed of light in the S system (i.e. $u = c$), then it would also be moving at the speed of light in the S' system. Also, if both u and v are small with respect to the speed of light, we will recover the intuitive Galilean transformation of velocities

$$u' \approx u - v.$$

The usual example given is that of a train (frame S' above) traveling due east with a velocity v with respect to the tracks (frame S). A child inside the train throws a baseball due east with a velocity u' with respect to the train. In nonrelativistic physics, an observer at rest on the tracks will measure the velocity of the baseball (due east) as $u = u' + v$, while in special relativity this is no longer true; instead the velocity of the baseball (due east) is given by the second equation: $u = (u' + v)/(1 + u'v/c^2)$. Again, there is nothing special about the x or east directions. This formalism applies to any direction by considering parallel and perpendicular components of motion to the direction of relative velocity v.

Other Consequences

Thomas Rotation

The orientation of an object (i.e. the alignment of its axes with the observer's axes) may be different for different observers. Unlike other relativistic effects, this effect becomes quite significant at fairly low velocities as can be seen in the spin of moving particles.

Equivalence of Mass and Energy

As an object's speed approaches the speed of light from an observer's point of view, its relativistic mass increases thereby making it more and more difficult to accelerate it from within the observer's frame of reference.

The energy content of an object at rest with mass m equals mc^2. Conservation of energy implies that, in any reaction, a decrease of the sum of the masses of particles must be accompanied by an increase in kinetic energies of the particles after the reaction. Similarly, the mass of an object can be increased by taking in kinetic energies.

In addition to the papers referenced above—which give derivations of the Lorentz transformation and describe the foundations of special relativity—Einstein also wrote at least four papers giving heuristic arguments for the equivalence (and transmutability) of mass and energy, for $E = mc^2$.

Mass–energy equivalence is a consequence of special relativity. The energy and momentum, which are separate in Newtonian mechanics, form a four-vector in relativity, and this relates the time component (the energy) to the space components (the momentum) in a non-trivial way. For an object at rest, the energy–momentum four-vector is $(E/c, 0, 0, 0)$: it has a time component which is the energy, and three space components which are zero. By changing frames with a Lorentz transformation in the x direction with a small value of the velocity v, the energy momentum four-vector becomes $(E/c, Ev/c^2, 0, 0)$. The momentum is equal to the energy multiplied by the velocity divided by c^2. As such, the Newtonian mass of an object, which is the ratio of the momentum to the velocity for slow velocities, is equal to E/c^2.

The energy and momentum are properties of matter and radiation, and it is impossible to deduce that they form a four-vector just from the two basic postulates of special relativity by themselves, because these don't talk about matter or radiation, they only talk about space and time. The derivation therefore requires some additional physical reasoning. In his 1905 paper, Einstein used the additional principles that Newtonian mechanics should hold for slow velocities, so that there is one energy scalar and one three-vector momentum at slow velocities, and that the conservation law for energy and momentum is exactly true in relativity. Furthermore, he assumed that the energy of light is transformed by the same Doppler-shift factor as its frequency, which he had previously shown to be true based on Maxwell's equations. The first of Einstein's papers on this subject was "Does the Inertia of a Body Depend upon its Energy Content?" in 1905. Although Einstein's argument in this paper is nearly universally accepted by physicists as correct, even self-evident, many authors over the years have suggested that it is wrong. Other authors suggest that the argument was merely inconclusive because it relied on some implicit assumptions.

Einstein acknowledged the controversy over his derivation in his 1907 survey paper on special relativity. There he notes that it is problematic to rely on Maxwell's equations for the heuristic mass–energy argument. The argument in his 1905 paper can be carried out with the emission of any massless particles, but the Maxwell equations are implicitly used to make it obvious that the emission of light in particular can be achieved only by doing work. To emit electromagnetic waves, all you have to do is shake a charged particle, and this is clearly doing work, so that the emission is of energy.

How far can one Travel from the Earth?

Since one can not travel faster than light, one might conclude that a human can never travel farther from Earth than 40 light years if the traveller is active between the ages of 20 and 60. One would easily think that a traveller would never be able to reach more than the very few solar systems which exist within the limit of 20–40 light years from the earth. But that would be a mistaken conclusion. Because of time dilation, a hypothetical spaceship can travel thousands of light years during the pilot's 40 active years. If a spaceship could be built that accelerates at a constant 1 g, it will, after a little less than a year, be travelling at almost the speed of light as seen from Earth. This is described by:

$$v(t) = \frac{at}{\sqrt{1 + \frac{a^2 t^2}{c^2}}}$$

where v(t) is the velocity at a time, t, a is the acceleration of 1g and t is the time as measured by people on Earth. Therefore, after 1 year of accelerating at 9.81 m/s², the spaceship will be travelling at v = 0.77c relative to Earth. Time dilation will increase the travellers life span as seen from the reference frame of the Earth to 2.7 years, but his lifespan measured by a clock travelling with him will not change. During his journey, people on Earth will experience more time than he does. A 5-year round trip for him will take 6½ Earth years and cover a distance of over 6 light-years. A 20-year round trip for him (5 years accelerating, 5 decelerating, twice each) will land him back on Earth having travelled for 335 Earth years and a distance of 331 light years. A full 40-year trip at 1 *g* will appear on Earth to last 58,000 years and cover a distance of 55,000 light years. A 40-year trip at 1.1 *g* will take 148,000 Earth years and cover about 140,000 light years. A one-way 28 year (14 years accelerating, 14 decelerating as measured with the cosmonaut's clock) trip at 1 *g* acceleration could reach 2,000,000 light-years to the Andromeda Galaxy. This same time dilation is why a muon travelling close to *c* is observed to travel much further than *c* times its half-life (when at rest).

Causality and Prohibition of Motion Faster than Light

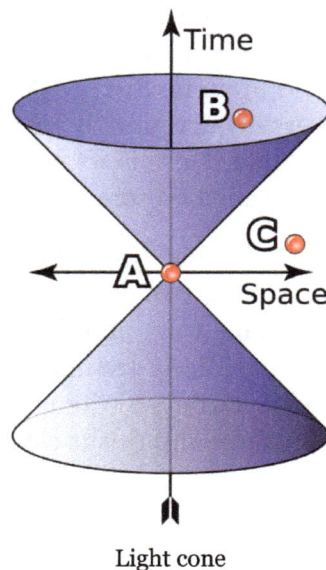

Light cone

In diagram 2 the interval AB is 'time-like'; i.e., there is a frame of reference in which events A and B occur at the same location in space, separated only by occurring at different times. If A precedes B in that frame, then A precedes B in all frames. It is hypothetically possible for matter (or information) to travel from A to B, so there can be a causal relationship (with A the cause and B the effect).

The interval AC in the diagram is 'space-like'; i.e., there is a frame of reference in which events A and C occur simultaneously, separated only in space. There are also frames in which A precedes C (as shown) and frames in which C precedes A. If it were possible for a cause-and-effect relationship to exist between events A and C, then paradoxes of causality would result. For example, if A was the cause, and C the effect, then there would be frames of reference in which the effect preceded the cause. Although this in itself won't give rise to a paradox, one can show that faster than light signals can be sent back into one's own past. A causal paradox can then be constructed by sending the signal if and only if no signal was received previously.

Therefore, if causality is to be preserved, one of the consequences of special relativity is that no information signal or material object can travel faster than light in vacuum. However, some "things" can still move faster than light. For example, the location where the beam of a search light hits the bottom of a cloud can move faster than light when the search light is turned rapidly.

Even without considerations of causality, there are other strong reasons why faster-than-light travel is forbidden by special relativity. For example, if a constant force is applied to an object for a limitless amount of time, then integrating $F = dp/dt$ gives a momentum that grows without bound, but this is simply because $p = m\gamma v$ approaches infinity as v approaches c. To an observer who is not accelerating, it appears as though the object's inertia is increasing, so as to produce a smaller acceleration in response to the same force. This behavior is observed in particle accelerators, where each charged particle is accelerated by the electromagnetic force.

Geometry of Spacetime

Comparison between Flat Euclidean Space and Minkowski Space

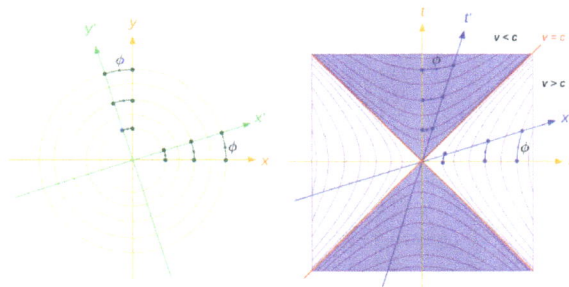

Orthogonality and rotation of coordinate systems compared between **left:** Euclidean space through circular angle φ, **right:** in Minkowski spacetime through hyperbolic angle φ (red lines labelled c denote the worldlines of a light signal, a vector is orthogonal to itself if it lies on this line).

Special relativity uses a 'flat' 4-dimensional Minkowski space – an example of a spacetime. Minkowski spacetime appears to be very similar to the standard 3-dimensional Euclidean space, but there is a crucial difference with respect to time.

In 3D space, the differential of distance (line element) ds is defined by

$$ds^2 = d\mathbf{x} \cdot d\mathbf{x} = dx_1^2 + dx_2^2 + dx_3^2,$$

where $d\mathbf{x} = (dx_1, dx_2, dx_3)$ are the differentials of the three spatial dimensions. In Minkowski geometry, there is an extra dimension with coordinate X^0 derived from time, such that the distance differential fulfills

$$ds^2 = -dX_0^2 + dX_1^2 + dX_2^2 + dX_3^2,$$

where $dX = (dX_0, dX_1, dX_2, dX_3)$ are the differentials of the four spacetime dimensions. This suggests a deep theoretical insight: special relativity is simply a rotational symmetry of our spacetime, analogous to the rotational symmetry of Euclidean space. Just as Euclidean space uses a Euclidean metric, so spacetime uses a Minkowski metric. Basically, special relativity can be stated as the *invariance of any spacetime interval* (that is the 4D distance between any two events) when viewed from *any inertial reference frame*. All equations and effects of special relativity can be derived from this rotational symmetry (the Poincaré group) of Minkowski spacetime.

The actual form of *ds* above depends on the metric and on the choices for the X^0 coordinate. To make the time coordinate look like the space coordinates, it can be treated as imaginary: $X_0 = ict$ (this is called a Wick rotation). According to Misner, Thorne and Wheeler (1971, §2.3), ultimately the deeper understanding of both special and general relativity will come from the study of the Minkowski metric and to take $X^0 = ct$, rather than a "disguised" Euclidean metric using *ict* as the time coordinate.

Some authors use $X^0 = t$, with factors of *c* elsewhere to compensate; for instance, spatial coordinates are divided by *c* or factors of $c^{\pm 2}$ are included in the metric tensor. These numerous conventions can be superseded by using natural units where $c = 1$. Then space and time have equivalent units, and no factors of *c* appear anywhere.

3D Spacetime

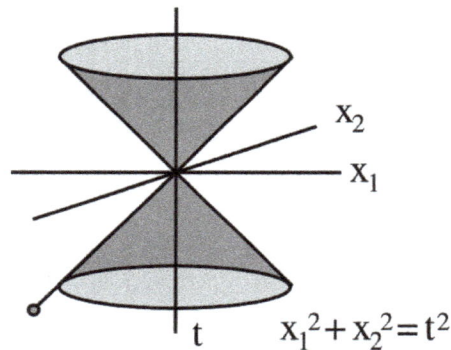

Three-dimensional dual-cone.

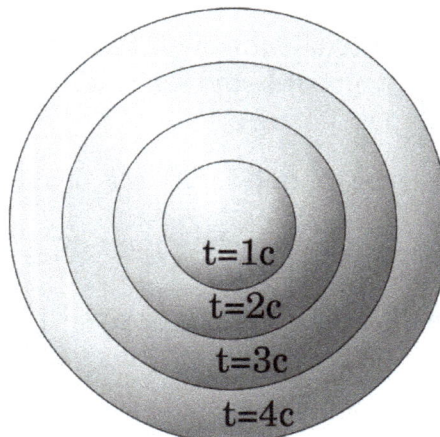

Null spherical space.

If we reduce the spatial dimensions to 2, so that we can represent the physics in a 3D space

$$ds^2 = dx_1^2 + dx_2^2 - c^2 dt^2,$$

we see that the null geodesics lie along a dual-cone defined by the equation;

$$ds^2 = 0 = dx_1^2 + dx_2^2 - c^2 dt^2$$

or simply

$$dx_1^2 + dx_2^2 = c^2 dt^2,$$

which is the equation of a circle of radius $c\,dt$.

4D Spacetime

If we extend this to three spatial dimensions, the null geodesics are the 4-dimensional cone:

$$ds^2 = 0 = dx_1^2 + dx_2^2 + dx_3^2 - c^2 dt^2$$

so

$$dx_1^2 + dx_2^2 + dx_3^2 = c^2 dt^2.$$

This null dual-cone represents the "line of sight" of a point in space. That is, when we look at the stars and say "The light from that star which I am receiving is X years old", we are looking down this line of sight: a null geodesic. We are looking at an event a distance $d = \sqrt{x_1^2 + x_2^2 + x_3^2}$ away and a time d/c in the past. For this reason the null dual cone is also known as the 'light cone'. (The point in the lower left of the picture above right represents the star, the origin represents the observer, and the line represents the null geodesic "line of sight".)

The cone in the $-t$ region is the information that the point is 'receiving', while the cone in the $+t$ section is the information that the point is 'sending'.

The geometry of Minkowski space can be depicted using Minkowski diagrams, which are useful also in understanding many of the thought-experiments in special relativity.

Note that, in 4d spacetime, the concept of the center of mass becomes more complicated.

Physics in Spacetime

Transformations of Physical Quantities between Reference Frames

Above, the Lorentz transformation for the time coordinate and three space coordinates illustrates that they are intertwined. This is true more generally: certain pairs of "timelike" and "spacelike" quantities naturally combine on equal footing under the same Lorentz transformation.

The Lorentz transformation in standard configuration above, i.e. for a boost in the x direction, can be recast into matrix form as follows:

$$\begin{pmatrix} ct' \\ x' \\ y' \\ z' \end{pmatrix} = \begin{pmatrix} \gamma & -\beta\gamma & 0 & 0 \\ -\beta\gamma & \gamma & 0 & 0 \\ 0 & 0 & 1 & 0 \\ 0 & 0 & 0 & 1 \end{pmatrix} \begin{pmatrix} ct \\ x \\ y \\ z \end{pmatrix} = \begin{pmatrix} \gamma ct - \gamma\beta x \\ \gamma x - \beta\gamma ct \\ y \\ z \end{pmatrix}.$$

In Newtonian mechanics, quantities which have magnitude and direction are mathematically described as 3d vectors in Euclidean space, and in general they are parametrized by time. In special relativity, this notion is extended by adding the appropriate timelike quantity to a spacelike vector quantity, and we have 4d vectors, or "four vectors", in Minkowski spacetime. The components of vectors are written using tensor index notation, as this has numerous advantages. The notation makes it clear the equations are manifestly covariant under the Poincaré group, thus bypassing the tedious calculations to check this fact. In constructing such equations, we often find that equations previously thought to be unrelated are, in fact, closely connected being part of the same tensor equation. Recognizing other physical quantities as tensors simplifies their transformation laws. Throughout, upper indices (superscripts) are contravariant indices rather than exponents except when they indicate a square (this is should be clear from the context), and lower indices (subscripts) are covariant indices. For simplicity and consistency with the earlier equations, Cartesian coordinates will be used.

The simplest example of a four-vector is the position of an event in spacetime, which constitutes a timelike component ct and spacelike component x = (x, y, z), in a contravariant position four vector with components:

$$X^\nu = (X^0, X^1, X^2, X^3) = (ct, x, y, z) = (ct, \mathbf{x}).$$

where we define $X^0 = ct$ so that the time coordinate has the same dimension of distance as the other spatial dimensions; so that space and time are treated equally. Now the transformation of the contravariant components of the position 4-vector can be compactly written as:

$$X^{\mu'} = \Lambda^{\mu'}{}_\nu X^\nu$$

where there is an implied summation on ν from 0 to 3, and $\Lambda^{\mu'}{}_\nu$ is a matrix.

More generally, all contravariant components of a four-vector T^ν transform from one frame to another frame by a Lorentz transformation:

$$T^{\mu'} = \Lambda^{\mu'}{}_\nu T^\nu$$

Examples of other 4-vectors include the four-velocity U^μ, defined as the derivative of the position 4-vector with respect to proper time:

$$U^\mu = \frac{dX^\mu}{d\tau} = \gamma(v)(c, v_x, v_y, v_z) = \gamma(v)(c, \mathbf{v}).$$

where the Lorentz factor is:

$$\gamma(v) = \frac{1}{\sqrt{1-(v/c)^2}}, \quad v^2 = v_x^2 + v_y^2 + v_z^2.$$

The relativistic energy $E = \gamma(v)mc^2$ and relativistic momentum $\mathbf{p} = \gamma(v)m\mathbf{v}$ of an object are respectively the timelike and spacelike components of a contravariant four momentum vector:

$$P^\mu = mU^\mu = m\gamma(v)(c, v_x, v_y, v_z) = (E/c, p_x, p_y, p_z) = (E/c, \mathbf{p}).$$

where m is the invariant mass.

The four-acceleration is the proper time derivative of 4-velocity:

$$A^\mu = \frac{dU^\mu}{d\tau}.$$

The transformation rules for *three*-dimensional velocities and accelerations are very awkward; even above in standard configuration the velocity equations are quite complicated owing to their non-linearity. On the other hand, the transformation of *four*-velocity and *four*-acceleration are simpler by means of the Lorentz transformation matrix.

The four-gradient of a scalar field φ transforms covariantly rather than contravariantly:

$$\left(\frac{1}{c}\frac{\partial\phi}{\partial t'} \quad \frac{\partial\phi}{\partial x'} \quad \frac{\partial\phi}{\partial y'} \quad \frac{\partial\phi}{\partial z'} \right) = \left(\frac{1}{c}\frac{\partial\phi}{\partial t} \quad \frac{\partial\phi}{\partial x} \quad \frac{\partial\phi}{\partial y} \quad \frac{\partial\phi}{\partial z} \right) \begin{pmatrix} \gamma & -\beta\gamma & 0 & 0 \\ -\beta\gamma & \gamma & 0 & 0 \\ 0 & 0 & 1 & 0 \\ 0 & 0 & 0 & 1 \end{pmatrix}.$$

that is:

$$(\bar{\partial}_{\mu'}\phi) = \Lambda_{\mu'}{}^\nu(\partial_\nu\phi), \quad \partial_\mu \equiv \frac{\partial}{\partial x^\mu}.$$

only in Cartesian coordinates. It's the covariant derivative which transforms in manifest covariance, in Cartesian coordinates this happens to reduce to the partial derivatives, but not in other coordinates.

More generally, the *covariant* components of a 4-vector transform according to the *inverse* Lorentz transformation:

$$\Lambda_{\mu'}{}^\nu T^{\mu'} = T^\nu$$

where $\Lambda_{\mu'}{}^\nu$ is the reciprocal matrix of $\Lambda^{\mu'}{}_\nu$.

The postulates of special relativity constrain the exact form the Lorentz transformation matrices take.

More generally, most physical quantities are best described as (components of) tensors. So to transform from one frame to another, we use the well-known tensor transformation law

$$T^{\alpha'\beta'\cdots\zeta'}_{\theta'\iota'\cdots\kappa'} = \Lambda^{\alpha'}{}_{\mu}\Lambda^{\beta'}{}_{\nu}\cdots\Lambda^{\zeta'}{}_{\rho}\Lambda_{\theta'}{}^{\sigma}\Lambda_{\iota'}{}^{\upsilon}\cdots\Lambda_{\kappa'}{}^{\phi}T^{\mu\nu\cdots\rho}_{\sigma\upsilon\cdots\phi}$$

where $\Lambda_{\chi'}{}^{\psi}$ is the reciprocal matrix of $\Lambda^{\chi'}{}_{\psi'}$. All tensors transform by this rule.

An example of a four dimensional second order antisymmetric tensor is the relativistic angular momentum, which has six components: three are the classical angular momentum, and the other three are related to the boost of the center of mass of the system. The derivative of the relativistic angular momentum with respect to proper time is the relativistic torque, also second order antisymmetric tensor.

The electromagnetic field tensor is another second order antisymmetric tensor field, with six components: three for the electric field and another three for the magnetic field. There is also the stress–energy tensor for the electromagnetic field, namely the electromagnetic stress–energy tensor.

Metric

The metric tensor allows one to define the inner product of two vectors, which in turn allows one to assign a magnitude to the vector. Given the four-dimensional nature of spacetime the Minkowski metric η has components (valid in any inertial reference frame) which can be arranged in a 4×4 matrix:

$$\eta_{\alpha\beta} = \begin{pmatrix} -1 & 0 & 0 & 0 \\ 0 & 1 & 0 & 0 \\ 0 & 0 & 1 & 0 \\ 0 & 0 & 0 & 1 \end{pmatrix}$$

which is equal to its reciprocal, $\eta^{\alpha\beta}$, in those frames. Throughout we use the signs as above, different authors use different conventions.

The Poincaré group is the most general group of transformations which preserves the Minkowski metric:

$$\eta_{\alpha\beta} = \eta_{\mu'\nu'}\Lambda^{\mu'}{}_{\alpha}\Lambda^{\nu'}{}_{\beta}$$

and this is the physical symmetry underlying special relativity.

The metric can be used for raising and lowering indices on vectors and tensors. Invariants can be constructed using the metric, the inner product of a 4-vector T with another 4-vector S is:

$$T^{\alpha}S_{\alpha} = T^{\alpha}\eta_{\alpha\beta}S^{\beta} = T_{\alpha}\eta^{\alpha\beta}S_{\beta} = \text{invariant scalar}$$

Invariant means that it takes the same value in all inertial frames, because it is a scalar (0 rank tensor), and so no Λ appears in its trivial transformation. The magnitude of the 4-vector T is the positive square root of the inner product with itself:

$$|\mathbf{T}| = \sqrt{T^{\alpha}T_{\alpha}}$$

One can extend this idea to tensors of higher order, for a second order tensor we can form the invariants:

$$T^{\alpha}{}_{\alpha}, T^{\alpha}{}_{\beta}T^{\beta}{}_{\alpha}, T^{\alpha}{}_{\beta}T^{\beta}{}_{\gamma}T^{\gamma}{}_{\alpha} = \text{invariant scalars},$$

similarly for higher order tensors. Invariant expressions, particularly inner products of 4-vectors with themselves, provide equations that are useful for calculations, because one doesn't need to perform Lorentz transformations to determine the invariants.

Relativistic Kinematics and Invariance

The coordinate differentials transform also contravariantly:

$$dX^{\mu'} = \Lambda^{\mu'}{}_{\nu} dX^{\nu}$$

so the squared length of the differential of the position four-vector dX^{μ} constructed using

$$d\mathbf{X}^2 = dX^{\mu} dX_{\mu} = \eta_{\mu\nu} dX^{\mu} dX^{\nu} = -(cdt)^2 + (dx)^2 + (dy)^2 + (dz)^2$$

is an invariant. Notice that when the line element $d\mathbf{X}^2$ is negative that $\sqrt{-d\mathbf{X}^2}$ is the differential of proper time, while when $d\mathbf{X}^2$ is positive, $\sqrt{d\mathbf{X}^2}$ is differential of the proper distance.

The 4-velocity U^{μ} has an invariant form:

$$\mathbf{U}^2 = \eta_{\nu\mu} U^{\nu} U^{\mu} = -c^2,$$

which means all velocity four-vectors have a magnitude of c. This is an expression of the fact that there is no such thing as being at coordinate rest in relativity: at the least, you are always moving forward through time. Differentiating the above equation by τ produces:

$$2\eta_{\mu\nu} A^{\mu} U^{\nu} = 0.$$

So in special relativity, the acceleration four-vector and the velocity four-vector are orthogonal.

Relativistic Dynamics and Invariance

The invariant magnitude of the momentum 4-vector generates the energy–momentum relation:

$$\mathbf{P}^2 = \eta^{\mu\nu} P_{\mu} P_{\nu} = -(E/c)^2 + p^2.$$

We can work out what this invariant is by first arguing that, since it is a scalar, it doesn't matter in which reference frame we calculate it, and then by transforming to a frame where the total momentum is zero.

$$\mathbf{P}^2 = -(E_{\text{rest}}/c)^2 = -(mc)^2.$$

We see that the rest energy is an independent invariant. A rest energy can be calculated even for particles and systems in motion, by translating to a frame in which momentum is zero.

The rest energy is related to the mass according to the celebrated equation discussed above:

$$E_{\text{rest}} = mc^2.$$

Note that the mass of systems measured in their center of momentum frame (where total momentum is zero) is given by the total energy of the system in this frame. It may not be equal to the sum of individual system masses measured in other frames.

To use Newton's third law of motion, both forces must be defined as the rate of change of momentum with respect to the same time coordinate. That is, it requires the 3D force defined above. Unfortunately, there is no tensor in 4D which contains the components of the 3D force vector among its components.

If a particle is not traveling at c, one can transform the 3D force from the particle's co-moving reference frame into the observer's reference frame. This yields a 4-vector called the four-force. It is the rate of change of the above energy momentum four-vector with respect to proper time. The covariant version of the four-force is:

$$F_v = \frac{dP_v}{d\tau} = mA_v$$

In the rest frame of the object, the time component of the four force is zero unless the "invariant mass" of the object is changing (this requires a non-closed system in which energy/mass is being directly added or removed from the object) in which case it is the negative of that rate of change of mass, times c. In general, though, the components of the four force are not equal to the components of the three-force, because the three force is defined by the rate of change of momentum with respect to coordinate time, i.e. dp/dt while the four force is defined by the rate of change of momentum with respect to proper time, i.e. $dp/d\tau$.

In a continuous medium, the 3D *density of force* combines with the *density of power* to form a covariant 4-vector. The spatial part is the result of dividing the force on a small cell (in 3-space) by the volume of that cell. The time component is $-1/c$ times the power transferred to that cell divided by the volume of the cell. This will be used below in the section on electromagnetism.

Relativity and Unifying Electromagnetism

Theoretical investigation in classical electromagnetism led to the discovery of wave propagation. Equations generalizing the electromagnetic effects found that finite propagation speed of the E and B fields required certain behaviors on charged particles. The general study of moving charges forms the Liénard–Wiechert potential, which is a step towards special relativity.

The Lorentz transformation of the electric field of a moving charge into a non-moving observer's reference frame results in the appearance of a mathematical term commonly called the magnetic field. Conversely, the *magnetic* field generated by a moving charge disappears and becomes a purely *electrostatic* field in a comoving frame of reference. Maxwell's equations are thus simply an empirical fit to special relativistic effects in a classical model of the Universe. As electric and magnetic fields are reference frame dependent and thus intertwined, one speaks of *electromagnetic* fields. Special relativity provides the transformation rules for how an electromagnetic field in one inertial frame appears in another inertial frame.

Maxwell's equations in the 3D form are already consistent with the physical content of special relativity, although they are easier to manipulate in a manifestly covariant form, i.e. in the language of tensor calculus.

Status

Special relativity in its Minkowski spacetime is accurate only when the absolute value of the gravitational potential is much less than c^2 in the region of interest. In a strong gravitational field, one must use general relativity. General relativity becomes special relativity at the limit of a weak field. At very small scales, such as at the Planck length and below, quantum effects must be taken into consideration resulting in quantum gravity. However, at macroscopic scales and in the absence of strong gravitational fields, special relativity is experimentally tested to extremely high degree of accuracy (10^{-20}) and thus accepted by the physics community. Experimental results which appear to contradict it are not reproducible and are thus widely believed to be due to experimental errors.

Special relativity is mathematically self-consistent, and it is an organic part of all modern physical theories, most notably quantum field theory, string theory, and general relativity (in the limiting case of negligible gravitational fields).

Newtonian mechanics mathematically follows from special relativity at small velocities (compared to the speed of light) – thus Newtonian mechanics can be considered as a special relativity of slow moving bodies.

Several experiments predating Einstein's 1905 paper are now interpreted as evidence for relativity. Of these it is known Einstein was aware of the Fizeau experiment before 1905, and historians have concluded that Einstein was at least aware of the Michelson–Morley experiment as early as 1899 despite claims he made in his later years that it played no role in his development of the theory.

- The Fizeau experiment (1851, repeated by Michelson and Morley in 1886) measured the speed of light in moving media, with results that are consistent with relativistic addition of colinear velocities.

- The famous Michelson–Morley experiment (1881, 1887) gave further support to the postulate that detecting an absolute reference velocity was not achievable. It should be stated here that, contrary to many alternative claims, it said little about the invariance of the speed of light with respect to the source and observer's velocity, as both source and observer were travelling together at the same velocity at all times.

- The Trouton–Noble experiment (1903) showed that the torque on a capacitor is independent of position and inertial reference frame.

- The Experiments of Rayleigh and Brace (1902, 1904) showed that length contraction doesn't lead to birefringence for a co-moving observer, in accordance with the relativity principle.

Particle accelerators routinely accelerate and measure the properties of particles moving at near the speed of light, where their behavior is completely consistent with relativity theory and inconsistent with the earlier Newtonian mechanics. These machines would simply not work if they were not engineered according to relativistic principles. In addition, a considerable number of modern experiments have been conducted to test special relativity. Some examples:

- Tests of relativistic energy and momentum – testing the limiting speed of particles

- Ives–Stilwell experiment – testing relativistic Doppler effect and time dilation

- Time dilation of moving particles – relativistic effects on a fast-moving particle's half-life

- Kennedy–Thorndike experiment – time dilation in accordance with Lorentz transformations

- Hughes–Drever experiment – testing isotropy of space and mass

- Modern searches for Lorentz violation – various modern tests

- Experiments to test emission theory demonstrated that the speed of light is independent of the speed of the emitter.

- Experiments to test the aether drag hypothesis – no "aether flow obstruction".

Theories of Relativity and Quantum Mechanics

Special relativity can be combined with quantum mechanics to form relativistic quantum mechanics. It is an unsolved problem in physics how *general* relativity and quantum mechanics can be unified; quantum gravity and a "theory of everything", which require such a unification, are active and ongoing areas in theoretical research.

The early Bohr–Sommerfeld atomic model explained the fine structure of alkali metal atoms using both special relativity and the preliminary knowledge on quantum mechanics of the time.

In 1928, Paul Dirac constructed an influential relativistic wave equation, now known as the Dirac equation in his honour, that is fully compatible both with special relativity and with the final version of quantum theory existing after 1926. This equation explained not only the intrinsic angular momentum of the electrons called *spin*, it also led to the prediction of the antiparticle of the electron (the positron), and fine structure could only be fully explained with special relativity. It was the first foundation of *relativistic quantum mechanics*. In non-relativistic quantum mechanics, spin is phenomenological and cannot be explained.

On the other hand, the existence of antiparticles leads to the conclusion that relativistic quantum mechanics is not enough for a more accurate and complete theory of particle interactions. Instead, a theory of particles interpreted as quantized fields, called *quantum field theory*, becomes necessary; in which particles can be created and destroyed throughout space and time.

Lorentz Transformation

In physics, the Lorentz transformation (or transformations) are coordinate transformations between two coordinate frames that move at constant velocity relative to each other.

Frames of reference can be divided into two groups: inertial (relative motion with constant velocity) and non-inertial (accelerating in curved paths, rotational motion with constant angular velocity, etc.). The term "Lorentz transformations" only refers to transformations between *inertial* frames, usually in the context of special relativity.

In each reference frame, an observer can use a local coordinate system (most exclusively Cartesian coordinates in this context) to measure lengths, and a clock to measure time intervals. An observer is a real or imaginary entity that can take measurements, say humans, or any other living organism—or even robots and computers. An event is something that happens at a point in space at an instant of time, or more formally a point in spacetime. The transformations connect the space and time coordinates of an event as measured by an observer in each frame.

They supersede the Galilean transformation of Newtonian physics, which assumes an absolute space and time. The Galilean transformation is a good approximation only at relative speeds much smaller than the speed of light. Lorentz transformations have a number of unintuitive features that do not appear in Galilean transformations. For example, they reflect the fact that observers moving at different velocities may measure different distances, elapsed times, and even different orderings of events, but always such that the speed of light is the same in all inertial reference frames. The invariance of light speed is one of the postulates of special relativity.

Historically, the transformations were the result of attempts by Lorentz and others to explain how the speed of light was observed to be independent of the reference frame, and to understand the symmetries of the laws of electromagnetism. The Lorentz transformation is in accordance with special relativity, but was derived before special relativity. The transformations are named after the Dutch physicist Hendrik Lorentz.

The Lorentz transformation is a linear transformation. It may include a rotation of space; a rotation-free Lorentz transformation is called a Lorentz boost. In Minkowski space, the mathematical model of spacetime in special relativity, the Lorentz transformations preserve the spacetime interval between any two events. This property is the defining property of a Lorentz transformation. They describe only the transformations in which the spacetime event at the origin is left fixed. They can be considered as a hyperbolic rotation of Minkowski space. The more general set of transformations that also includes translations is known as the Poincaré group.

History

Many physicists—including Woldemar Voigt, George FitzGerald, Joseph Larmor, and Hendrik Lorentz himself—had been discussing the physics implied by these equations since 1887. Early in 1889, Oliver Heaviside had shown from Maxwell's equations that the electric field surrounding a spherical distribution of charge should cease to have spherical symmetry once the charge is in motion relative to the ether. FitzGerald then conjectured that Heaviside's distortion result might be applied to a theory of intermolecular forces. Some months later, FitzGerald published the conjecture that bodies in motion are being contracted, in order to explain the baffling outcome of the 1887 ether-wind experiment of Michelson and Morley. In 1892, Lorentz independently presented the same idea in a more detailed manner, which was subsequently called FitzGerald–Lorentz contraction hypothesis. Their explanation was widely known before 1905.

Lorentz (1892–1904) and Larmor (1897–1900), who believed the luminiferous ether hypothesis, also looked for the transformation under which Maxwell's equations are invariant when transformed from the ether to a moving frame. They extended the FitzGerald–Lorentz contraction hypothesis and found out that the time coordinate has to be modified as well ("local time"). Henri Poincaré gave a physical interpretation to local time (to first order in v/c) as the consequence of clock synchronization, under the assumption that the speed of light is constant in moving frames. Larmor is credited to have been the first to understand the crucial time dilation property inherent in his equations.

In 1905, Poincaré was the first to recognize that the transformation has the properties of a mathematical group, and named it after Lorentz. Later in the same year Albert Einstein published what is now called special relativity, by deriving the Lorentz transformation under the assumptions of the principle of relativity and the constancy of the speed of light in any inertial reference frame, and by abandoning the mechanical ether.

Derivation

An *event* is something that happens at a certain point in spacetime, or more generally, the point in spacetime itself. In any inertial frame an event is specified by a time coordinate t and a set of Cartesian coordinates x, y, z to specify position in space in that frame. Subscripts label individual events.

From Einstein's second postulate of relativity follows immediately

$$c^2(t_2 - t_1)^2 - (x_2 - x_1)^2 - (y_2 - y_1)^2 - (z_2 - z_1)^2 = 0$$

in all inertial frames for events connected by *light signals*. The quantity on the left is called the *spacetime interval* between events (t_1, x_1, y_1, z_1) and (t_2, x_2, y_2, z_2). The interval between *any two* events, not necessarily separated by light signals, is in fact invariant, i.e., independent of the state of relative motion of observers in different inertial frames, as is shown here (where one can also find several more explicit derivations than presently given) using homogeneity and isotropy of space. The transformation sought after thus must possess the property that

$$c^2(t_2 - t_1)^2 - (x_2 - x_1)^2 - (y_2 - y_1)^2 - (z_2 - z_1)^2 = c^2(t_2' - t_1')^2 - (x_2' - x_1')^2 - (y_2' - y_1')^2 - (z_2' - z_1')^2.$$

where t, x, y, z are the spacetime coordinates used to define events in one frame, and t', x', y', z' are the coordinates in another frame. Now one observes that a *linear* solution to the simpler problem

$$c^2 t^2 - x^2 - y^2 - z^2 = c^2 t'^2 - x'^2 - y'^2 - z'^2$$

solves the general problem too. Finding the solution to the simpler problem is just a matter of look-up in the theory of classical groups that preserve bilinear forms of various signature.[nb 2] The Lorentz transformation is thus an element of the group O(3, 1) or, for those that prefer the other metric signature, O(1, 3).

Generalities

The relations between the primed and unprimed spacetime coordinates are the Lorentz transformations, each coordinate in one frame is a linear function of all the coordinates in the other frame, and the inverse functions are the inverse transformation. Depending on how the frames move relative to each other, and how they are oriented in space relative to each other, other parameters that describe direction, speed, and orientation enter the transformation equations.

Transformations describing relative motion with constant (uniform) velocity and without rotation of the space coordinate axes are called *boosts*, and the relative velocity between the frames is the parameter of the transformation. The other basic type of Lorentz transformations is rotations in the spatial coordinates only, these are also inertial frames since there is no relative motion, the frames are simply tilted (and not continuously rotating), and in this case quantities defining the rotation are the parameters of the transformation (e.g., axis–angle representation, or Euler angles, etc.). A combination of a rotation and boost is a *homogenous transformation*, which transforms the origin back to the origin.

The full Lorentz group O(3, 1) also contains special transformations that are neither rotations nor boosts, but rather reflections in a plane through the origin. Two of these can be singled out; spatial inversion in which the spatial coordinates of all events are reversed in sign and temporal inversion in which the time coordinate for each event gets its sign reversed.

Boosts should not be conflated with mere displacements in spacetime; in this case, the coordinate systems are simply shifted and there is no relative motion. However, these also count as symmetries forced by special relativity since they leave the spacetime interval invariant. A combination of a rotation with a boost, followed by a shift in spacetime, is an *inhomogenous Lorentz transformation*, an element of the Poincaré group, which is also called the inhomogeneous Lorentz group.

Physical Formulation of Lorentz Boosts

Coordinate Transformation

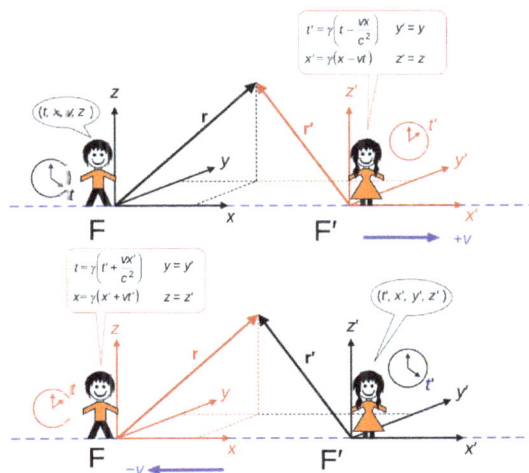

The spacetime coordinates of an event, as measured by each observer in their inertial reference frame (in standard configuration) are shown in the speech bubbles.
Top: frame F' moves at velocity v along the x-axis of frame F.
Bottom: frame F moves at velocity $-v$ along the x'-axis of frame F'.

A "stationary" observer in frame F defines events with coordinates t, x, y, z. Another frame F' moves with velocity v relative to F, and an observer in this "moving" frame F' defines events using the coordinates t', x', y', z'.

The coordinate axes in each frame are parallel (the x and x' axes are parallel, the y and y' axes are parallel, and the z and z' axes are parallel), remain mutually perpendicular, and relative motion is along the coincident xx' axes. At $t = t' = 0$, the origins of both coordinate systems are the same, $(x,$

$y, z) = (x', y', z') = (0, 0, 0)$. In other words, the times and positions are coincident at this event. If all these hold, then the coordinate systems are said to be in standard configuration, or synchronized.

If an observer in F records an event t, x, y, z, then an observer in F' records the *same* event with coordinates

$$
\boxed{
\begin{array}{c}
\textbf{Lorentz boost } (x \text{ } direction) \\[8pt]
t' = \gamma\left(t - \dfrac{vx}{c^2}\right) \\[8pt]
x' = \gamma\left(x - vt\right) \\[6pt]
y' = y \\[4pt]
z' = z
\end{array}
}
$$

where v is the relative velocity between frames in the x-direction, c is the speed of light, and

$$
\gamma = \dfrac{1}{\sqrt{1 - \dfrac{v^2}{c^2}}}
$$

(lowercase gamma) is the Lorentz factor.

Here, v is the *parameter* of the transformation, for a given boost it is a constant number, but can take a continuous range of values. In the setup used here, positive relative velocity $v > 0$ is motion along the positive directions of the xx' axes, zero relative velocity $v = 0$ is no relative motion, while negative relative velocity $v < 0$ is relative motion along the negative directions of the xx' axes. The magnitude of relative velocity v cannot equal or exceed c, so only subluminal speeds $-c < v < c$ are allowed. The corresponding range of γ is $1 \leq \gamma < \infty$.

The transformations are not defined if v is outside these limits. At the speed of light ($v = c$) γ is infinite, and faster than light ($v > c$) γ is a complex number, each of which make the transformations unphysical. The space and time coordinates are measurable quantities and numerically must be real numbers, not complex.

As an active transformation, an observer in F$'$ notices the coordinates of the event to be "boosted" in the negative directions of the xx' axes, because of the $-v$ in the transformations. This has the equivalent effect of the *coordinate system* F$'$ boosted in the positive directions of the xx' axes, while the event does not change and is simply represented in another coordinate system, a passive transformation.

The inverse relations (t, x, y, z in terms of t', x', y', z') can be found by algebraically solving the original set of equations. A more efficient way is to use physical principles. Here F' is the "stationary" frame while F is the "moving" frame. According to the principle of relativity, there is no privileged frame of reference, so the transformations from F' to F must take exactly the same form as the transformations from F to F'. The only difference is F' moves with velocity $-v$ relative to F (i.e., the relative velocity has the same magnitude but is oppositely directed). Thus if an observer in F' notes an event t', x', y', z', then an observer in F notes the *same* event with coordinates

---

Inverse Lorentz boost

(*x direction*)

$$t = \gamma \left(t' + \frac{vx'}{c^2} \right)$$

$$x = \gamma \left(x' + vt' \right)$$

$$y = y'$$

$$z = z',$$

---

and the value of γ remains unchanged. This "trick" of simply reversing the direction of relative velocity while preserving its magnitude, and exchanging primed and unprimed variables, always applies to finding the inverse transformation of every boost in any direction.

Sometimes it is more convenient to use $\beta = v/c$ (lowercase beta) instead of v, so that

$$ct' = \gamma \left(ct - \beta x \right),$$
$$x' = \gamma \left(x - \beta ct \right),$$

which shows much more clearly the symmetry in the transformation. From the allowed ranges of v and the definition of β, it follows $-1 < \beta < 1$. The use of β and γ is standard throughout the literature.

The Lorentz transformations can also be derived in a way that resembles circular rotations in 3d space using the hyperbolic functions. For the boost in the x direction, the results are

---

Lorentz boost (*x direction with rapidity* ζ)

$$ct' = ct \cosh \zeta - x \sinh \zeta$$
$$x' = x \cosh \zeta - ct \sinh \zeta$$
$$y' = y$$
$$z' = z$$

---

where ζ (lowercase zeta) is a parameter called *rapidity* (many other symbols are used, including θ, ϕ, φ, η, ψ, ξ). Given the strong resemblance to rotations of spatial coordinates in 3d space in the Cartesian xy, yz, and zx planes, a Lorentz boost can be thought of as a hyperbolic rotation of space-time coordinates in the xt, yt, and zt Cartesian-time planes of 4d Minkowski space. The parameter ζ is the hyperbolic angle of rotation, analogous to the ordinary angle for circular rotations. This transformation can be illustrated with a Minkowski diagram.

The hyperbolic functions arise from the *difference* between the squares of the time and spatial coordinates in the spacetime interval, rather than a sum. The geometric significance of the hyperbolic functions can be visualized by taking $x = 0$ or $ct = 0$ in the transformations. Squaring and subtracting the results, one can derive hyperbolic curves of constant coordinate values but varying ζ, which parametrizes the curves according to the identity

$$\cosh^2 \zeta - \sinh^2 \zeta = 1.$$

Conversely the ct and x axes can be constructed for varying coordinates but constant ζ. The definition

$$\tanh \zeta = \frac{\sinh \zeta}{\cosh \zeta},$$

provides the link between a constant value of rapidity, and the slope of the ct axis in spacetime. A consequence these two hyperbolic formulae is an identity that matches the Lorentz factor

$$\cosh \zeta = \frac{1}{\sqrt{1 - \tanh^2 \zeta}}.$$

Comparing the Lorentz transformations in terms of the relative velocity and rapidity, or using the above formulae, the connections between β, γ, and ζ are

$$\beta = \tanh \zeta,$$
$$\gamma = \cosh \zeta,$$
$$\beta\gamma = \sinh \zeta.$$

Taking the inverse hyperbolic tangent gives the rapidity

$$\zeta = \tanh^{-1} \beta.$$

Since $-1 < \beta < 1$, it follows $-\infty < \zeta < \infty$. From the relation between ζ and β, positive rapidity $\zeta > 0$ is motion along the positive directions of the xx' axes, zero rapidity $\zeta = 0$ is no relative motion, while negative rapidity $\zeta < 0$ is relative motion along the negative directions of the xx' axes.

The inverse transformations are obtained by exchanging primed and unprimed quantities to switch the coordinate frames, and negating rapidity $\zeta \to -\zeta$ since this is equivalent to negating the relative velocity. Therefore,

---

Inverse Lorentz boost (x *direction with rapidity* ζ)

$$ct = ct' \cosh \zeta + x' \sinh \zeta$$
$$x = x' \cosh \zeta + ct' \sinh \zeta$$
$$y = y'$$
$$z = z'$$

The inverse transformations can be similarly visualized by considering the cases when $x' = 0$ and $ct' = 0$.

So far the Lorentz transformations have been applied to *one event*. If there are two events, there is a spatial separation and time interval between them. It follows from the linearity of the Lorentz transformations that two values of space and time coordinates can be chosen, the Lorentz transformations can be applied to each, then subtracted to get the Lorentz transformations of the differences;

$$\Delta t' = \gamma \left(\Delta t - \frac{v \Delta x}{c^2} \right),$$

$$\Delta x' = \gamma \left(\Delta x - v \Delta t \right),$$

with inverse relations

$$\Delta t = \gamma \left(\Delta t' + \frac{v \Delta x'}{c^2} \right),$$

$$\Delta x = \gamma \left(\Delta x' + v \Delta t' \right).$$

where Δ (uppercase delta) indicates a difference of quantities; e.g., $\Delta x = x_2 - x_1$ for two values of x coordinates, and so on.

These transformations on *differences* rather than spatial points or instants of time are useful for a number of reasons:

- in calculations and experiments, it is lengths between two points or time intervals that are measured or of interest (e.g., the length of a moving vehicle, or time duration it takes to travel from one place to another),

- the transformations of velocity can be readily derived by making the difference infinitesimally small and dividing the equations, and the process repeated for the transformation of acceleration,

- if the coordinate systems are never coincident (i.e., not in standard configuration), and if both observers can agree on an event t_0, x_0, y_0, z_0 in F and t_0', x_0', y_0', z_0' in F', then they can use that event as the origin, and the spacetime coordinate differences are the differences between their coordinates and this origin, e.g., $\Delta x = x - x_0$, $\Delta x' = x' - x_0'$, etc.

Physical Implications

A critical requirement of the Lorentz transformations is the invariance of the speed of light, a fact used in their derivation, and contained in the transformations themselves. If in F the equation for a pulse of light along the x direction is $x = ct$, then in F' the Lorentz transformations give $x' = ct'$, and vice versa, for any $-c < v < c$.

For relative speeds much less than the speed of light, the Lorentz transformations reduce to the Galilean transformation

$$t' \approx t$$
$$x' \approx x - vt$$

in accordance with the correspondence principle. It is sometimes said that nonrelativistic physics is a physics of "instantaneous action at a distance".

Three unintuitive, but correct, predictions of the transformations are:

Time dilation

> Suppose there is a clock at rest in F. If a time interval (say a "tick") is measured at the same point so that $\Delta x = 0$, then the transformations give this tick in F' by $\Delta t' = \gamma \Delta t$. Conversely, suppose there is a clock at rest in F'. If a tick is measured at the same point so that $\Delta x' = 0$, then the transformations give this tick in F by $\Delta t = \gamma \Delta t'$. Either way, the boosted observer measures longer time intervals than the observer in the other frame.

Relativity of simultaneity

> Suppose two events occur simultaneously ($\Delta t = 0$) along the x axis, but separated by a non-zero displacement Δx. Then in F', we find that $\Delta t' = \gamma \dfrac{-v\Delta x}{c^2}$, so the events are no longer simultaneous according to a moving observer.

Length contraction

> Suppose there is a rod at rest in F aligned along the x axis, with length Δx. In F', the rod moves with velocity $-v$, so its length must be measured by taking two simultaneous ($\Delta t' = 0$) measurements at opposite ends. Under these conditions, the inverse Lorentz transform shows that $\Delta x = \gamma \Delta x'$. In F the two measurements are no longer simultaneous, but this does not matter because the rod is at rest in F. We conclude that the boosted observer measures a shorter length, by a factor of γ, than the observer in the rest frame of the rod. Length contraction affects any geometric quantity related to lengths, so from the perspective of a moving observer, areas and volumes will also appear to shrink along the direction of motion.

Vector Transformations

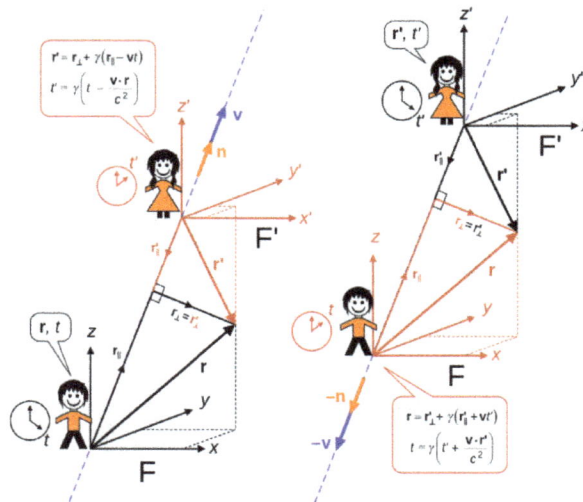

An observer in frame F observes F' to move with velocity $\mathbf{v}$, while F' observes F to move with velocity $-\mathbf{v}$. The coordinate axes of each frame are still parallel and orthogonal. The position vector as measured in each frame is split into components parallel and perpendicular to the relative velocity vector $\mathbf{v}$. **Left:** Standard configuration. **Right:** Inverse configuration.

The use of vectors allows positions and velocities to be expressed in arbitrary directions compactly. A single boost in any direction depends on the full relative velocity vector v with a magnitude $|\mathbf{v}| = v$ that cannot equal or exceed c, so that $0 \leq v < c$.

Only time and the coordinates parallel to the direction of relative motion change, while those coordinates perpendicular do not. With this in mind, split the spatial position vector r as measured in F, and r′ as measured in F', each into components perpendicular ($\perp$) and parallel ($\parallel$) to v,

$$\mathbf{r} = \mathbf{r}_\perp + \mathbf{r}_\parallel, \quad \mathbf{r}' = \mathbf{r}_{\perp'} + \mathbf{r}_{\parallel'},$$

then the transformations are

$$t' = \gamma\left(t - \frac{\mathbf{r}_\parallel \cdot \mathbf{v}}{c^2}\right)$$

$$\mathbf{r}_\parallel' = \gamma(\mathbf{r}_\parallel - \mathbf{v}t)$$

$$\mathbf{r}_\perp' = \mathbf{r}_\perp$$

where $\cdot$ is the dot product. The Lorentz factor γ retains its definition for a boost in any direction, since it depends only on the magnitude of the relative velocity. The definition $\beta = v/c$ with magnitude $0 \leq \beta < 1$ is also used by some authors.

Introducing a unit vector $\mathbf{n} = \mathbf{v}/v = \boldsymbol{\beta}/\beta$ in the direction of relative motion, the relative velocity is $\mathbf{v} = v\mathbf{n}$ with magnitude v and direction n, and vector projection and rejection give respectively

$$\mathbf{r}_\parallel = (\mathbf{r} \cdot \mathbf{n})\mathbf{n}, \quad \mathbf{r}_\perp = \mathbf{r} - (\mathbf{r} \cdot \mathbf{n})\mathbf{n}$$

Accumulating the results gives the full transformations,

> **Lorentz boost** (*in direction* **n** *with magnitude* v)
>
> $$t' = \gamma\left(t - \frac{v\mathbf{n} \cdot \mathbf{r}}{c^2}\right),$$
>
> $$\mathbf{r}' = \mathbf{r} + (\gamma - 1)(\mathbf{r} \cdot \mathbf{n})\mathbf{n} - \gamma t v\mathbf{n}.$$

The projection and rejection also applies to r′. For the inverse transformations, exchange r and r′ to switch observed coordinates, and negate the relative velocity $\mathbf{v} \to -\mathbf{v}$ (or simply the unit vector n $\to$ −n since the magnitude v is always positive) to obtain

> **Inverse Lorentz boost** (*in direction* **n** *with magnitude* v)
>
> $$t = \gamma\left(t' + \frac{\mathbf{r}' \cdot v\mathbf{n}}{c^2}\right),$$
>
> $$\mathbf{r} = \mathbf{r}' + (\gamma - 1)(\mathbf{r}' \cdot \mathbf{n})\mathbf{n} + \gamma t' v\mathbf{n},$$

The unit vector has the advantage of simplifying equations for a single boost, allows either v or β to be reinstated when convenient, and the rapidity parametrization is immediately obtained by replacing β and βγ. It is not convenient for multiple boosts.

The vectorial relation between relative velocity and rapidity is

$$\beta = \beta \mathbf{n} = \mathbf{n} \tanh \zeta,$$

and the "rapidity vector" can be defined as

$$\zeta = \zeta \mathrm{n} = \mathrm{n} \tanh^{-1} \beta,$$

each of which serves as a useful abbreviation in some contexts. The magnitude of ζ is the absolute value of the rapidity scalar confined to $0 \leq \zeta < \infty$, which agrees with the range $0 \leq \beta < 1$.

Transformation of Velocities

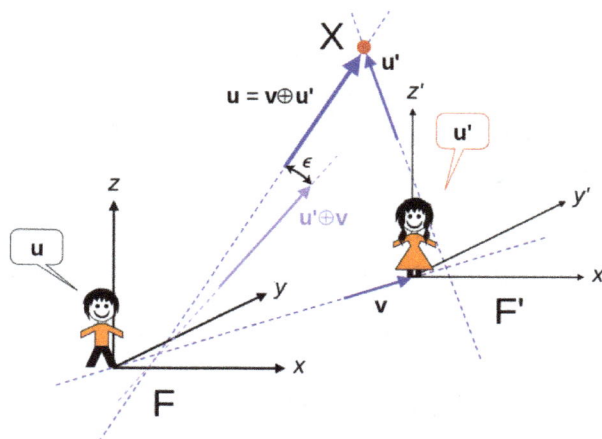

The transformation of velocities provides the definition relativistic velocity addition ⊕, the ordering of vectors is chosen to reflect the ordering of the addition of velocities; first **v** (the velocity of F′ relative to F) then **u′** (the velocity of X relative to F′) to obtain **u** = **v** ⊕ **u′** (the velocity of X relative to F).

Defining the coordinate velocities and Lorentz factor by

$$\mathbf{u} = \frac{d\mathbf{r}}{dt}, \quad \mathbf{u}' = \frac{d\mathbf{r}'}{dt'}, \quad \gamma_v = \frac{1}{\sqrt{1 - \dfrac{\mathbf{v} \cdot \mathbf{v}}{c^2}}}$$

taking the differentials in the coordinates and time of the vector transformations, then dividing equations, leads to

$$\mathbf{u}' = \frac{1}{1 - \dfrac{\mathbf{v} \cdot \mathbf{u}}{c^2}} \left[\frac{\mathbf{u}}{\gamma_v} - \mathbf{v} + \frac{1}{c^2} \frac{\gamma_v}{\gamma_v + 1} (\mathbf{u} \cdot \mathbf{v}) \mathbf{v} \right]$$

The velocities u and u′ are the velocity of some massive object. They can also be for a third inertial frame (say F''), in which case they must be *constant*. Denote either entity by X. Then X moves with velocity u relative to F, or equivalently with velocity u' relative to F', in turn F' moves with velocity v relative to F. The inverse transformations can be obtained in a similar way, or as with position coordinates exchange u and u', and change v to –v.

The transformation of velocity is useful in stellar aberration, the Fizeau experiment, and the relativistic Doppler effect.

The Lorentz transformations of acceleration can be similarly obtained by taking differentials in the velocity vectors, and dividing these by the time differential.

Transformation of other Quantities

In general, given four quantities A and $Z = (Z_x, Z_y, Z_z)$ and their Lorentz-boosted counterparts A' and $Z' = (Z_x', Z_y', Z_z')$, a relation of the form

$$A^2 - \mathbf{Z} \cdot \mathbf{Z} = A'^2 - \mathbf{Z}' \cdot \mathbf{Z}'$$

implies the quantities transform under Lorentz transformations similar to the transformation of spacetime coordinates;

$$A' = \gamma \left(A - \frac{v\mathbf{n} \cdot \mathbf{Z}}{c} \right),$$

$$\mathbf{Z}' = \mathbf{Z} + (\gamma - 1)(\mathbf{Z} \cdot \mathbf{n})\mathbf{n} - \frac{\gamma A v \mathbf{n}}{c}.$$

The decomposition of Z (and Z') into components perpendicular and parallel to v is exactly the same as for the position vector, as is the process of obtaining the inverse transformations (exchange (A, Z) and (A', Z') to switch observed quantities, and reverse the direction of relative motion by n → –n).

The quantities (A, Z) collectively make up a *four vector*, where A is the "timelike component", and Z the "spacelike component". Examples of A and Z are the following:

Four vector	A	Z
Position four vector	Time (multiplied by c), ct	Position vector, r
Four momentum	Energy (divided by c), E/c	Momentum, p
Wave four vector	angular frequency (divided by c), ω/c	wave vector, k
Four spin	(No name), s_t	Spin, s
Four current	Charge density (multiplied by c), ρc	Current density, j
Electromagnetic four potential	Electric potential (divided by c), φ/c	Magnetic potential, A

For a given object (e.g., particle, fluid, field, material), if A or Z correspond to properties specific to the object like its charge density, mass density, spin, etc., its properties can be fixed in the rest frame of that object. Then the Lorentz transformations give the corresponding properties in a frame moving relative to the object with constant velocity. This breaks some notions taken for granted in non-relativistic physics. For example, the energy E of an object is a scalar in non-relativistic mechanics, but not in relativistic mechanics because energy changes under Lorentz transformations; its value is different for various inertial frames. In the rest frame of an object, it has a rest energy and zero momentum. In a boosted frame its energy is different and it appears to have a momentum. Similarly, in non-relativistic quantum mechanics the spin of a particle is a constant vector, but in relativistic quantum mechanics spin s depends on relative motion. In the rest frame of the particle, the spin pseudovector can be fixed to be its ordinary non-relativistic spin with a zero timelike quantity s_t, however a boosted observer will perceive a nonzero timelike component and an altered spin.

Not all quantities are invariant in the form as shown above, for example orbital angular momentum L does not have a timelike quantity, and neither does the electric field E nor the magnetic field B. The definition of angular momentum is L = r × p, and in a boosted frame the altered angular momentum is L′ = r′ × p′. Applying this definition using the transformations of coordinates and momentum leads to the transformation of angular momentum. It turns out L transforms with another vector quantity N = (E/c^2)r − tp related to boosts. For the case of the E and B fields, the transformations cannot be obtained as directly using vector algebra. The Lorentz force is the definition of these fields, and in F it is F = q(E + v×B) while in $F′$ it is F′ = q(E′ + v′×B′). A method of deriving the EM field transformations in an efficient way which also illustrates the unit of the electromagnetic field uses tensor algebra, given below.

Mathematical Formulation

Throughout, italic non-bold capital letters are 4×4 matrices, while non-italic bold letters are 3×3 matrices.

Homogeneous Lorentz Group

Writing the coordinates in column vectors and the Minkowski metric η as a square matrix

$$X' = \begin{bmatrix} ct' \\ x' \\ y' \\ z' \end{bmatrix}, \quad \eta = \begin{bmatrix} -1 & 0 & 0 & 0 \\ 0 & 1 & 0 & 0 \\ 0 & 0 & 1 & 0 \\ 0 & 0 & 0 & 1 \end{bmatrix}, \quad X = \begin{bmatrix} ct \\ x \\ y \\ z \end{bmatrix}$$

the spacetime interval takes the form (T denotes transpose)

$$X \cdot X = X^{\mathrm{T}} \eta X = X'^{\mathrm{T}} \eta X'$$

and is invariant under a Lorentz transformation

$$X' = \Lambda X$$

where Λ is a square matrix which can depend on parameters.

The set of all Lorentz transformations Λ in this article is denoted $\mathcal{L}$. This set together with matrix multiplication forms a group, in this context known as the *Lorentz group*. Also, the above expression $X \cdot X$ is a quadratic form of signature (3,1) on spacetime, and the group of transformations which leaves this quadratic form invariant is the indefinite orthogonal group O(3,1), a Lie group. In other words, the Lorentz group is O(3,1). As presented in this article, any Lie groups mentioned are matrix Lie groups. In this context the operation of composition amounts to matrix multiplication.

From the invariance of the spacetime interval it follows

$$\eta = \Lambda^{T} \eta \Lambda$$

and this matrix equation contains the general conditions on the Lorentz transformation to ensure invariance of the spacetime interval. Taking the determinant of the equation using the product rule gives immediately

$$[\det(\Lambda)]^2 = 1 \quad \Rightarrow \quad \det(\Lambda) = \pm 1$$

Writing the Minkowski metric as a block matrix, and the Lorentz transformation in the most general form,

$$\eta = \begin{bmatrix} -1 & 0 \\ 0 & \mathbf{I} \end{bmatrix}, \quad \Lambda = \begin{bmatrix} \Gamma & -\mathbf{a}^{T} \\ -\mathbf{b} & \mathbf{M} \end{bmatrix},$$

carrying out the block matrix multiplications obtains general conditions on Γ, a, b, M to ensure relativistic invariance. Not much information can be directly extracted from all the conditions, however one of the results

$$\Gamma^2 = 1 + \mathbf{b}^{T} \mathbf{b}$$

is useful; $b^{T}b \geq 0$ always so it follows that

$$\Gamma^2 \geq 1 \quad \Rightarrow \quad \Gamma \leq -1, \quad \Gamma \geq 1$$

The negative inequality may be unexpected, because Γ multiplies the time coordinate and this has an effect on time symmetry. If the positive equality holds, then Γ is the Lorentz factor.

The determinant and inequality provide four ways to classify Lorentz transformations (herein LTs for brevity). Any particular LT has only one determinant sign *and* only one inequality. There are four sets which include every possible pair given by the intersections ("n"-shaped symbol meaning "and") of these classifying sets.

Intersection, $\cap$	Antichronous (or non-orthochronous) LTs $\mathcal{L}^{\downarrow} = \{\Lambda : \Gamma \leq -1\}$	Orthochronous LTs $\mathcal{L}^{\uparrow} = \{\Lambda : \Gamma \geq 1\}$
Proper LTs $\mathcal{L}_{+} = \{\Lambda : \det(\Lambda) = +1\}$	Proper antichronous LTs $\mathcal{L}_{+}^{\downarrow} = \mathcal{L}_{+} \cap \mathcal{L}^{\downarrow}$	Proper orthochronous LTs $\mathcal{L}_{+}^{\uparrow} = \mathcal{L}_{+} \cap \mathcal{L}^{\uparrow}$
Improper LTs $\mathcal{L}_{-} = \{\Lambda : \det(\Lambda) = -1\}$	Improper antichronous LTs $\mathcal{L}_{-}^{\downarrow} = \mathcal{L}_{-} \cap \mathcal{L}^{\downarrow}$	Improper orthochronous LTs $\mathcal{L}_{-}^{\uparrow} = \mathcal{L}_{-} \cap \mathcal{L}^{\uparrow}$

where "+" and "−" indicate the determinant sign, while "↑" for ≥ and "↓" for ≤ denote the inequalities.

The full Lorentz group splits into the union ("∪"-shaped symbol meaning "or") of four disjoint sets

$$\mathcal{L} = \mathcal{L}_+^\uparrow \cup \mathcal{L}_-^\uparrow \cup \mathcal{L}_+^\downarrow \cup \mathcal{L}_-^\downarrow$$

A subgroup of a group must be closed under the same operation of the group (here matrix multiplication). In other words, for two Lorentz transformations Λ and L from a particular set, the composite Lorentz transformations ΛL and $L\Lambda$ must be in the same set as Λ and L. This will not always be the case; it can be shown that the composition of *any* two Lorentz transformations always has the positive determinant and positive inequality, a proper orthochronous transformation. The sets $\mathcal{L}_+^\uparrow$, $\mathcal{L}_+$, $\mathcal{L}^\uparrow$, and $\mathcal{L}_0 = \mathcal{L}_+^\uparrow \cup \mathcal{L}_+^\downarrow$ all form subgroups. The other sets involving the improper and/or antichronous properties (i.e. $\mathcal{L}_+^\downarrow$, $\mathcal{L}_-^\downarrow$, $\mathcal{L}_-^\uparrow$) do not form subgroups, because the composite transformation always has a positive determinant or inequality, whereas the original separate transformations will have negative determinants and/or inequalities.

Proper Transformations

The Lorentz boost is

$$X' = B(\mathbf{v})X$$

where the boost matrix is

$$B(\mathbf{v}) = \begin{bmatrix} \gamma & -\gamma\beta n_x & -\gamma\beta n_y & -\gamma\beta n_z \\ -\gamma\beta n_x & 1+(\gamma-1)n_x^2 & (\gamma-1)n_x n_y & (\gamma-1)n_x n_z \\ -\gamma\beta n_y & (\gamma-1)n_y n_x & 1+(\gamma-1)n_y^2 & (\gamma-1)n_y n_z \\ -\gamma\beta n_z & (\gamma-1)n_z n_x & (\gamma-1)n_z n_y & 1+(\gamma-1)n_z^2 \end{bmatrix}.$$

The boosts along the Cartesian directions can be readily obtained, for example the unit vector in the x direction has components $n_x = 1$ and $n_y = n_z = 0$.

The matrices make one or more successive transformations easier to handle, rather than rotely iterating the transformations to obtain the result of more than one transformation. If a frame F' is boosted with velocity u relative to frame F, and another frame F'' is boosted with velocity v relative to F', the separate boosts are

$$X'' = B(\mathbf{v})X', \quad X' = B(\mathbf{u})X$$

and the composition of the two boosts connects the coordinates in F'' and F,

$$X'' = B(\mathbf{v})B(\mathbf{u})X.$$

Successive transformations act on the left. If u and v are collinear (parallel or antiparallel along the same line of relative motion), the boost matrices commute: $B(v)B(u) = B(u)B(v)$ and this composite transformation happens to be another boost.

If u and v are not collinear but in different directions, the situation is considerably more complicated. Lorentz boosts along different directions do not commute: $B(v)B(u)$ and $B(u)B(v)$ are not equal. Also, each of these compositions is *not* a single boost, but still a Lorentz transformation as each boost still preserves invariance of the spacetime interval. It turns out the composition of any two Lorentz boosts is equivalent to a boost followed or preceded by a rotation on the spatial coordinates, in the form of $R(\rho)B(w)$ or $B(w)R(\rho)$. The w and w are composite velocities, while ρ and ρ are rotation parameters (e.g. axis-angle variables, Euler angles, etc.). The rotation in block matrix form is simply

$$R(\rho) = \begin{bmatrix} 1 & 0 \\ 0 & R(\rho) \end{bmatrix},$$

where $R(\rho)$ is a 3d rotation matrix, which rotates any 3d vector in one sense (active transformation), or equivalently the coordinate frame in the opposite sense (passive transformation). It is *not* simple to connect w and ρ (or w and ρ) to the original boost parameters u and v. In a composition of boosts, the R matrix is named the Wigner rotation, and gives rise to the Thomas precession. These articles give the explicit formulae for the composite transformation matrices, including expressions for w, ρ, $\overline{w}$, $\overline{\rho}$.

In this article the axis-angle representation is used for ρ. The rotation is about an axis in the direction of a unit vector e, through angle θ (positive anticlockwise, negative clockwise, according to the right-hand rule). The "axis-angle vector"

$$\theta = \theta e$$

will serve as a useful abbreviation.

Spatial rotations alone are also Lorentz transformations they leave the spacetime interval invariant. Like boosts, successive rotations about different axes do not commute. Unlike boosts, the composition of any two rotations is equivalent to a single rotation. Some other similarities and differences between the boost and rotation matrices include:

- inverses: $B(v)^{-1} = B(-v)$ (relative motion in the opposite direction), and $R(\theta)^{-1} = R(-\theta)$ (rotation in the opposite sense about the same axis)

- identity transformation for no relative motion/rotation: $B(0) = R(0) = I$

- unit determinant: $\det(B) = \det(R) = +1$. This property makes them proper transformations.

- matrix symmetry: B is symmetric (equals transpose), while R is nonsymmetric but orthogonal (transpose equals inverse, $R^T = R^{-1}$).

The most general proper Lorentz transformation $\Lambda(v, \theta)$ includes a boost and rotation together, and is a nonsymmetric matrix. As special cases, $\Lambda(0, \theta) = R(\theta)$ and $\Lambda(v, 0) = B(v)$. An explicit form of the general Lorentz transformation is cumbersome to write down and will not be given here. Nevertheless, closed form expressions for the transformation matrices will be given below using group theoretical arguments. It will be easier to use the rapidity parametrization for boosts, in which case one writes $\Lambda(\zeta, \theta)$ and $B(\zeta)$.

The Lie Group SO⁺(3,1)

The set of transformations

$$\{B(\zeta), R(\theta), \Lambda(\zeta, \theta)\}$$

with matrix multiplication as the operation of composition forms a group, called the "restricted Lorentz group", and is the special indefinite orthogonal group SO⁺(3,1). (The plus sign indicates positive unit determinant).

For simplicity, look at the infinitesimal Lorentz boost in the x direction (examining a boost in any other direction, or rotation about any axis, follows an identical procedure). The infinitesimal boost is a small boost away from the identity, obtained by the Taylor expansion of the boost matrix to first order about $\zeta = 0$,

$$B_x = I + \zeta \left. \frac{\partial B_x}{\partial \zeta} \right|_{\zeta=0} + \cdots$$

where the higher order terms not shown are negligible because ζ is small, and B_x is simply the boost matrix in the x direction. The derivative of the matrix is the matrix of derivatives (of the entries, with respect to the same variable), and it is understood the derivatives are found first then evaluated at $\zeta = 0$,

$$\left. \frac{\partial B_x}{\partial \zeta} \right|_{\zeta=0} = -K_x.$$

For now, K_x is defined by this result (its significance will be explained shortly). In the limit of an infinite number of infinitely small steps, the finite boost transformation in the form of a matrix exponential is obtained

$$B_x = \lim_{N \to \infty} \left(I - \frac{\zeta}{N} K_x \right)^N = e^{-\zeta K_x}$$

where the limit definition of the exponential has been used. More generally

$$B(\zeta) = e^{-\zeta \cdot K}, \quad R(\theta) = e^{\theta \cdot J}.$$

The axis-angle vector θ and rapidity vector ζ are altogether six continuous variables which make up the group parameters (in this particular representation), and the generators of the group are K = (K_x, K_y, K_z) and J = (J_x, J_y, J_z), each vectors of matrices with the explicit forms

$$K_x = \begin{bmatrix} 0 & 1 & 0 & 0 \\ 1 & 0 & 0 & 0 \\ 0 & 0 & 0 & 0 \\ 0 & 0 & 0 & 0 \end{bmatrix}, \quad K_y = \begin{bmatrix} 0 & 0 & 1 & 0 \\ 0 & 0 & 0 & 0 \\ 1 & 0 & 0 & 0 \\ 0 & 0 & 0 & 0 \end{bmatrix}, \quad K_z = \begin{bmatrix} 0 & 0 & 0 & 1 \\ 0 & 0 & 0 & 0 \\ 0 & 0 & 0 & 0 \\ 1 & 0 & 0 & 0 \end{bmatrix}$$

$$J_x = \begin{bmatrix} 0 & 0 & 0 & 0 \\ 0 & 0 & 0 & 0 \\ 0 & 0 & 0 & -1 \\ 0 & 0 & 1 & 0 \end{bmatrix}, \quad J_y = \begin{bmatrix} 0 & 0 & 0 & 0 \\ 0 & 0 & 0 & 1 \\ 0 & 0 & 0 & 0 \\ 0 & -1 & 0 & 0 \end{bmatrix}, \quad J_z = \begin{bmatrix} 0 & 0 & 0 & 0 \\ 0 & 0 & -1 & 0 \\ 0 & 1 & 0 & 0 \\ 0 & 0 & 0 & 0 \end{bmatrix}$$

These are all defined in an analogous way to K_x above, although the minus signs in the boost generators are conventional. Physically, the generators of the Lorentz group correspond to important symmetries in spacetime: J are the *rotation generators* which correspond to angular momentum, and K are the *boost generators* which correspond to the motion of the system in spacetime. The derivative of any smooth curve $C(t)$ with $C(0) = I$ in the group depending on some group parameter t with respect to that group parameter, evaluated at $t = 0$, serves as a definition of a corresponding group generator G, and this reflects an infinitesimal transformation away from the identity. The smooth curve can always be taken as an exponential as the exponential will always map G smoothly back into the group via $t \to \exp(tG)$ for all t; this curve will yield G again when differentiated at $t = 0$.

Expanding the exponentials in their Taylor series obtains

$$B(\zeta) = I - \sinh \zeta \, (n \cdot K) + (\cosh \zeta - 1)(n \cdot K)^2$$

$$R(\theta) = I + \sin \theta (e \cdot J) + (1 - \cos \theta)(e \cdot J)^2.$$

which compactly reproduce the boost and rotation matrices as given in the previous section.

It has been stated that the general proper Lorentz transformation is a product of a boost and rotation. At the *infinitesimal* level the product

$$\begin{aligned} \Lambda &= (I - \zeta \cdot K + \cdots)(I + \theta \cdot J + \cdots) \\ &= (I + \theta \cdot J + \cdots)(I - \zeta \cdot K + \cdots) \\ &= I - \zeta \cdot K + \theta \cdot J + \cdots \end{aligned}$$

is commutative because only linear terms are required (products like $(\theta \cdot J)(\zeta \cdot K)$ and $(\zeta \cdot K)(\theta \cdot J)$ count as higher order terms and are negligible). Taking the limit as before leads to the finite transformation in the form of an exponential

$$\Lambda(\zeta, \theta) = e^{-\zeta \cdot k + \theta \cdot j}.$$

The converse is also true, but the decomposition of a finite general Lorentz transformation into such factors is nontrivial. In particular,

$$e^{-\zeta \cdot k + \theta \cdot j} \neq e^{-\zeta \cdot k} e^{\theta \cdot J}$$

because the generators do not commute. For a description of how to find the factors of a general Lorentz transformation in terms of a boost and a rotation *in principle* (this usually does not yield an intelligible expression in terms of generators J and K). If, on the other hand, *the decomposition is given* in terms of the generators, and one wants to find the product in terms of the generators, then the Baker–Campbell–Hausdorff formula applies.

The Lie Algebra so(3,1)

Lorentz generators can be added together, or multiplied by real numbers, to obtain more Lorentz generators. In other words the set of all Lorentz generators

$$V = \{\zeta \cdot K + \theta \cdot J\}$$

together with the operations of ordinary matrix addition and multiplication of a matrix by a number, forms a vector space over the real numbers. The generators $J_x, J_y, J_z, K_x, K_y, K_z$ form a basis set of V, and the components of the axis-angle and rapidity vectors, $\theta_x, \theta_y, \theta_z, \zeta_x, \zeta_y, \zeta_z$, are the coordinates of a Lorentz generator with respect to this basis.

Three of the commutation relations of the Lorentz generators are

$$[J_x, J_y] = J_z, \quad [K_x, K_y] = -J_z, \quad [J_x, K_y] = K_z,$$

where the bracket $[A, B] = AB - BA$ is known as the *commutator*, and the other relations can be found by taking cyclic permutations of x, y, z components (i.e. change x to y, y to z, and z to x, repeat).

These commutation relations, and the vector space of generators, fulfill the definition of the Lie algebra $\mathfrak{so}(3,1)$. In summary, a Lie algebra is defined as a vector space V over a field of numbers, and with a binary operation [,] (called a Lie bracket in this context) on the elements of the vector space, satisfying the axioms of bilinearity, alternatization, and the Jacobi identity. Here the operation [,] is the commutator which satisfies all of these axioms, the vector space is the set of Lorentz generators V as given previously, and the field is the set of real numbers.

Linking terminology used in mathematics and physics: A group generator is any element of the Lie algebra. A group parameter is a component of a coordinate vector representing an arbitrary element of the Lie algebra with respect to some basis. A basis, then, is a set of generators being a basis of the Lie algebra in the usual vector space sense.

The exponential map (Lie theory) from the Lie algebra to the Lie group,

$$\exp:\mathfrak{so}(3,1) \to SO(3,1),$$

provides a one-to-one correspondence between small enough neighborhoods of the origin of the Lie algebra and neighborhoods of the identity element of the Lie group. It the case of the Lorentz group, the exponential map is just the matrix exponential. Globally, the exponential map is not one-to-one, but in the case of the Lorentz group, it is surjective (onto). Hence any group element can be expressed as an exponential of an element of the Lie algebra.

Improper Transformations

Lorentz transformations also include parity inversion

$$P = \begin{bmatrix} 1 & 0 \\ 0 & -\mathbf{I} \end{bmatrix}$$

which negates all the spatial coordinates only, and time reversal

$$T = \begin{bmatrix} -1 & 0 \\ 0 & \mathbf{I} \end{bmatrix}$$

which negates the time coordinate only, because these transformations leave the spacetime interval invariant. Here I is the 3d identity matrix. These are both symmetric, they are their own inverses, and each have determinant −1. This latter property makes them improper transformations.

If Λ is a proper orthochronous Lorentz transformation, then $T\Lambda$ is improper antichronous, $P\Lambda$ is improper orthochronous, and $TP\Lambda = PT\Lambda$ is proper antichronous.

Inhomogeneous Lorentz Group

Two other spacetime symmetries have not been accounted for. For the spacetime interval to be invariant, it can be shown that it is necessary and sufficient for the coordinate transformation to be of the form

$$X' = \Lambda X + C$$

where C is a constant column containing translations in time and space. If $C \neq 0$, this is an inhomogenous Lorentz transformation or Poincaré transformation. If $C = 0$, this is a homogeneous Lorentz transformation. Poincaré transformations are not dealt further in this article.

Tensor Formulation

Contravariant Vectors

Writing the general matrix transformation of coordinates as the matrix equation

$$\begin{bmatrix} x'^0 \\ x'^1 \\ x'^2 \\ x'^3 \end{bmatrix} = \begin{bmatrix} \Lambda^0{}_0 & \Lambda^0{}_1 & \Lambda^0{}_2 & \Lambda^0{}_3 \\ \Lambda^1{}_0 & \Lambda^1{}_1 & \Lambda^1{}_2 & \Lambda^1{}_3 \\ \Lambda^2{}_0 & \Lambda^2{}_1 & \Lambda^2{}_2 & \Lambda^2{}_3 \\ \Lambda^3{}_0 & \Lambda^3{}_1 & \Lambda^3{}_2 & \Lambda^3{}_3 \end{bmatrix} \begin{bmatrix} x^0 \\ x^1 \\ x^2 \\ x^3 \end{bmatrix}$$

allows the transformation of other physical quantities that cannot be expressed as four-vectors; e.g., tensors or spinors of any order in 4d spacetime, to be defined. In the corresponding tensor index notation, the above matrix expression is

$$x'^\nu = \Lambda^\nu{}_\mu x^\mu,$$

where lower and upper indices label covariant and contravariant components respectively, and the summation convention is applied. It is a standard convention to use Greek indices that take the value 0 for time components, and 1, 2, 3 for space components, while Latin indices simply take the values 1, 2, 3, for spatial components. Note that the first index (reading left to right) corresponds in the matrix notation to a *row index*. The second index corresponds to the column index.

The transformation matrix is universal for all four-vectors, not just 4-dimensional spacetime coordinates. If A is any four-vector, then in tensor index notation

$$A'^{\nu} = \Lambda^{\nu}{}_{\mu} A^{\mu}.$$

Alternatively, one writes

$$A^{\nu'} = \Lambda^{\nu'}{}_{\mu} A^{\mu}.$$

in which the primed indices denote the indices of A in the primed frame. This notation cuts risk of exhausting the Greek alphabet roughly in half.

For a general n-component object one may write

$$X'^{\alpha} = \Pi(\Lambda)^{\alpha}{}_{\beta} X^{\beta},$$

where Π is the appropriate representation of the Lorentz group, an $n \times n$ matrix for every Λ. In this case, the indices should *not* be thought of as spacetime indices (sometimes called Lorentz indices), and they run from 1 to n. E.g., if X is a bispinor, then the indices are called *Dirac indices*.

Covariant Vectors

There are also vector quantities with covariant indices. They are generally obtained from their corresponding objects with contravariant indices by the operation of *lowering an index*; e.g.,

$$x_{\nu} = \eta_{\mu\nu} x^{\mu},$$

where η is the metric tensor. The inverse of this transformation is given by

$$x^{\mu} = \eta^{\nu\mu} x_{\nu},$$

where, when viewed as matrices, $\eta^{\mu\nu}$ is the inverse of $\eta_{\mu\nu}$. As it happens, $\eta^{\mu\nu} = \eta_{\mu\nu}$. This is referred to as *raising an index*. To transform a covariant vector A_{μ}, first raise its index, then transform it according to the same rule as for contravariant 4-vectors, then finally lower the index;

$$A'_{\nu} = \eta_{\rho\nu} \Lambda^{\rho}{}_{\sigma} \eta^{\mu\sigma} A_{\mu}.$$

But

$$\eta_{\rho\nu} \Lambda^{\rho}{}_{\sigma} \eta^{\mu\sigma} = \left(\Lambda^{-1}\right)^{\mu}{}_{\nu},$$

I. e., it is the (μ, ν)-component of the *inverse* Lorentz transformation. One defines (as a matter of notation),

$$\Lambda_\nu{}^\mu \equiv \left(\Lambda^{-1}\right)^\mu{}_\nu,$$

and may in this notation write

$$A'_\nu = \Lambda_\nu{}^\mu A_\mu.$$

Now for a subtlety. The implied summation on the right hand side of

$$A'_\nu = \Lambda_\nu{}^\mu A_\mu = \left(\Lambda^{-1}\right)^\mu{}_\nu A_\mu$$

is running over *a row index* of the matrix representing Λ^{-1}. Thus, in terms of matrices, this transformation should be thought of as the *inverse transpose* of Λ acting on the column vector A_μ. That is, in pure matrix notation,

$$A' = \left(\Lambda^{-1}\right)^{\mathrm{T}} A.$$

This means exactly that covariant vectors (thought of as column matrices) transform according to the dual representation of the standard representation of the Lorentz group. This notion generalizes to general representations, simply replace Λ with $\Pi(\Lambda)$.

Tensors

If A and B are linear operators on vector spaces U and V, then a linear operator $A \otimes B$ may be defined on the tensor product of U and V, denoted $U \otimes V$ according to

$$(A \otimes B)(u \otimes v) = Au \otimes Bv, \qquad u \in U, v \in V, u \otimes v \in U \otimes V. \tag{T1}$$

From this it is immediately clear that if u and v are a four-vectors in V, then $u \otimes v \in T_2 V \equiv V \otimes V$ transforms as

$$u \otimes v \to \Lambda u \otimes \Lambda v = \Lambda^\mu{}_\nu u^\nu \otimes \Lambda^\rho{}_\sigma v^\sigma = \Lambda^\mu{}_\nu \Lambda^\rho{}_\sigma u^\nu \otimes v^\sigma \equiv \Lambda^\mu{}_\nu \Lambda^\rho{}_\sigma w^{\nu\sigma}. \tag{T2}$$

The second step uses the bilinearity of the tensor product and the last step defines a 2-tensor on component form, or rather, it just renames the tensor $u \otimes v$.

These observations generalize in an obvious way to more factors, and using the fact that a general tensor on a vector space V can be written as a sum of a coefficient (component!) times tensor products of basis vectors and basis covectors, one arrives at the transformation law for any tensor quantity T. It is given by

$$T^{\alpha'\beta'\cdots\zeta'}_{\theta't'\cdots\kappa'} = \Lambda^{\alpha'}{}_\mu \Lambda^{\beta'}{}_\nu \cdots \Lambda^{\zeta'}{}_\rho \Lambda_{\theta'}{}^\sigma \Lambda_{t'}{}^\upsilon \cdots \Lambda_{\kappa'}{}^\zeta T^{\mu\nu\cdots\rho}_{\sigma\upsilon\cdots\zeta}, \tag{T3}$$

where $\Lambda_{x'}{}^\psi$ is defined above. This form can generally be reduced to the form for general n-component objects given above with a single matrix $(\Pi(\Lambda))$ operating on column vectors. This latter form is sometimes preferred; e.g., for the electromagnetic field tensor.

Transformation of the Electromagnetic Field

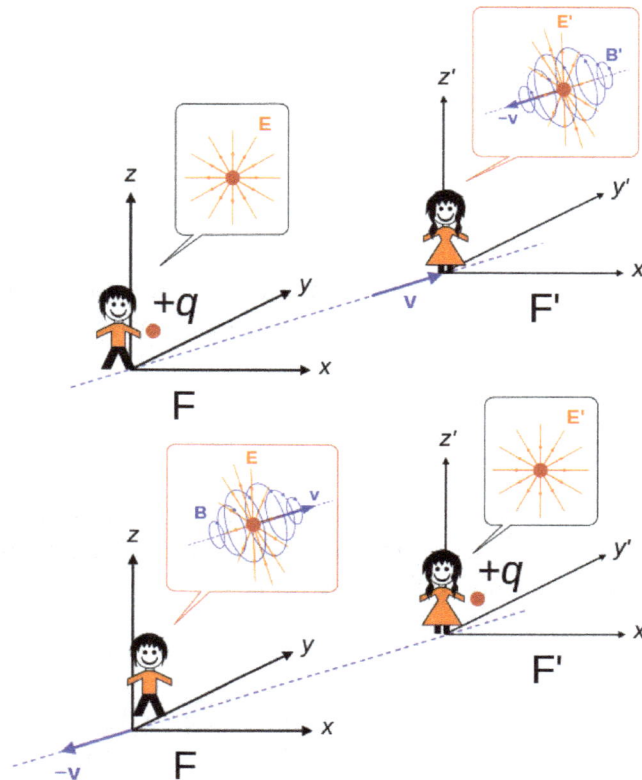

Lorentz boost of an electric charge, the charge is at rest in one frame or the other.

Lorentz transformations can also be used to illustrate that the magnetic field B and electric field E are simply different aspects of the same force — the electromagnetic force, as a consequence of relative motion between electric charges and observers. The fact that the electromagnetic field shows relativistic effects becomes clear by carrying out a simple thought experiment.

- An observer measures a charge at rest in frame F. The observer will detect a static electric field. As the charge is stationary in this frame, there is no electric current, so the observer does not observe any magnetic field.

- The other observer in frame F′ moves at velocity v relative to F and the charge. *This* observer sees a different electric field because the charge moves at velocity −v in their rest frame. The motion of the charge corresponds to an electric current, and thus the observer in frame F′ also sees a magnetic field.

The electric and magnetic fields transform differently from space and time, but exactly the same way as relativistic angular momentum and the boost vector.

The electromagnetic field strength tensor is given by

$$
F^{\mu\nu} = \begin{bmatrix}
0 & -\dfrac{1}{c}E_x & -\dfrac{1}{c}E_y & -\dfrac{1}{c}E_z \\[2mm]
\dfrac{1}{c}E_x & 0 & -B_z & B_y \\[2mm]
\dfrac{1}{c}E_y & B_z & 0 & -B_x \\[2mm]
\dfrac{1}{c}E_z & -B_y & B_x & 0
\end{bmatrix} \quad \text{(SI units, signature } (+,-,-,-)).
$$

in SI units. In relativity, the Gaussian system of units is often preferred over SI units, even in texts whose main choice of units is SI units, because in it the electric field E and the magnetic induction B have the same units making the appearance of the electromagnetic field tensor more natural. Consider a Lorentz boost in the x-direction. It is given by

$$
\Lambda^{\mu}{}_{\nu} = \begin{bmatrix}
\gamma & -\gamma\beta & 0 & 0 \\
-\gamma\beta & \gamma & 0 & 0 \\
0 & 0 & 1 & 0 \\
0 & 0 & 0 & 1
\end{bmatrix}, \qquad
F^{\mu\nu} = \begin{bmatrix}
0 & E_x & E_y & E_z \\
-E_x & 0 & B_z & -B_y \\
-E_y & -B_z & 0 & B_x \\
-E_z & B_y & -B_x & 0
\end{bmatrix} \quad \text{(Gaussian units, signature } (-,+,+,+)),
$$

where the field tensor is displayed side by side for easiest possible reference in the manipulations below.

The general transformation law (T3) becomes

$$
F^{\mu'\nu'} = \Lambda^{\mu'}{}_{\mu}\Lambda^{\nu'}{}_{\nu}F^{\mu\nu}.
$$

For the magnetic field one obtains

$$
\begin{aligned}
B_{x'} &= F^{2'3'} = \Lambda^2{}_{\mu}\Lambda^3{}_{\nu}F^{\mu\nu} = \Lambda^2{}_2\Lambda^3{}_3 F^{23} = 1\times 1\times B_x \\
&= B_x,
\end{aligned}
$$

$$
\begin{aligned}
B_{y'} &= F^{3'1'} = \Lambda^3{}_{\mu}\Lambda^1{}_{\nu}F^{\mu\nu} = \Lambda^3{}_3\Lambda^1{}_{\nu}F^{3\nu} = \Lambda^3{}_3\Lambda^1{}_0 F^{30} + \Lambda^3{}_3\Lambda^1{}_1 F^{31} \\
&= 1\times(-\beta\gamma)(-E_z) + 1\times\gamma B_y = \gamma B_y + \beta\gamma E_z \\
&= \left(\mathbf{B}-\beta\times\mathbf{E}\right)_y
\end{aligned}
$$

$$
\begin{aligned}
B_{z'} &= F^{1'2'} = \Lambda^1{}_{\mu}\Lambda^2{}_{\nu}F^{\mu\nu} = \Lambda^1{}_{\mu}\Lambda^2{}_2 F^{\mu 2} = \Lambda^1{}_0\Lambda^2{}_2 F^{02} + \Lambda^1{}_1\Lambda^2{}_2 F^{12} \\
&= (-\gamma\beta)\times 1\times E_y + \gamma\times 1\times B_z = \gamma B_z - \beta\gamma E \\
&= \left(\mathbf{B}-\beta\times\mathbf{E}\right)_z
\end{aligned}
$$

For the electric field results

$$E_{x'} = F^{0'1'} = \Lambda^0{}_\mu \Lambda^1{}_\nu F^{\mu\nu} = \Lambda^0{}_1 \Lambda^1{}_0 F^{10} + \Lambda^0{}_0 \Lambda^1{}_1 F^{01} = (-\gamma\beta)(-\gamma\beta)(-E_x) + \gamma\gamma E_x = -\gamma^2\beta^2(E_x) + \gamma^2 E_x = E_x(1-\beta^2)\gamma^2$$

$$= E_x,$$

$$E_{y'} = F^{0'2'} = \Lambda^0{}_\mu \Lambda^2{}_\nu F^{\mu\nu} = \Lambda^0{}_\mu \Lambda^2{}_2 F^{\mu2} = \Lambda^0{}_0 \Lambda^2{}_2 F^{02} + \Lambda^0{}_1 \Lambda^2{}_2 F^{12} = \gamma \times 1 \times E_y + (-\beta\gamma) \times 1 \times B_z = \gamma E_y - \beta\gamma B_z$$

$$= \gamma(\mathbf{E} + \boldsymbol{\beta} \times \mathbf{B})_y$$

$$E_{z'} = F^{0'3'} = \Lambda^0{}_\mu \Lambda^3{}_\nu F^{\mu\nu} = \Lambda^0{}_\mu \Lambda^3{}_3 F^{\mu3} = \Lambda^0{}_0 \Lambda^3{}_3 F^{03} + \Lambda^0{}_1 \Lambda^3{}_3 F^{13}$$

$$= \gamma \times 1 \times E_z - \beta\gamma \times 1 \times (-B_y) = \gamma E_z + \beta\gamma B_y$$

$$= \gamma(\mathbf{E} + \boldsymbol{\beta} \times \mathbf{B})_z.$$

Here, $\boldsymbol{\beta} = (\beta, 0, 0)$ is used. These results can be summarized by

$$\mathbf{E}_{\parallel'} = \mathbf{E}_\parallel$$
$$\mathbf{B}_{\parallel'} = \mathbf{B}_\parallel$$
$$\mathbf{E}_{\perp'} = \gamma\left(\mathbf{E}_\perp + \boldsymbol{\beta} \times \mathbf{B}_\perp\right) = \gamma\left(\mathbf{E} + \boldsymbol{\beta} \times \mathbf{B}\right)_\perp,$$
$$\mathbf{B}_{\perp'} = \gamma\left(\mathbf{B}_\perp - \boldsymbol{\beta} \times \mathbf{E}_\perp\right) = \gamma\left(\mathbf{B} - \boldsymbol{\beta} \times \mathbf{E}\right)_\perp,$$

and are independent of the metric signature. For SI units, substitute $E \rightarrow E/c$. Misner, Thorne & Wheeler (1973) refer to this last form as the 3 + 1 view as opposed to the *geometric view* represented by the tensor expression

$$F^{\mu'\nu'} = \Lambda^{\mu'}{}_\mu \Lambda^{\nu'}{}_\nu F^{\mu\nu},$$

and make a strong point of the ease with which results that are difficult to achieve using the 3 + 1 view can be obtained and understood. Only objects that have well defined Lorentz transformation properties (in fact under *any* smooth coordinate transformation) are geometric objects. In the geometric view, the electromagnetic field is a six-dimensional geometric object in *spacetime* as opposed to two interdependent, but separate, 3-vector fields in *space* and *time*. The fields E (alone) and B (alone) do not have well defined Lorentz transformation properties. The mathematical underpinnings are equations (T1) and (T2) that immediately yield (T3). One should note that the primed and unprimed tensors refer to the *same event in spacetime*. Thus the complete equation with spacetime dependence is

$$F^{\mu'\nu'}\left(x'\right) = \Lambda^{\mu'}{}_\mu \Lambda^{\nu'}{}_\nu F^{\mu\nu}\left(\Lambda^{-1}x'\right) = \Lambda^{\mu'}{}_\mu \Lambda^{\nu'}{}_\nu F^{\mu\nu}(x).$$

Length contraction has an effect on charge density ρ and current density J, and time dilation has an effect on the rate of flow of charge (current), so charge and current distributions must transform in a related way under a boost. It turns out they transform exactly like the space-time and energy-momentum four-vectors,

$$\mathbf{j}' = \mathbf{j} - \gamma\rho v\mathbf{n} + (\gamma - 1)(\mathbf{j} \cdot \mathbf{n})\mathbf{n}$$
$$\rho' = \gamma\left(\rho - \mathbf{j} \cdot \frac{v\mathbf{n}}{c^2}\right),$$

or, in the simpler geometric view,

$$j^{\mu'} = \Lambda^{\mu'}{}_{\mu} j^{\mu}.$$

One says that charge density transforms as the time component of a four-vector. It is a rotational scalar. The current density is a 3-vector.

The Maxwell equations are invariant under Lorentz transformations.

Spinors

Equation (T1) hold unmodified for any representation of the Lorentz group, including the bispinor representation. In (T2) one simply replaces all occurrences of Λ by the bispinor representation $\Pi(\Lambda)$,

$$u \otimes v \rightarrow \Pi(\Lambda)u \otimes \Pi(\Lambda)v = \Pi(\Lambda)^{\alpha}{}_{\beta} u^{\beta} \otimes \Pi(\Lambda)^{\rho}{}_{\sigma} v^{\sigma} = \Pi(\Lambda)^{\alpha}{}_{\beta} \Pi(\Lambda)^{\rho}{}_{\sigma} u^{\beta} \otimes v^{\sigma} \equiv \Pi(\Lambda)^{\alpha}{}_{\beta} \Pi(\Lambda)^{\rho}{}_{\sigma} w^{\alpha\beta}. \qquad \textbf{(T4)}$$

The above equation could, for instance, be the transformation of a state in Fock space describing two free electrons.

Transformation of General Fields

A general *noninteracting* multi-particle state (Fock space state) in quantum field theory transforms according to the rule

$$U(\Lambda, a)\Psi_{p_1\sigma_1 n_1; p_2\sigma_2 n_2; \cdots}$$

$$= e^{-ia_{\mu}\left[(\Lambda p_1)^{\mu} + (\Lambda p_2)^{\mu} + \cdots\right]} \sqrt{\frac{(\Lambda p_1)^0 (\Lambda p_2)^0 \cdots}{p_1^0 p_2^0 \cdots}} \left(\sum_{\sigma_1'\sigma_2'\cdots} D^{(j_1)}_{\sigma_1'\sigma_1}[W(\Lambda, p_1)] D^{(j_2)}_{\sigma_2'\sigma_2}[W(\Lambda, p_2)] \cdots \right) \Psi_{\Lambda p_1\sigma_1' n_1; \Lambda p_2\sigma_2' n_2; \cdots}, \qquad (1)$$

where $W(\Lambda, p)$ is the Wigner rotation and $D^{(j)}$ is the $(2j + 1)$-dimensional representation of SO(3).

Ladder Paradox

The ladder paradox (or barn-pole paradox) is a thought experiment in special relativity. It involves a ladder, parallel to the ground, travelling horizontally and therefore undergoing a Lorentz length contraction. As a result, the ladder fits inside a garage which would normally be too small to contain it. On the other hand, from the point of view of an observer moving with the ladder, it is the garage that is moving, so it is the garage which will be contracted to an even smaller size, thus being unable to contain the ladder. This apparent paradox results from the mistaken assumption of absolute simultaneity. The ladder fits into the garage only if both of its ends are simultaneously inside the garage. In relativity, simultaneity is relative to each observer, and so the question of whether the ladder fits inside the garage is relative to each observer, and the paradox is resolved.

Paradox

The simplest version of the problem involves a garage, with a front and back door which are open, and a ladder which, when at rest with respect to the garage, is too long to fit inside. We now move the ladder at a high horizontal velocity through the stationary garage. Because of its high velocity, the ladder undergoes the relativistic effect of length contraction, and becomes significantly shorter. As a result, as the ladder passes through the garage, it is, for a time, completely contained inside it. We could, if we liked, simultaneously close both doors for a brief time, to demonstrate that the ladder fits.

So far, this is consistent. The apparent paradox comes when we consider the symmetry of the situation. As an observer moving with the ladder is travelling at constant velocity in the inertial reference frame of the garage, this observer also occupies an inertial frame, where, by the principle of relativity, the same laws of physics apply. From this perspective, it is the ladder which is now stationary, and the garage which is moving with high velocity. It is therefore the garage which is length contracted, and we now conclude that it is far too small to have ever fully contained the ladder as it passed through: the ladder does not fit, and we can't close both doors on either side of the ladder without hitting it. This apparent contradiction is the paradox.

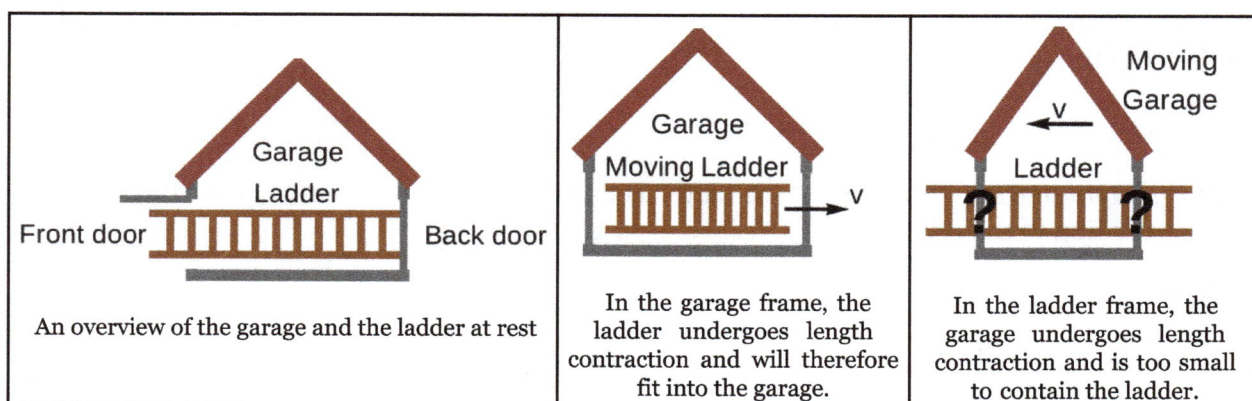

| An overview of the garage and the ladder at rest | In the garage frame, the ladder undergoes length contraction and will therefore fit into the garage. | In the ladder frame, the garage undergoes length contraction and is too small to contain the ladder. |

Resolution

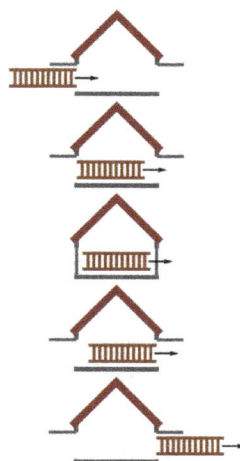

Scenario in the garage frame: a length contracted ladder passing through the garage

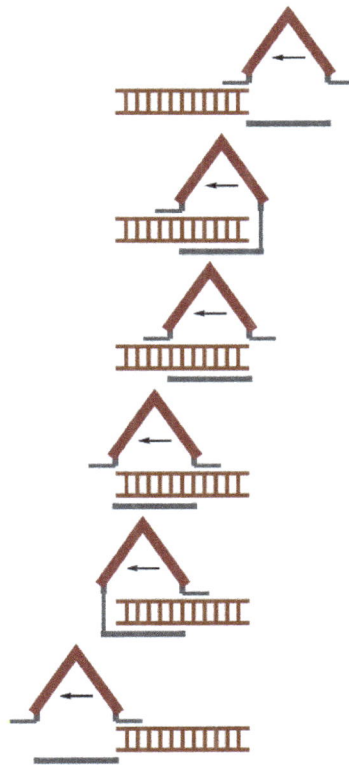

Scenario in the ladder frame: a length contracted garage passing over the ladder

The solution to the apparent paradox lies in the relativity of simultaneity: what one observer (e.g. with the garage) considers to be two simultaneous events may not in fact be simultaneous to another observer (e.g. with the ladder). When we say the ladder "fits" inside the garage, what we mean precisely is that, at some specific time, the position of the back of the ladder and the position of the front of the ladder were both inside the garage; in other words, the front and back of the ladder were inside the garage simultaneously. As simultaneity is relative, then, two observers can disagree without contradiction on whether the ladder fits. To the observer with the garage, the back end of the ladder was in the garage at the same time that the front end of the ladder was, and so the ladder fit; but to the observer with the ladder, these two events were not simultaneous, and the ladder did not fit.

A clear way of seeing this is to consider the doors, which, in the frame of the garage, close for the brief period that the ladder is fully inside. We now look at these events in the frame of the ladder. The first event is the front of the ladder approaching the exit door of the garage. The door closes, and then opens again to let the front of the ladder pass through. At a later time, the back of the ladder passes through the entrance door, which closes and then opens. We see that, as simultaneity is relative, the two doors did not need to be shut at the same time, and the ladder did not need to fit inside the garage.

The situation can be further illustrated by the Minkowski diagram below. The diagram is in the rest frame of the garage. The vertical light-blue band shows the garage in space-time, and the light-red band shows the ladder in space-time. The x and t axes are the garage space and time axes, respectively, and x′ and t′ are the ladder space and time axes, respectively.

In the frame of the garage, the ladder at any specific time is represented by a horizontal set of points, parallel to the x axis, in the red band. One example is the bold blue line segment, which lies inside the blue band representing the garage, and which represents the ladder at a time when it is fully inside the garage. In the frame of the ladder, however, sets of simultaneous events lie on lines parallel to the x' axis; the ladder at any specific time is therefore represented by a cross section of such a line with the red band. One such example is the bold red line segment. We see that such line segments never lie fully inside the blue band; that is, the ladder never lies fully inside the garage.

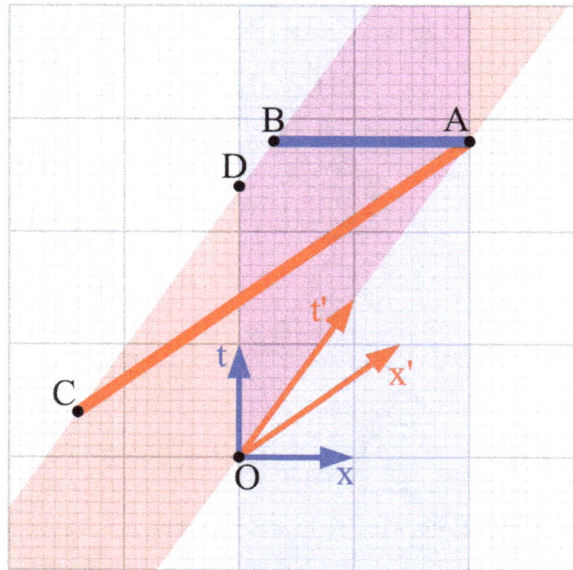

A Minkowski diagram of ladder paradox. The garage is shown in light blue, the ladder in light red. The diagram is in the rest frame of the garage, with x and t being the garage space and time axes, respectively. The ladder frame is for a person sitting on the front of the ladder, with x' and t' being the ladder space and time axes respectively.

Shutting the Ladder in the Garage

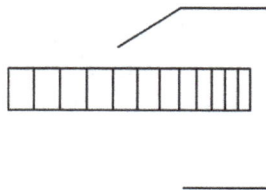

A ladder contracting under acceleration to fit into a length contracted garage

In a more complicated version of the paradox, we can physically trap the ladder once it is fully inside the garage. This could be done, for instance, by not opening the exit door again after we close it. In the frame of the garage, we assume the exit door is immovable, and so when the ladder hits it, we say that it instantaneously stops. By this time, the entrance door has also closed, and so the ladder is stuck inside the garage. As its relative velocity is now zero, it is not length contracted, and is now longer than the garage; it will have to bend, snap, or explode.

Again, the puzzle comes from considering the situation from the frame of the ladder. In the above analysis, in its own frame, the ladder was always longer than the garage. So how did we ever close the doors and trap it inside?

It is worth noting here a general feature of relativity: we have deduced, by considering the frame of the garage, that we do indeed trap the ladder inside the garage. This must therefore be true in any frame - it cannot be the case that the ladder snaps in one frame but not in another. From the ladder's frame, then, we know that there must be some explanation for how the ladder came to be trapped; we must simply find the explanation.

The explanation is that, although all parts of the ladder simultaneously decelerate to zero in the garage's frame, because simultaneity is relative, the corresponding decelerations in the frame of the ladder are not simultaneous. Instead, each part of the ladder decelerates sequentially, from front to back, until finally the back of the ladder decelerates, by which time it is already within the garage.

As length contraction and time dilation are both controlled by the Lorentz transformations, the ladder paradox can be seen as a physical correlate of the twin paradox, in which instance one of a set of twins leaves earth, travels at speed for a period, and returns to earth a bit younger than the earthbound twin. As in the case of the ladder trapped inside the barn, if neither frame of reference is privileged — each is moving only relative to the other — how can it be that it's the traveling twin and not the stationary one who is younger (just as it's the ladder rather than the barn which is shorter)? In both instances it is the acceleration-deceleration that differentiates the phenomena: it's the twin, not the earth (or the ladder, not the barn) that undergoes the force of deceleration in returning to the temporal (or physical, in the case of the ladder-barn) inertial frame.

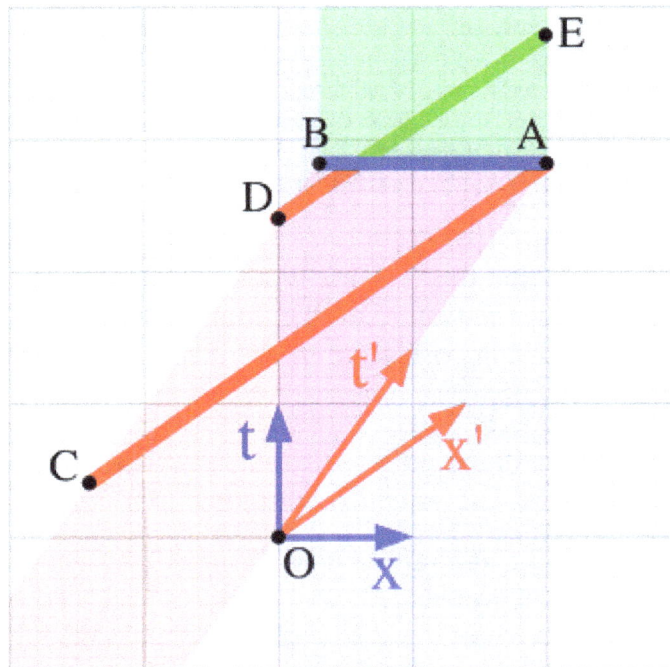

A Minkowski diagram of the case where the ladder is stopped all along its length, simultaneously in the garage frame. When this occurs, the garage frame sees the ladder as AB, but the ladder frame sees the ladder as AC. When the back of the ladder enters the garage at point D, it has not yet felt the effects of the acceleration of its front end. At this time, according to someone at rest with respect to the back of the ladder, the front of the ladder will be at point E and will see the ladder as DE. It is seen that this length in the ladder frame is not the same as CA, the rest length of the ladder before the deceleration.

Ladder Paradox and Transmission of Force

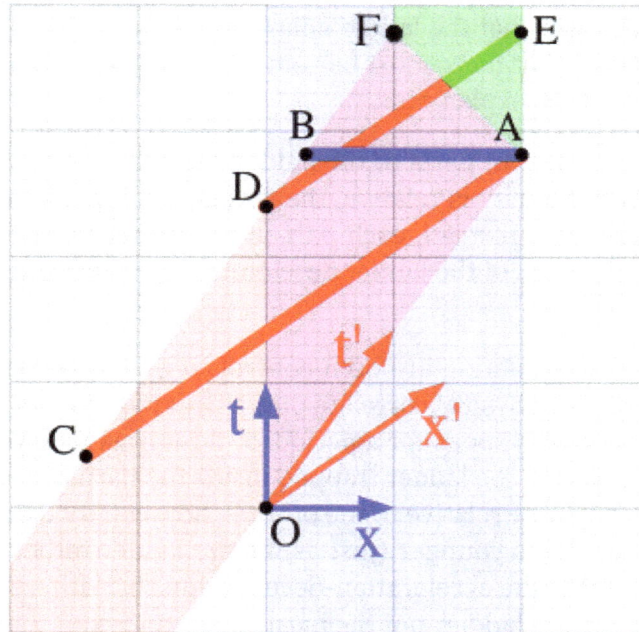

A Minkowski diagram of the case where the ladder is stopped by impact with the back wall of the garage. The impact is event A. At impact, the garage frame sees the ladder as AB, but the ladder frame sees the ladder as AC. The ladder does not move out of the garage, so its front end now goes directly upward, through point E. The back of the ladder will not change its trajectory in space-time until it feels the effects of the impact. The effect of the impact can propagate outward from A no faster than the speed of light, so the back of the ladder will never feel the effects of the impact until point F or later, at which time the ladder is well within the garage in both frames. Note that when the diagram is drawn in the frame of the ladder, the speed of light is the same, but the ladder is longer, so it takes more time for the force to reach the back end; this gives enough time for the back of the ladder to move inside the garage.

What if the back door (the door the ladder exits out of) is closed permanently and does not open? Suppose that the door is so solid that the ladder will not penetrate it when it collides, so it must stop. Then, as in the scenario described above, in the frame of reference of the garage, there is a moment when the ladder is completely within the garage (i.e., the back of the ladder is inside the front door), before it collides with the back door and stops. However, from the frame of reference of the ladder, the ladder is too big to fit in the garage, so by the time it collides with the back door and stops, the back of the ladder still has not reached the front door. This seems to be a paradox. The question is, does the back of the ladder cross the front door or not?

The difficulty arises mostly from the assumption that the ladder is rigid (i.e., maintains the same shape). Ladders seem pretty rigid in everyday life. But being rigid requires that it can transfer force at infinite speed (i.e., when you push one end the other end must react immediately, otherwise the ladder will deform). This contradicts special relativity, which states that information can only travel at most the speed of light (which is too fast for us to notice in real life, but is significant in the ladder scenario). So objects cannot be perfectly rigid under special relativity.

In this case, by the time the front of the ladder collides with the back door, the back of the ladder does not know it yet, so it keeps moving forwards (and the ladder "compresses"). In both the frame of the garage and the inertial frame of the ladder, the back end keeps moving at the time of the collision, until at least the point where the back of the ladder comes into the light cone of the collision (i.e., a point where force moving backwards at the speed of light from the point of the collision will reach it). At this point the ladder is actually shorter than the original contracted length, so the back end is well inside the garage. Calculations in both frames of reference will show this to be the case.

What happens after the force reaches the back of the ladder (the "green" zone in the diagram) is not specified. Depending on the physics, the ladder could break into a million pieces; or, if it were sufficiently elastic, it could re-expand to its original length and push the back end out of the garage. Any realistic material would violently explode into a plasma.

Man Falling into Grate Variation

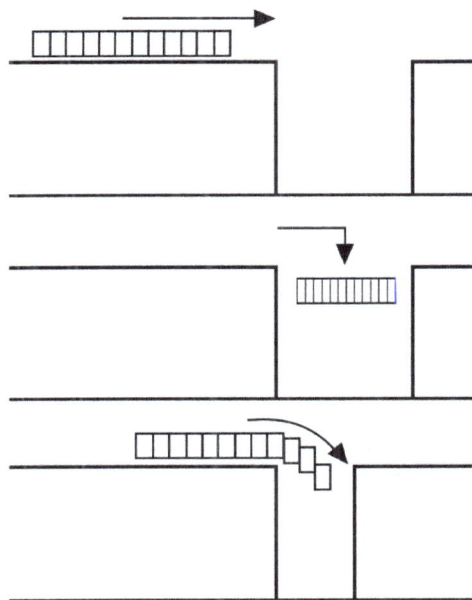

A man (represented by a segmented rod) falling into a grate

This paradox was originally proposed and solved by Wolfgang Rindler and involved a fast walking man, represented by a rod, falling into a grate. It is assumed that the rod is entirely over the grate in the grate frame of reference before the downward acceleration begins simultaneously and equally applied to each point in the rod.

From the perspective of the grate, the rod undergoes a length contraction and fits into the grate. However, from the perspective of the rod, it is the *grate* undergoing a length contraction, through which it seems the rod is then too long to fall.

In fact, the downward acceleration of the rod, which is simultaneous in the grate's frame of reference, is not simultaneous in the rod's frame of reference. In the rod's frame of reference, the bottom of the front of the rod is first accelerated downward, and as time goes by, more and more of the rod is subjected to the downward acceleration, until finally the back of the rod is accelerated downward. This results in a bending of the rod in the rod's frame of reference. It should be stressed that, since this bending occurs in the rod's rest frame, it is a true physical distortion of the rod which will cause stresses to occur in the rod.

Bar and Ring Paradox

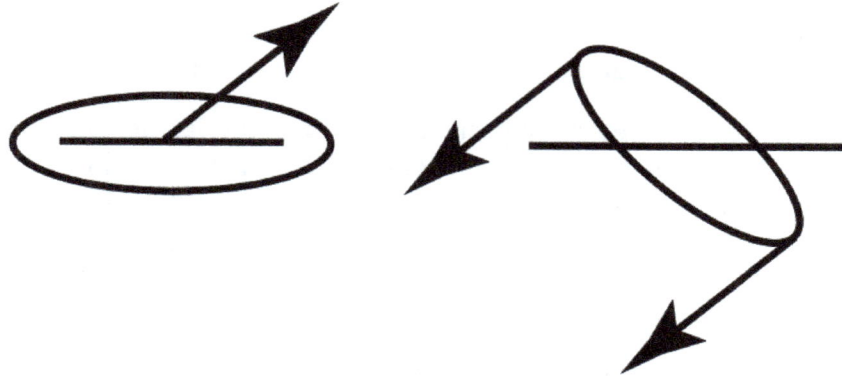

The diagram on the left illustrates a bar and a ring in the rest frame of the ring at the instant that their centers coincide. The bar is Lorentz-contracted and moving upward and to the right while the ring is stationary and uncontracted. The diagram on the right illustrates the situation at the same instant, but in the rest frame of the bar. The ring is now Lorentz-contracted and rotated with respect to the bar, and the bar is uncontracted. Again, the ring passes over the bar without touching it.

The above paradox is complicated: It involves non-inertial frames of reference since at one moment the man is walking horizontally, and a moment later he is falling downward. It involves a physical deformation of the man (or segmented rod), since the rod is bent in one frame of reference and straight in another. These aspects of the problem introduce complications involving the stiffness of the bar which tends to obscure the real nature of the "paradox". A very similar but simpler problem involving only inertial frames is the "bar and ring" paradox (Ferraro 2007) in which a bar which is slightly larger in length than the diameter of a ring is moving upward and to the right with its long axis horizontal, while the ring is stationary and the plane of the ring is also horizontal. If the motion of the bar is such that the center of the bar coincides with the center of the ring at some point in time, then the bar will be Lorentz-contracted due to the forward component of its motion, and it will pass through the ring. The paradox occurs when the problem is considered in the rest frame of the bar. The ring is now moving downward and to the left, and will be Lorentz-contracted along its horizontal length, while the bar will not be contracted at all. How can the bar pass through the ring?

The resolution of the paradox again lies in the relativity of simultaneity (Ferraro 2007). The length of a physical object is defined as the distance between two *simultaneous* events occurring at each end of the body, and since simultaneity is relative, so is this length. This variability in length is just the Lorentz contraction. Similarly, a physical angle is defined as the angle formed by three *simultaneous* events, and this angle will also be a relative quantity. In the above paradox, although the rod and the plane of the ring are parallel in the rest frame of the ring, they are not parallel in the rest frame of the rod. The uncontracted rod passes through the Lorentz-contracted ring because the plane of the ring is rotated relative to the rod by an amount sufficient to let the rod pass through.

In mathematical terms, a Lorentz transformation can be separated into the product of a spatial rotation and a "proper" Lorentz transformation which involves no spatial rotation. The mathematical resolution of the bar and ring paradox is based on the fact that the product of two proper Lorentz transformations may produce a Lorentz transformation which is not proper, but rather includes a spatial rotation component.

Time Dilation

Time dilation explains why two working clocks will report different times after different accelerations. For example, ISS astronauts return from missions having aged slightly less than had they remained on Earth, and GPS satellites work because they adjust for similar bending of spacetime to coordinate with systems on Earth.

In the theory of relativity, time dilation is a difference of elapsed time between two events as measured by observers either moving relative to each other or differently situated from a gravitational mass or masses.

A clock at rest with respect to one observer may be measured to tick at a different rate when compared to a second observer's clock. This effect arises neither from technical aspects of the clocks nor from the propagation time of signals, but from the nature of spacetime.

Overview

Clocks on the Space Shuttle run slightly slower than reference clocks on Earth, while clocks on GPS and Galileo satellites run slightly faster. Such time dilation has been repeatedly demonstrated, for instance by small disparities in atomic clocks on Earth and in space, even though both clocks work perfectly (it is not a mechanical malfunction). The nature of spacetime is such that time measured along different trajectories is affected by differences in either gravity or velocity – each of which affects time in different ways.

In theory, and to make a clearer example, time dilation could affect planned meetings for astronauts with advanced technologies and greater travel speeds. The astronauts would have to set their clocks to count exactly 80 years, whereas mission control – back on Earth – might need to count 81 years. The astronauts would return to Earth, after their mission, having aged one year less than the people staying on Earth. What is more, the local experience of time passing never actually changes for anyone. In other words, the astronauts on the ship as well as the mission control crew on Earth each feel normal, despite the effects of time dilation (i.e. to the traveling party, those stationary are living "faster"; while to those who stood still, their counterparts in motion live "slower" at any given moment).

With technology limiting the velocities of astronauts, these differences are minuscule: after 6 months on the International Space Station (ISS), the astronaut crew has indeed aged less than those on Earth, but only by about 0.005 seconds (nowhere near the 1 year disparity from the theoretical example). The effects would be greater if the astronauts were traveling nearer to the speed of light (299,792,458 m/s), instead of their actual speed – which is the speed of the orbiting ISS, about 7,700 m/s.

Time dilation is caused by differences in either gravity or relative velocity. In the case of ISS, time is slower due to the velocity in circular orbit; this effect is slightly reduced by the opposing effect of less gravitational potential.

Relative Velocity Time Dilation

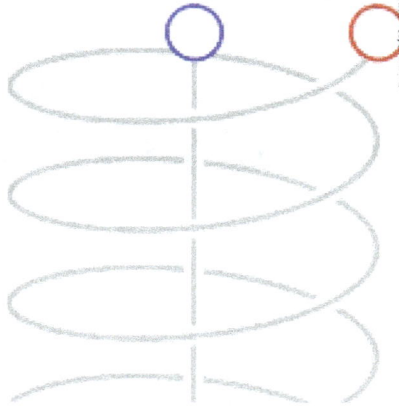

From the local frame of reference (the blue clock), the relatively accelerated red clock moves slower

When two observers are in relative uniform motion and uninfluenced by any gravitational mass, the point of view of each will be that the other's (moving) clock is ticking at a *slower* rate than the local clock. The faster the relative velocity, the greater the magnitude of time dilation. This case is sometimes called special relativistic time dilation.

For instance, two rocket ships (A and B) speeding past one another in space would experience time dilation. If they somehow had a clear view into each other's ships, each crew would see the others' clocks and movement as going more slowly. That is, inside the frame of reference of Ship A, everything is moving normally, but everything over on Ship B appears to be moving more slowly (and vice versa).

From a local perspective, time registered by clocks that are at rest with respect to the local frame of reference (and far from any gravitational mass) always appears to pass at the same rate. In other words, if a new ship, Ship C, travels alongside Ship A, it is "at rest" relative to Ship A. From the point of view of Ship A, new Ship C's time would appear normal too.

A question arises: If Ship A and Ship B both think each other's time is moving slower, who will have aged more if they decided to meet up? With a more sophisticated understanding of relative velocity time dilation, this seeming twin paradox turns out not to be a paradox at all (the resolution of the paradox involves a jump in time, as a result of the accelerated observer turning around). Similarly, understanding the twin paradox would help explain why astronauts on the ISS age slower (e.g. 0.007 seconds behind for every six months) even though they are experiencing relative velocity time dilation.

Gravitational Time Dilation

Time passes more quickly further from a center of gravity, as is witnessed with massive objects (like the Earth)

The key is that both observers are differently situated in their distance from a significant gravitational mass. The general theory of relativity describes how, for both observers, the clock that is closer to the gravitational mass, i.e. deeper in its "gravity well", appears to go more slowly than the clock that is more distant from the mass. In the case of a satellite orbiting a planet, it has the opposite effect of the relative velocity time dilation.

Gravitational time dilation is at play e.g. for ISS astronauts. With respect to ground observers the ISS astronauts's relative velocity slows down their time, whereas the reduced gravitational influence at their location speeds it up. The two opposing effects are not equally strong. At the ISS altitude the net effect is a slowing down of clocks, whereas in much higher orbits clocks run faster than on the ground.

This effect is not restricted to astronauts in space; a climber's time is passing slightly faster at the top of a mountain (a high altitude, farther from the Earth's center of gravity) compared to people at sea level. It has also been calculated that due to time dilation, the core of the Earth is 2.5 years younger than the crust.

As with all time dilation, the local experience of time is normal (nobody notices a difference within their own frame of reference). In the situations of velocity time dilation, both observers saw the other as moving slower (a reciprocal effect). Now, with gravitational time dilation, both observers – those at sea level, versus the climber – agree that the clock nearer the mass is slower in rate, and they agree on the ratio of the difference (time dilation from gravity is therefore not reciprocal). That is, the climber sees the sea level clocks as moving more slowly, and those living at sea level see the climber's clock as moving faster.

Time Dilation: Special vs. General Theories of Relativity

In Albert Einstein's theory of relativity, time dilation in these two circumstances can be summarized:

- In special relativity (or, hypothetically far from all gravitational mass), clocks that are moving with respect to an inertial system of observation are measured to be running more slowly. This effect is described precisely by the Lorentz transformation.

- In general relativity, clocks at a position with lower gravitational potential – such as in closer proximity to a planet – are found to be running more slowly. The articles on gravitational time dilation and gravitational redshift give a more detailed discussion.

Special and general relativistic effects can combine.

In special relativity, the time dilation effect is reciprocal: as observed from the point of view of either of two clocks which are in motion with respect to each other, it will be the other clock that is time dilated. (This presumes that the relative motion of both parties is uniform; that is, they do not accelerate with respect to one another during the course of the observations.) In contrast, gravitational time dilation (as treated in general relativity) is not reciprocal: an observer at the top of a tower will observe that clocks at ground level tick slower, and observers on the ground will agree about the direction and the magnitude of the difference. There is still some disagreement in a sense, because all the observers believe their own local clocks are correct, but the direction and ratio of gravitational time dilation is agreed by all observers, independent of their altitude.

Science Fiction Implications

Science fiction enthusiasts have noted the implications time dilation has on forward time travel, technically making it possible. The Hafele and Keating experiment involved flying planes around the world with atomic clocks on board. Upon the trips' completion the clocks were compared to a static, ground based atomic clock. It was found that 273 ± 7 nanoseconds had been gained on the planes' clocks. The current human time travel record holder is Russian cosmonaut Sergei Krikalev, who beat the previous record of about 20 milliseconds by cosmonaut Sergei Avdeyev.

Simple Inference of Time Dilation Due to Relative Velocity

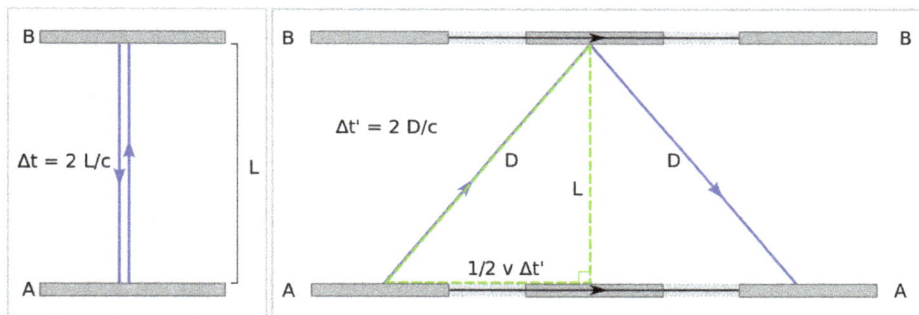

Left: Observer at rest measures time $2L/c$ between co-local events of light signal generation at A and arrival at A.
Right: Events according to an observer moving to the left of the setup: bottom mirror A when signal is generated at time $t'=0$, top mirror B when signal gets reflected at time $t'=D/c$, bottom mirror A when signal returns at time $t'=2D/c$

Time dilation can be inferred from the observed constancy of the speed of light in all reference frames.

This constancy of the speed of light means, counter to intuition, that speeds of material objects and light are not additive. It is not possible to make the speed of light appear greater by approaching at speed towards the material source that is emitting light. It is not possible to make the speed of light appear less by receding from the source at speed. From one point of view, it is the implications of this unexpected constancy that take away from constancies expected elsewhere.

Consider a simple clock consisting of two mirrors A and B, between which a light pulse is bouncing. The separation of the mirrors is L and the clock ticks once each time the light pulse hits a given mirror.

In the frame where the clock is at rest (diagram at left), the light pulse traces out a path of length $2L$ and the period of the clock is $2L$ divided by the speed of light

$$\Delta t = \frac{2L}{c}.$$

From the frame of reference of a moving observer traveling at the speed v relative to the rest frame of the clock (diagram at lower right), the light pulse traces out a *longer*, angled path. The second postulate of special relativity states that the speed of light in free space is constant for all inertial observers, which implies a lengthening of the period of this clock from the moving observer's perspective. That is to say, in a frame moving relative to the clock, the clock appears to be running more slowly. Straightforward application of the Pythagorean theorem leads to the well-known prediction of special relativity:

The total time for the light pulse to trace its path is given by

$$\Delta t' = \frac{2D}{c}.$$

The length of the half path can be calculated as a function of known quantities as

$$D = \sqrt{\left(\frac{1}{2}v\Delta t'\right)^2 + L^2}.$$

Substituting D from this equation into the previous and solving for $\Delta t'$ gives:

$$\Delta t' = \frac{\sqrt{(v\Delta t')^2 + (2L)^2}}{c}$$

$$(\Delta t')^2 = \frac{v^2}{c^2}(\Delta t')^2 + \left(\frac{2L}{c}\right)^2$$

$$(1 - \frac{v^2}{c^2})(\Delta t')^2 = \left(\frac{2L}{c}\right)^2$$

$$(\Delta t')^2 = \frac{\left(\frac{2L}{c}\right)^2}{(1 - \frac{v^2}{c^2})}$$

$$\Delta t' = \frac{\frac{2L}{c}}{\sqrt{1 - \frac{v^2}{c^2}}}$$

and thus, with the definition of Δt:

$$\Delta t' = \frac{\Delta t}{\sqrt{1 - \dfrac{v^2}{c^2}}}$$

which expresses the fact that for the moving observer the period of the clock is longer than in the frame of the clock itself.

Due to Relative Velocity Symmetric between Observers

Minkowski diagram

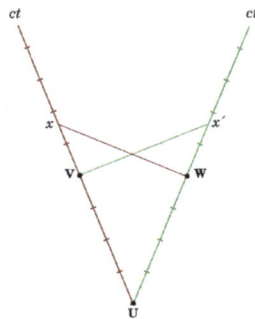

Time UV of a clock in S is shorter compared to Ux′ in S′, and time UW of a clock in S′ is shorter compared to Ux in S.

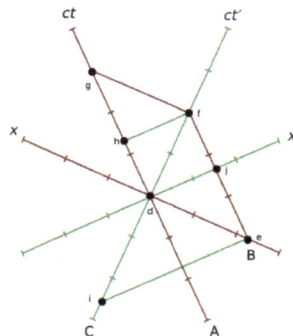

Clock C in relative motion between two synchronized clocks A and B. C meets A at *d*, and B at *f*.

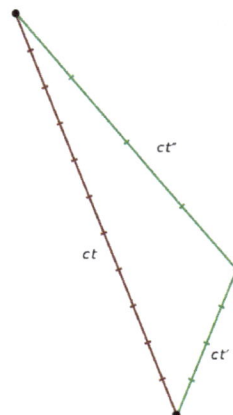

Twin paradox. One twin has to change frames, leading to different proper times in the twin's world lines.

Common sense would dictate that if time passage has slowed for a moving object, the moving object would observe the external world to be correspondingly "sped up". Counterintuitively, special relativity predicts the opposite.

A similar oddity occurs in everyday life. If Sam sees Abigail at a distance, she appears small to him; and at the same time, Sam appears small to Abigail. Being very familiar with the effects of perspective, we see no mystery or a hint of a paradox in this situation.

One is accustomed to the notion of relativity with respect to distance: the distance from Los Angeles to New York is by convention the same as the distance from New York to Los Angeles. On the other hand, when speeds are considered, one thinks of an object as "actually" moving, overlooking that its motion is always relative to something else – to the stars, the ground or to oneself. If one object is moving with respect to another, the latter is moving with respect to the former and with equal relative speed.

In the special theory of relativity, a moving clock is found to be ticking slowly with respect to the observer's clock. If Sam and Abigail are on different trains in near-lightspeed relative motion, Sam measures (by all methods of measurement) clocks on Abigail's train to be running slowly and similarly, Abigail measures clocks on Sam's train to be running slowly.

Note that in all such attempts to establish "synchronization" within the reference system, the question of whether something happening at one location is in fact happening simultaneously with something happening elsewhere, is of key importance. Calculations are ultimately based on determining which events are simultaneous. Furthermore, establishing simultaneity of events separated in space necessarily requires transmission of information between locations, which by itself is an indication that the speed of light will enter the determination of simultaneity.

It is a natural and legitimate question to ask how, in detail, special relativity can be self-consistent if clock C is time-dilated with respect to clock B and clock B is also time-dilated with respect to clock C. It is by challenging the assumptions built into the common notion of simultaneity that logical consistency can be restored. Simultaneity is a relationship between an observer in a particular frame of reference and a set of events. By analogy, left and right are accepted to vary with the position of the observer, because they apply to a relationship. In a similar vein, Plato explained that up and down describe a relationship to the earth and one would not fall off at the antipodes.

In relativity, temporal coordinate systems are set up using a procedure for synchronizing clocks. It is now usually called the *Poincaré-Einstein synchronization procedure*. An observer with a clock sends a light signal out at time t_1 according to his clock. At a distant event, that light signal is reflected back, and arrives back at the observer at time t_2 according to his clock. Since the light travels the same path at the same rate going both out and back for the observer in this scenario, the coordinate time of the event of the light signal being reflected for the observer t_E is $t_E = (t_1 + t_2) / 2$. In this way, a single observer's clock can be used to define temporal coordinates which are good anywhere in the universe.

However, since those clocks are in motion in all other inertial frames, these clock indications are thus not synchronous in those frames, which is the basis of relativity of simultaneity. Because the pairs of putatively simultaneous moments are identified differently by different observers, each can treat the other clock as being the slow one without relativity being self-contradictory. Symmetric time dilation occurs with respect to coordinate systems set up in this manner. It is an effect where another clock is measured to run more slowly than one's own clock. Observers do not consider their own clock time to be affected, but may find that it is observed to be affected in another coordinate system.

Proper Time and Minkowski Diagram

This symmetry can be demonstrated in a Minkowski diagram (second image on the right). Clock C resting in inertial frame S' meets clock A at d and clock B at f (both resting in S). All three clocks simultaneously start to tick in S. The worldline of A is the ct-axis, the worldline of B intersecting f is parallel to the ct-axis, and the worldline of C is the ct'-axis. All events simultaneous with d in S are on the x-axis, in S' on the x'-axis.

The proper time between two events is indicated by a clock present at both events. It is invariant, i.e., in all inertial frames it is agreed that this time is indicated by that clock. Interval df is therefore the proper time of clock C, and is shorter with respect to the coordinate times $ef=dg$ of clocks B and A in S. Conversely, also proper time ef of B is shorter with respect to time if in S', because event e was measured in S' already at time i due to relativity of simultaneity, long before C started to tick.

From that it can be seen, that the proper time between two events indicated by an unaccelerated clock present at both events, compared with the synchronized coordinate time measured in all other inertial frames, is always the *minimal* time interval between those events. However, the interval between two events can also correspond to the proper time of accelerated clocks present at both events. Under all possible proper times between two events, the proper time of the unaccelerated clock is *maximal*, which is the solution to the twin paradox.

Overview of Formulae

Time Dilation Due to Relative Velocity

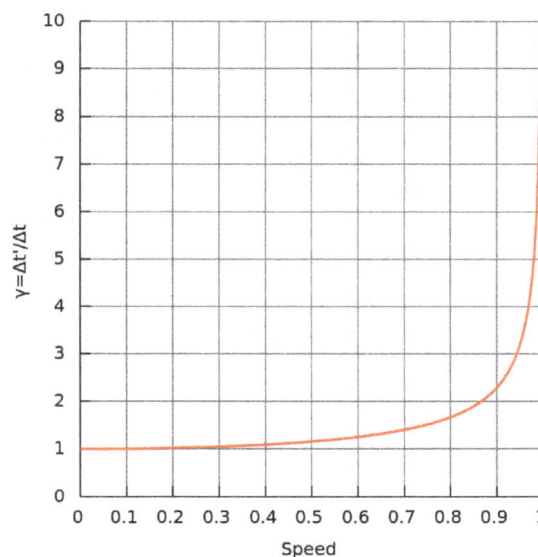

Lorentz factor as a function of speed (in natural units where $c = 1$). Notice that for small speeds (less than 0.1), γ is approximately 1

The formula for determining time dilation in special relativity is:

$$\Delta t' = \gamma \Delta t = \frac{\Delta t}{\sqrt{1 - \dfrac{v^2}{c^2}}}$$

where Δt is the time interval between *two co-local events* (i.e. happening at the same place) for an observer in some inertial frame (e.g. ticks on his clock), known as the *proper time*, $\Delta t'$ is the time interval between those same events, as measured by another observer, inertially moving with velocity v with respect to the former observer, v is the relative velocity between the observer and the moving clock, c is the speed of light, and the Lorentz factor (conventionally denoted by the Greek letter gamma or γ) is

$$\gamma = \frac{1}{\sqrt{1 - \dfrac{v^2}{c^2}}}.$$

Thus the duration of the clock cycle of a moving clock is found to be increased: it is measured to be "running slow". The range of such variances in ordinary life, where $v \ll c$, even considering space travel, are not great enough to produce easily detectable time dilation effects and such vanishingly small effects can be safely ignored for most purposes. It is only when an object approaches speeds on the order of 30,000 km/s (1/10 the speed of light) that time dilation becomes important.

Time dilation by the Lorentz factor was predicted by Joseph Larmor (1897), at least for electrons orbiting a nucleus. Thus "... individual electrons describe corresponding parts of their orbits in

times shorter for the [rest] system in the ratio : $\sqrt{1 - \dfrac{v^2}{c^2}}$ " (Larmor 1897). Time dilation of magnitude corresponding to this (Lorentz) factor has been experimentally confirmed, as described below.

Time Dilation Due to Gravitation and Motion Together

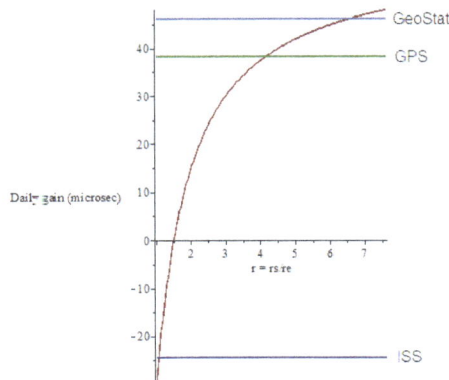

Daily time dilation (gain or loss if negative) in microseconds as a function of (circular) orbit radius $r = rs/re$, where rs is satellite orbit radius and re is the equatorial Earth radius, calculated using the Schwarzschild metric. At $r \approx 1.497$ there is no time dilation. Here the effects of motion and reduced gravity cancel. ISS astronauts fly below, whereas GPS and Geostationary satellites fly above.

High accuracy timekeeping, low earth orbit satellite tracking, and pulsar timing are applications that require the consideration of the combined effects of mass and motion in producing time dilation. Practical examples include the International Atomic Time standard and its relationship with the Barycentric Coordinate Time standard used for interplanetary objects.

Relativistic time dilation effects for the solar system and the earth can be modeled very precisely by the Schwarzschild solution to the Einstein field equations. In the Schwarzschild metric, the interval dt_E is given by

$$dt_E^2 = \left(1 - \frac{2GM_i}{r_i c^2}\right)dt_c^2 - \left(1 - \frac{2GM_i}{r_i c^2}\right)^{-1}\frac{dx^2 + dy^2 + dz^2}{c^2}$$

where:

dt_E is a small increment of proper time t_E (an interval that could be recorded on an atomic clock);

dt_c is a small increment in the coordinate t_c (coordinate time);

dx, dy and dz are small increments in the three coordinates x, y, z of the clock's position; and

GM_i/r_i represents the sum of the Newtonian gravitational potentials due to the masses in the neighborhood, based on their distances r_i from the clock. This sum GM_i/r_i includes any tidal potentials, and is represented as U (using the positive astronomical sign convention for gravitational potentials). The coordinate velocity of the clock is given by

$$v^2 = \frac{dx^2 + dy^2 + dz^2}{dt_c^2}.$$

The coordinate time t_c is the time that would be read on a hypothetical "coordinate clock" situated infinitely far from all gravitational masses ($U = 0$), and stationary in the system of coordinates ($v = 0$). The exact relation between the rate of proper time and the rate of coordinate time for a clock with a radial component of velocity is

$$\frac{dt_E}{dt_c} = \sqrt{1 - \frac{2U}{c^2} - \frac{v^2}{c^2} - \left(\frac{c^2}{2U} - 1\right)^{-1}\frac{v_{\parallel}^2}{c^2}}$$

where:

$v_{\parallel}$ is the radial velocity, and

$U = GM_i/r_i$ is the Newtonian potential, equivalent to half of the escape velocity squared.

The above equation is exact under the assumptions of the Schwarzschild solution.

Experimental Confirmation

Time dilation has been tested a number of times. The routine work carried on in particle accelerators since the 1950s, such as those at CERN, is a continuously running test of the time dilation of special relativity. The specific experiments include:

Velocity Time Dilation Tests

- Ives and Stilwell (1938, 1941). The stated purpose of these experiments was to verify the time dilation effect, predicted by Larmor–Lorentz ether theory, due to motion through the ether using Einstein's suggestion that Doppler effect in canal rays would provide a suitable experiment. These experiments measured the Doppler shift of the radiation emitted from cathode rays, when viewed from directly in front and from directly behind. The high and low frequencies detected were not the classically predicted values

$$\frac{f_0}{1-v/c} \quad \text{and} \quad \frac{f_0}{1+v/c}.$$

The high and low frequencies of the radiation from the moving sources were measured as

$$\sqrt{\frac{1+v/c}{1-v/c}}f_0 = \gamma(1+v/c)f_0 \quad \text{and} \quad \sqrt{\frac{1-v/c}{1+v/c}}f_0 = \gamma(1-v/c)f_0,$$

as deduced by Einstein (1905) from the Lorentz transformation, when the source is running slow by the Lorentz factor.

- Rossi and Hall (1941) compared the population of cosmic-ray-produced muons at the top of a mountain to that observed at sea level. Although the travel time for the muons from the top of the mountain to the base is several muon half-lives, the muon sample at the base was only moderately reduced. This is explained by the time dilation attributed to their high speed relative to the experimenters. That is to say, the muons were decaying about 10 times slower than if they were at rest with respect to the experimenters.

- Hasselkamp, Mondry, and Scharmann (1979) measured the Doppler shift from a source moving at right angles to the line of sight. The most general relationship between frequencies of the radiation from the moving sources is given by:

$$f_{\text{detected}} = f_{\text{rest}}\left(1-\frac{v}{c}\cos\phi\right)/\sqrt{1-v^2/c^2}$$

as deduced by Einstein (1905). For $\phi = 90°$ ($\cos\phi = 0$) this reduces to $f_{\text{detected}} = f_{\text{rest}}\gamma$. This lower frequency from the moving source can be attributed to the time dilation effect and is often called the transverse Doppler effect and was predicted by relativity.

- In 2010 time dilation was observed at speeds of less than 10 meters per second using optical atomic clocks connected by 75 meters of optical fiber.

Gravitational Time Dilation Tests

- In 1959 Robert Pound and Glen A. Rebka measured the very slight gravitational red shift in the frequency of light emitted at a lower height, where Earth's gravitational field is relatively more intense. The results were within 10% of the predictions of general relativity. In 1964, Pound and J. L. Snider measured a result within 1% of the value predicted by gravitational time dilation.

- In 2010 gravitational time dilation was measured at the earth's surface with a height difference of only one meter, using optical atomic clocks.

Velocity and Gravitational Time Dilation Combined-Effect Tests

- Hafele and Keating, in 1971, flew caesium atomic clocks east and west around the earth in commercial airliners, to compare the elapsed time against that of a clock that remained at the U.S. Naval Observatory. Two opposite effects came into play. The clocks were expected to age more quickly than the reference clock, since they were in a higher (weaker) gravitational potential for most of the trip (c.f. Pound–Rebka experiment). But also, contrastingly, the moving clocks were expected to age more slowly because of the speed of their travel. From the actual flight paths of each trip, the theory predicted that the flying clocks, compared with reference clocks at the U.S. Naval Observatory, should have lost 40±23 nanoseconds during the eastward trip and should have gained 275±21 nanoseconds during the westward trip. Relative to the atomic time scale of the U.S. Naval Observatory, the flying clocks lost 59±10 nanoseconds during the eastward trip and gained 273±7 nanoseconds during the westward trip (where the error bars represent standard deviation). In 2005, the National Physical Laboratory in the United Kingdom reported their limited replication of this experiment. The NPL experiment differed from the original in that the caesium clocks were sent on a shorter trip (London–Washington, D.C. return), but the clocks were more accurate. The reported results are within 4% of the predictions of relativity, within the uncertainty of the measurements.

- The Global Positioning System can be considered a continuously operating experiment in both special and general relativity. The in-orbit clocks are corrected for both special and general relativistic time dilation effects as described above, so that (as observed from the earth's surface) they run at the same rate as clocks on the surface of the Earth.

Muon Lifetime

A comparison of muon lifetimes at different speeds is possible. In the laboratory, slow muons are produced; and in the atmosphere, very fast moving muons are introduced by cosmic rays. Taking the muon lifetime at rest as the laboratory value of 2.197 µs, the lifetime of a cosmic ray produced muon traveling at 98% of the speed of light is about five times longer, in agreement with observations. In the muon storage ring at CERN the lifetime of muons circulating with $\gamma = 29.327$ was found to be dilated to 64.378 µs, confirming time dilation to an accuracy of 0.9 ± 0.4 parts per thousand. In this experiment the "clock" is the time taken by processes leading to muon decay, and these processes take place in the moving muon at its own "clock rate", which is much slower than the laboratory clock.

Space Flight

Time dilation would make it possible for passengers in a fast-moving vehicle to travel further into the future while aging very little, in that their great speed slows down the passage of on-board time relative to that of an observer. That is, the ship's clock (and according to relativity, any human traveling with it) shows less elapsed time than the clocks of observers on earth. For sufficiently high speeds the effect is dramatic. For example, one year of travel might correspond to ten years at home. Indeed, a constant 1 g acceleration would permit humans to travel through the entire known Universe in one human lifetime. The space travelers could return to Earth billions of years in the future. A scenario based on this idea was presented in the novel *Planet of the Apes* by Pierre Boulle.

A more likely use of this effect would be to enable humans to travel to nearby stars without spending their entire lives aboard a ship. However, any such application of time dilation during interstellar travel would require the use of some new, advanced method of propulsion. The Orion Project has been the only major attempt toward this idea.

Current space flight technology has fundamental theoretical limits based on the practical problem that an increasing amount of energy is required for propulsion as a craft approaches the speed of light. The likelihood of collision with small space debris and other particulate material is another practical limitation. At the velocities presently attained, however, time dilation occurs but is too small to be a factor in space travel. Travel to regions of spacetime where gravitational time dilation is taking place, such as within the gravitational field of a black hole but outside the event horizon (perhaps on a hyperbolic trajectory exiting the field), could also yield results consistent with present theory.

Time Dilation at Constant Force

In special relativity, time dilation is most simply described in circumstances where relative velocity is unchanging. Nevertheless, the Lorentz equations allow one to calculate proper time and movement in space for the simple case of a spaceship which is applied with a force per unit mass, relative to some reference object in uniform (i.e. constant velocity) motion, equal to g throughout the period of measurement.

Let t be the time in an inertial frame subsequently called the rest frame. Let x be a spatial coordinate, and let the direction of the constant acceleration as well as the spaceship's velocity (relative to the rest frame) be parallel to the x-axis. Assuming the spaceship's position at time $t = 0$ being $x = 0$ and the velocity being v_0 and defining the following abbreviation

$$\gamma_0 = \frac{1}{\sqrt{1 - v_0^2 / c^2}},$$

the following formulas hold:

Position:

$$x(t) = \frac{c^2}{g}\left(\sqrt{1 + \frac{(gt + v_0\gamma_0)^2}{c^2}} - \gamma_0 \right).$$

Velocity:

$$v(t) = \frac{gt + v_0\gamma_0}{\sqrt{1 + \dfrac{(gt + v_0\gamma_0)^2}{c^2}}}.$$

Proper time:

$$\tau(t) = \tau_0 + \int_0^t \sqrt{1 - \left(\frac{v(t')}{c}\right)^2}\, dt'.$$

In the case where $v(0) = v_0 = 0$ and $\tau(0) = \tau_0 = 0$ the integral can be expressed as a logarithmic function or, equivalently, as an inverse hyperbolic function:

$$\tau(t) = \frac{c}{g} \ln\left(\frac{gt}{c} + \sqrt{1 + \left(\frac{gt}{c}\right)^2}\right) = \frac{c}{g} \operatorname{arsinh}\left(\frac{gt}{c}\right).$$

Spacetime Geometry of Velocity Time Dilation

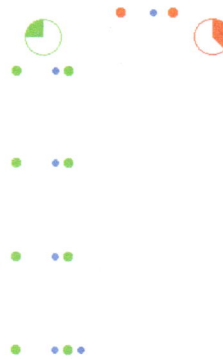

Time dilation in transverse motion

The green dots and red dots in the animation represent spaceships. The ships of a given fleet (color) have no velocity relative to each other, so for the clocks on board the individual ships within a given fleet, the same amount of time elapses relative to each other. Therefore, ships within a given fleet can set up a procedure to maintain a synchronized standard fleet time. The ships of the red fleet are moving with a velocity of 0.866c with respect to the green fleet.

The blue dots represent pulses of light. One cycle of light-pulses between two green ships takes two seconds of "green time", one second for each leg.

As seen from the perspective of the reds, the transit time of the light pulses they exchange among each other is one second of "red time" for each leg. As seen from the perspective of the greens, the red ships' cycle of exchanging light pulses travels a diagonal path that is two light-seconds long. (As seen from the green perspective the reds travel 1.73 () light-seconds of distance for every two seconds of green time.)

One of the red ships emits a light pulse towards the greens every second of red time. These pulses are received by ships of the green fleet with two-second intervals as measured in green time. Not shown in the animation is that all aspects of physics are proportionally involved. The light pulses that are emitted by the reds at a particular frequency as measured in red time are received at a lower frequency as measured by the detectors of the green fleet that measure against green time, and vice versa.

The animation cycles between the green perspective and the red perspective, to emphasize the symmetry. As there is no such thing as absolute motion in relativity (as is also the case for Newtonian mechanics), both the green and the red fleet are entitled to consider themselves motionless *in their own frame of reference.*

Again, it is vital to understand that the results of these interactions and calculations reflect the real state of the ships as it emerges from their situation of relative motion. It is not a mere quirk of the method of measurement or communication.

Length Contraction

Length contraction is the phenomenon of a decrease in length of an object as measured by an observer who is traveling at any non-zero velocity relative to the object. This contraction (more formally called Lorentz contraction or Lorentz–FitzGerald contraction after Hendrik Lorentz and George Francis FitzGerald) is usually only noticeable at a substantial fraction of the speed of light. Length contraction is only in the direction parallel to the direction in which the observed body is travelling. This effect is negligible at everyday speeds, and can be ignored for all regular purposes. Only at greater speeds does it become relevant. At a speed of 13,400,000 m/s (30 million mph, $0.0447c$) contracted length is 99.9% of the length at rest; at a speed of 42,300,000 m/s (95 million mph, $0.141c$), the length is still 99%. As the magnitude of the velocity approaches the speed of light, the effect becomes dominant, as can be seen from the formula:

$$L = \frac{L_0}{\gamma(v)} = L_0\sqrt{1 - v^2/c^2}$$

where

L_0 is the proper length (the length of the object in its rest frame),

L is the length observed by an observer in relative motion with respect to the object,

v is the relative velocity between the observer and the moving object,

c is the speed of light,

and the *Lorentz factor, $\gamma(v)$*, is defined as

$$\gamma(v) \equiv \frac{1}{\sqrt{1 - v^2/c^2}}.$$

In this equation it is assumed that the object is parallel with its line of movement. For the observer in relative movement, the length of the object is measured by subtracting the simultaneously measured distances of both ends of the object. An observer at rest viewing an object travelling very close to the speed of light would observe the length of the object in the direction of motion as very near zero.

History

Length contraction was postulated by George FitzGerald (1889) and Hendrik Antoon Lorentz (1892) to explain the negative outcome of the Michelson–Morley experiment and to rescue the hypothesis of the stationary aether (Lorentz–FitzGerald contraction hypothesis). Although both FitzGerald and Lorentz alluded to the fact that electrostatic fields in motion were deformed ("Heaviside-Ellipsoid" after Oliver Heaviside, who derived this deformation from electromagnetic theory in 1888), it was considered an ad hoc hypothesis, because at this time there was no sufficient reason to assume that intermolecular forces behave the same way as electromagnetic ones. In 1897 Joseph Larmor developed a model in which all forces are considered to be of electromagnetic origin, and length contraction appeared to be a direct consequence of this model. Yet it was shown by Henri Poincaré (1905) that electromagnetic forces alone cannot explain the electron's stability. So he had to introduce another ad hoc hypothesis: non-electric binding forces (Poincaré stresses) that ensure the electron's stability, give a dynamical explanation for length contraction, and thus hide the motion of the stationary aether.

Eventually, Albert Einstein (1905) was the first to completely remove the ad hoc character from the contraction hypothesis, by demonstrating that this contraction did not require motion through a supposed aether, but could be explained using special relativity, which changed our notions of space, time, and simultaneity. Einstein's view was further elaborated by Hermann Minkowski, who demonstrated the geometrical interpretation of all relativistic effects by introducing his concept of four-dimensional spacetime.

Basis in Relativity

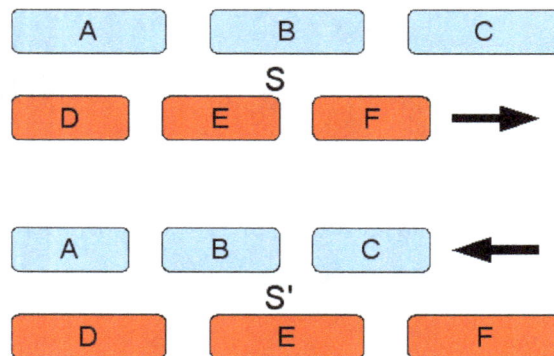

Length contraction: Three blue rods are at rest in S, and three red rods in S'. At the instant when the left ends of A and D attain the same position on the axis of x, the lengths of the rods shall be compared. In S the simultaneous positions of the left side of A and the right side of C are more distant than those of D and F. While in S' the simultaneous positions of the left side of D and the right side of F are more distant than those of A and C.

First it is necessary to carefully consider the methods for measuring the lengths of resting and moving objects. Here, "object" simply means a distance with endpoints that are always mutually at rest, *i.e.*, that are at rest in the same inertial frame of reference. If the relative velocity between an observer (or his measuring instruments) and the observed object is zero, then the proper length L_0 of the object can simply be determined by directly superposing a measuring rod. However, if the relative velocity > 0, then one can proceed as follows:

The observer installs a row of clocks that either are synchronized a) by exchanging light signals according to the Poincaré-Einstein synchronization, or b) by "slow clock transport", that is, one clock is transported along the row of clocks in the limit of vanishing transport velocity. Now, when the synchronization process is finished, the object is moved along the clock row and every clock stores the exact time when the left or the right end of the object passes by. After that, the observer only has to look after the position of a clock A that stored the time when the left end of the object was passing by, and a clock B at which the right end of the object was passing by *at the same time*. It's clear that distance AB is equal to length L of the moving object. Using this method, the definition of simultaneity is crucial for measuring the length of moving objects.

Another method is to use a clock indicating its proper time T_0, which is traveling from one endpoint of the rod to the other in time T as measured by clocks in the rod's rest frame. The length of the rod can be computed by multiplying its travel time by its velocity, thus $L_0 = T \cdot v$ in the rod's rest frame or $L = T_0 \cdot v$ in the clock's rest frame.

In Newtonian mechanics, simultaneity and time duration are absolute and therefore both methods lead to the equality of L and L_0. Yet in relativity theory the constancy of light velocity in all inertial frames in connection with relativity of simultaneity and time dilation destroys this equality. In the first method an observer in one frame claims to have measured the object's endpoints simultaneously, but the observers in all other inertial frames will argue that the object's endpoints were *not* measured simultaneously. In the second method, times T and T_0 are not equal due to time dilation, resulting in different lengths too.

The deviation between the measurements in all inertial frames is given by the formulas for Lorentz transformation and time dilation. It turns out, that the proper length remains unchanged and always denotes the greatest length of an object, yet the length of the same object as measured in another inertial frame is shorter than the proper length. This contraction only occurs in the line of motion, and can be represented by the following relation (where v is the relative velocity and c the speed of light)

$$L = L_0 / \gamma.$$

Symmetry

The principle of relativity (according to which the laws of nature must assume the same form in all inertial reference frames) requires that length contraction is symmetrical: If a rod rests in inertial frame S, it has its proper length in S and its length is contracted in S'. However, if a rod rests in S', it has its proper length in S' and its length is contracted in S. This can be vividly illustrated using symmetric Minkowski diagrams (or Loedel diagrams), because the Lorentz transformation geometrically corresponds to a rotation in four-dimensional spacetime.

Minkowski diagram

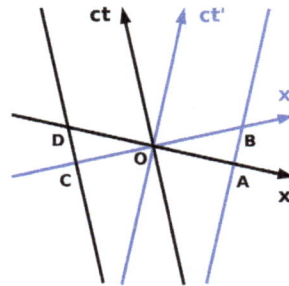

In S all events parallel to the axis of x are simultaneous, while in S' all events parallel to the axis of x' are simultaneous.

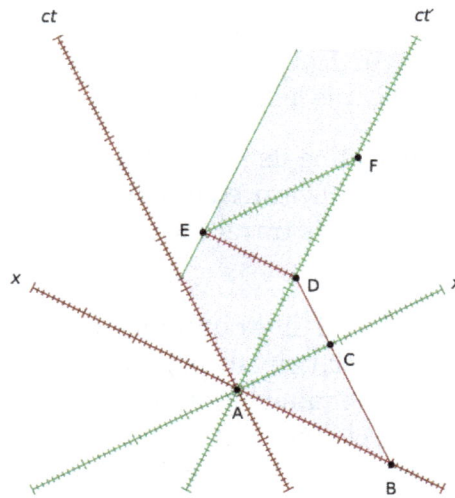

A rod is transported from S to S'

First image: If a rod at rest in S' is given, then its endpoints are located upon the ct' axis and the axis parallel to it. In this frame the simultaneous (parallel to the axis of x') positions of the endpoints are O and B, thus the *proper* length is given by OB. But in S the simultaneous (parallel to the axis of x) positions are O and A, thus the *contracted* length is given by OA.

On the other hand, if another rod is at rest in S, then its endpoints are located upon the ct axis and the axis parallel to it. In this frame the simultaneous (parallel to the axis of x) positions of the endpoints are O and D, thus the *proper* length is given by OD. But in S' the simultaneous (parallel to the axis of x') positions are O and C, thus the *contracted* length is given by OC.

Second image: A train at rest in S and a station at rest in S' with relative velocity of $v = 0.8c$ are given. In S a rod with proper length $L_0 = AB = 30$ cm is located, so its contracted length L' in S' is given by:

$$L' = AC = L_0 / \gamma = 18 \text{ cm}.$$

Then the rod will be thrown out of the train in S and will come to rest at the station in S'. Its length has to be measured again according to the methods given above, and now the proper length

$L_0' = \mathrm{EF} = 30$ cm will be measured in S' (the rod has become larger in that system), while in S the rod is in motion and therefore its length is contracted (the rod has become smaller in that system):

$$L = \mathrm{DE} = L_0' / \gamma = 18 \text{ cm}.$$

Experimental Verifications

Any observer co-moving with the observed object cannot measure the object's contraction, because he can judge himself and the object as at rest in the same inertial frame in accordance with the principle of relativity (as it was demonstrated by the Trouton-Rankine experiment). So length contraction cannot be measured in the object's rest frame, but only in a frame in which the observed object is in motion. In addition, even in such a non-co-moving frame, *direct* experimental confirmations of length contraction are hard to achieve, because at the current state of technology, objects of considerable extension cannot be accelerated to relativistic speeds. And the only objects traveling with the speed required are atomic particles, yet whose spatial extensions are too small to allow a direct measurement of contraction.

However, there are *indirect* confirmations of this effect in a non-co-moving frame:

- It was the negative result of a famous experiment, that required the introduction of length contraction: the Michelson-Morley experiment (and later also the Kennedy–Thorndike experiment). In special relativity its explanation is as follows: In its rest frame the interferometer can be regarded as at rest in accordance with the relativity principle, so the propagation time of light is the same in all directions. Although in a frame in which the interferometer is in motion, the transverse beam must traverse a longer, diagonal path with respect to the non-moving frame thus making its travel time longer, the factor by which the longitudinal beam would be delayed by taking times L/(c-v) & L/(c+v) for the forward and reverse trips respectively is even longer. Therefore, in the longitudinal direction the interferometer is supposed to be contracted, in order to restore the equality of both travel times in accordance with the negative experimental result(s). Thus the two-way speed of light remains constant and the round trip propagation time along perpendicular arms of the interferometer is independent of its motion & orientation.

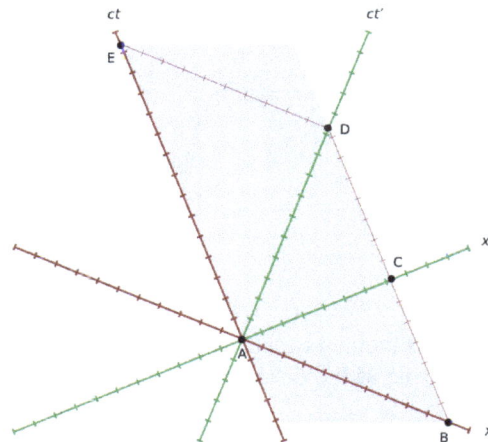

Muon-atmosphere-scenario

- The range of action of muons at high velocities is much higher than that of slower ones. The atmosphere has its proper length in the Earth frame, while the increased muon range is explained by their longer lifetimes due to time dilation. However, in the muon frame their lifetime is unchanged but the atmosphere is contracted so that even their small range is sufficient to reach the surface of earth.

- Heavy ions that are spherical when at rest should assume the form of "pancakes" or flat disks when traveling nearly at the speed of light. And in fact, the results obtained from particle collisions can only be explained when the increased nucleon density due to length contraction is considered.

- The ionization ability of electrically charged particles with large relative velocities is higher than expected. In pre-relativistic physics the ability should decrease at high velocities, because the time in which ionizing particles in motion can interact with the electrons of other atoms or molecules is diminished. Though in relativity, the higher-than-expected ionization ability can be explained by length contraction of the Coulomb field in frames in which the ionizing particles are moving, which increases their electrical field strength normal to the line of motion.

- In free-electron lasers, relativistic electrons were injected into an undulator, so that synchrotron radiation is generated. In the proper frame of the electrons, the undulator is contracted which leads to an increased radiation frequency. Additionally, to find out the frequency as measured in the laboratory frame, one has to apply the relativistic Doppler effect. So, only with the aid of length contraction and the relativistic Doppler effect, the extremely small wavelength of undulator radiation can be explained.

Reality of Length Contraction

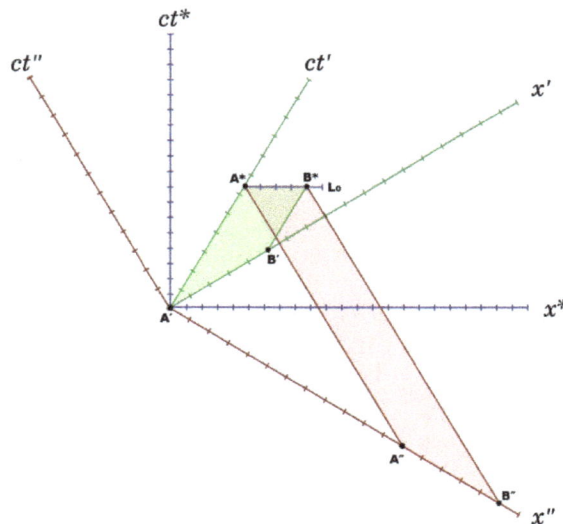

Minkowski diagram of Einstein's 1911 thought experiment on length contraction. Two rods of rest length $A'B' = A''B'' = L_0$ are moving with 0.6c in opposite direction, resulting in $A^*B^* < L_0$.

In 1911 Vladimir Varićak asserted that length contraction is "real" according to Lorentz, while it is "apparent or subjective" according to Einstein. Einstein replied:

The author unjustifiably stated a difference of Lorentz's view and that of mine *concerning the physical facts*. The question as to whether length contraction *really* exists or not is misleading. It doesn't "really" exist, in so far as it doesn't exist for a comoving observer; though it "really" exists, *i.e.* in such a way that it could be demonstrated in principle by physical means by a non-comoving observer.

—Albert Einstein, 1911

Einstein also argued in that paper, that length contraction is not simply the product of *arbitrary* definitions concerning the way clock regulations and length measurements are performed. He presented the following thought experiment: Let A'B' and A"B" be the endpoints of two rods of the same proper length. Let them move in opposite directions at the same speed with respect to a resting coordinate x-axis. Endpoints A'A" meet at point A*, and B'B" meet at point B*, both points being marked on that axis. Einstein pointed out that length A*B* is shorter than A'B' or A"B", which can also be demonstrated by one of the rods when brought to rest with respect to that axis.

Paradoxes

Due to superficial application of the contraction formula some paradoxes can occur. Examples are the ladder paradox and Bell's spaceship paradox. However, those paradoxes can simply be solved by a correct application of relativity of simultaneity. Another famous paradox is the Ehrenfest paradox, which proves that the concept of rigid bodies is not compatible with relativity, reducing the applicability of Born rigidity, and showing that for a co-rotating observer the geometry is in fact non-euclidean.

Visual Effects

Length contraction refers to measurements of position made at simultaneous times according to a coordinate system. This could suggest that if one could take a picture of a fast moving object, that the image would show the object contracted in the direction of motion. However, such visual effects are completely different measurements, as such a photograph is taken from a distance, while length contraction can only directly be measured at the exact location of the object's endpoints. It was shown by several authors such as Roger Penrose and James Terrell that moving objects generally do not appear length contracted on a photograph. For instance, for a small angular diameter, a moving sphere remains circular and is rotated. This kind of visual rotation effect is called Penrose-Terrell rotation.

Derivation

Lorentz Transformation

Length contraction can be derived from the Lorentz transformation in several ways:

$$x' = \gamma\left(x - vt\right),$$
$$t' = \gamma\left(t - vx/c^2\right).$$

Moving Length is Known

In an inertial reference frame S, x_1 and x_2 shall denote the endpoints of an object in motion in this frame. There, its length L was measured according to the above convention by determining the simultaneous positions of its endpoints at $t_1 = t_2$. Now, the proper length of this object in S' shall be calculated by using the Lorentz transformation. Transforming the time coordinates from S into S' results in different times, but this is not problematic, as the object is at rest in S' where it does not matter when the endpoints are measured. Therefore, the transformation of the spatial coordinates suffices, which gives:

$$x_1' = \gamma(x_1 - vt_1) \quad \text{and} \quad x_2' = \gamma(x_2 - vt_2).$$

Since $t_1 = t_2$, and by setting $L = x_2 - x_1$ and $L_0' = x_2' - x_1'$, the proper length in S' is given by

$$L_0' = L \cdot \gamma. \qquad (1),$$

with respect to which the measured length in S is contracted by

$$L = L_0' / \gamma. \qquad (2)$$

According to the relativity principle, objects that are at rest in S have to be contracted in S' as well. By exchanging the above signs and primes symmetrically, it follows:

$$L_0 = L' \cdot \gamma. \qquad (3)$$

Thus the contracted length as measured in S' is given by:

$$L' = L_0 / \gamma. \qquad (4)$$

Proper Length is Known

Conversely, if the object rests in S and its proper length is known, the simultaneity of the measurements at the object's endpoints has to be considered in another frame S', as the object constantly changes its position there. Therefore, both spatial and temporal coordinates must be transformed:

$$x_1' = \gamma(x_1 - vt_1) \quad \text{and} \quad x_2' = \gamma(x_2 - vt_2)$$
$$t_1' = \gamma(t_1 - vx_1 / c^2) \quad \text{and} \quad t_2' = \gamma(t_2 - vx_2 / c^2).$$

With $t_1 = t_2$ and $L_0 = x_2 - x_1$ this results in non-simultaneous differences:

$$\Delta x' = \gamma L_0$$
$$\Delta t' = \gamma v L_0 / c^2$$

In order to obtain the simultaneous positions of both endpoints, the distance traveled by the second endpoint with during $\Delta t'$ must be subtracted from $\Delta x'$:

$$L' = \Delta x' - v\Delta t'$$
$$= \gamma L_0 - \gamma v^2 L_0 / c^2$$
$$= L_0 / \gamma$$

So the moving length in S' is contracted. Likewise, the preceding calculation gives a symmetric result for an object at rest in S':

$$L = L_0' / \gamma .$$

Time Dilation

Length contraction can also be derived from time dilation, according to which the rate of a single "moving" clock (indicating its proper time T_0) is lower with respect to two synchronized "resting" clocks (indicating T). Time dilation was experimentally confirmed multiple times, and is represented by the relation:

$$T = T_0 \cdot \gamma .$$

Suppose a rod of proper length L_0 at rest in S and a clock at rest in S' are moving along each other. The respective travel times of the clock between the rod's endpoints are given by $T = L_0 / v$ in S and $T_0' = L' / v$ in S' , thus $L_0 = Tv$ and $L' = T_0' v$. By inserting the time dilation formula, the ratio between those lengths is:

$$\frac{L'}{L_0} = \frac{T_0' v}{Tv} = 1 / \gamma$$

Therefore, the length measured in S' is given by

$$L' = L_0 / \gamma .$$

So since the clock's travel time across the rod is longer in S than in S' (time dilation in S), the rod's length is also longer in S than in S' (length contraction in S'). Likewise, if the clock were at rest in S and the rod in S' , the above procedure would give

$$L = L_0' / \gamma .$$

Geometrical Considerations

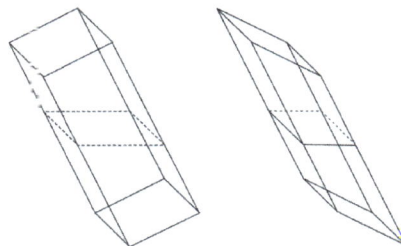

Cuboids in Euclidean and Minkowski spacetime

Additional geometrical considerations show, that length contraction can be regarded as a *trigono-metric* phenomenon, with analogy to parallel slices through a cuboid before and after a *rotation* in E³. This is the Euclidean analog of *boosting* a cuboid in E¹,². In the latter case, however, we can interpret the boosted cuboid as the *world slab* of a moving plate.

Image: Left: a *rotated cuboid* in three-dimensional euclidean space E³. The cross section is *longer* in the direction of the rotation than it was before the rotation. Right: the *world slab* of a moving thin plate in Minkowski spacetime (with one spatial dimension suppressed) E¹,², which is a *boosted cuboid*. The cross section is *thinner* in the direction of the boost than it was before the boost. In both cases, the transverse directions are unaffected and the three planes meeting at each corner of the cuboids are *mutually orthogonal* (in the sense of E¹,² at right, and in the sense of E³ at left).

In special relativity, Poincaré transformations are a class of affine transformations which can be characterized as the transformations between alternative Cartesian coordinate charts on Minkowski spacetime corresponding to alternative states of inertial motion (and different choices of an origin). Lorentz transformations are Poincaré transformations which are linear transformations (preserve the origin). Lorentz transformations play the same role in Minkowski geometry (the Lorentz group forms the *isotropy group* of the self-isometries of the spacetime) which are played by rotations in euclidean geometry. Indeed, special relativity largely comes down to studying a kind of noneuclidean trigonometry in Minkowski spacetime, as suggested by the following table:

Three plane trigonometries			
Trigonometry	Circular	Parabolic	Hyperbolic
Kleinian Geometry	euclidean plane	Galilean plane	Minkowski plane
Symbol	E^2	$E^{0,1}$	$E^{1,1}$
Quadratic form	positive definite	degenerate	non-degenerate but indefinite
Isometry group	E(2)	E(0,1)	E(1,1)
Isotropy group	SO(2)	SO(0,1)	SO(1,1)
type of isotropy	rotations	shears	boosts
Cayley algebra	complex numbers	dual numbers	split-complex numbers
ε^2	-1	0	1
Spacetime interpretation	none	Newtonian spacetime	Minkowski spacetime
slope	$\tan \varphi = m$	$\tan p\,\varphi = u$	$\tanh \varphi = v$
"cosine"	$\cos \varphi = (1+m^2)^{-1/2}$	$\cos p\,\varphi = 1$	$\cosh \varphi = (1-v^2)^{-1/2}$
"sine"	$\sin \varphi = m\,(1+m^2)^{-1/2}$	$\sin p\,\varphi = u$	$\sinh \varphi = v\,(1-v^2)^{-1/2}$
"secant"	$\sec \varphi = (1+m^2)^{1/2}$	$\sec p\,\varphi = 1$	$\text{sech}\,\varphi = (1-v^2)^{1/2}$
"cosecant"	$\csc \varphi = m^{-1}(1+m^2)^{1/2}$	$\csc p\,\varphi = u^{-1}$	$\text{csch}\,\varphi = v^{-1}(1-v^2)^{1/2}$

Mass–energy Equivalence

The four-meter tall sculpture of Einstein's 1905 formula $E = mc^2$ at the 2006 Walk of Ideas, Berlin, Germany.

value

energy | mass | speed of light

$$E = mc^2$$

J | kg | 299,792,458 m/s

units

$c^2 = 89,875,517,873,681,800 \ m^2/s^2$

$E = mc^2$ explained.

In physics, mass–energy equivalence is a concept formulated by Albert Einstein that explains the relationship between mass and energy. It expresses the law of equivalence of energy and mass using the formula

$$E = mc^2$$

where E is the energy of a physical system, m is the mass of the system, and c is the speed of light in a vacuum (about 3×10^8 m/s). In words, energy equals mass multiplied by the square of the speed of light. Because the speed of light is a very large number in everyday units, the formula implies that any small amount of matter contains a very large amount of energy. Some of this energy may be released as heat and light by chemical or nuclear transformations. This also serves to convert units of mass to units of energy, no matter what system of measurement units is used.

Mass–energy equivalence arose originally from special relativity as a paradox described by Henri Poincaré. Einstein proposed it in 1905, in the paper *Does the inertia of a body depend upon its energy-content?*, one of his *Annus Mirabilis (Miraculous Year) papers*. Einstein was the first to propose that the equivalence of mass and energy is a general principle and a consequence of the symmetries of space and time.

A consequence of the mass–energy equivalence is that if a body is stationary, it still has some internal or intrinsic energy, called its rest energy. Rest mass and rest energy are equivalent and remain proportional to each other. When the body is in motion (relative to an observer), its total energy is greater than its rest energy. The rest mass (or rest energy) remains an important quantity in this case because it remains the same regardless of this motion, even for the extreme speeds or gravity considered in special and general relativity; thus it is also called the invariant mass.

Nomenclature

The formula was initially written in many different notations, and its interpretation and justification was further developed in several steps.

In "*Does the inertia of a body depend upon its energy content?*" (1905), Einstein used V to mean the speed of light in a vacuum and L to mean the energy lost by a body in the form of radiation. Consequently, the equation $E = mc^2$ was not originally written as a formula but as a sentence in German saying that *if a body gives off the energy L in the form of radiation, its mass diminishes by L/V^2*. A remark placed above it informed that the equation was approximated by neglecting "magnitudes of fourth and higher orders" of a series expansion.

In May 1907, Einstein explained that the expression for energy ε of a moving mass point assumes the simplest form, when its expression for the state of rest is chosen to be $\varepsilon_0 = \mu V^2$ (where μ is the mass), which is in agreement with the "principle of the equivalence of mass and energy". In addition, Einstein used the formula $\mu = E_0/V^2$, with E_0 being the energy of a system of mass points to describe the energy and mass increase of that system when the velocity of the differently moving mass points is increased.

In June 1907, Max Planck rewrote Einstein's mass–energy relationship as $M = E_0 + pV_0/c^2$, where p is the pressure and V the volume to express the relation between mass, its *latent energy*, and thermodynamic energy within the body. Subsequently, in October 1907, this was rewritten as $M_0 = E_0/c^2$ and given a quantum interpretation by Johannes Stark, who assumed its validity and correctness (*Gültigkeit*).

In December 1907, Einstein expressed the equivalence in the form $M = \mu + E_0/c^2$ and concluded: *A mass μ is equivalent, as regards inertia, to a quantity of energy μc^2. [...] It appears far more natural to consider every inertial mass as a store of energy*.

In 1909, Gilbert N. Lewis and Richard C. Tolman used two variations of the formula: $m = E/c^2$ and $m_0 = E_0/c^2$, with E being the energy of a moving body, E_0 its rest energy, m the relativistic mass, and m_0 the invariant mass. The same relations in different notation were used by Hendrik Lorentz in 1913 (published 1914), though he placed the energy on the left-hand side: $\varepsilon = Mc^2$ and $\varepsilon_0 = mc^2$, with ε being the total energy (rest energy plus kinetic energy) of a moving material point, ε_0 its rest energy, M the relativistic mass, and m the invariant (or rest) mass.

In 1911, Max von Laue gave a more comprehensive proof of $M_0 = E_0/c^2$ from the stress–energy tensor, which was later (1918) generalized by Felix Klein.

Einstein returned to the topic once again after World War II and this time he wrote $E = mc^2$ in the title of his article intended as an explanation for a general reader by analogy.

Conservation of Mass and Energy

Mass and energy can be seen as two names (and two measurement units) for the same underlying, conserved physical quantity. Thus, the laws of conservation of energy and conservation of (total) mass are equivalent and both hold true. Einstein elaborated in a 1946 essay that "the principle of the conservation of mass [...] proved inadequate in the face of the special theory of relativity. It was therefore merged with the energy conservation principle—just as, about 60 years before, the principle of the conservation of mechanical energy had been combined with the principle of the conservation of heat [thermal energy]. We might say that the principle of the conservation of energy, having previously swallowed up that of the conservation of heat, now proceeded to swallow that of the conservation of mass—and holds the field alone."

If the conservation of mass law is interpreted as conservation of *rest* mass, it does not hold true in special relativity. The *rest* energy (equivalently, rest mass) of a particle can be converted, not "to energy" (it already *is* energy (mass)), but rather to *other* forms of energy (mass) that require motion, such as kinetic energy, thermal energy, or radiant energy. Similarly, kinetic or radiant energy can be converted to other kinds of particles that have rest energy (rest mass). In the transformation process, neither the total amount of mass nor the total amount of energy changes, since both properties are connected via a simple constant. This view requires that if either energy or (total) mass disappears from a system, it is always found that both have simply moved to another place, where they are both measurable as an increase of both energy and mass that corresponds to the loss in the first system.

Fast-moving Objects and Systems of Objects

When an object is pushed in the direction of motion, it gains momentum and energy, but when the object is already traveling near the speed of light, it cannot move much faster, no matter how much energy it absorbs. Its momentum and energy continue to increase without bounds, whereas its speed approaches a constant value—the speed of light. This implies that in relativity the momentum of an object cannot be a constant times the velocity, nor can the kinetic energy be a constant times the square of the velocity.

A property called the relativistic mass is defined as the ratio of the momentum of an object to its velocity. Relativistic mass depends on the motion of the object, so that different observers in relative motion see different values for it. If the object is moving slowly, the relativistic mass is nearly equal to the rest mass and both are nearly equal to the usual Newtonian mass. If the object is moving quickly, the relativistic mass is greater than the rest mass by an amount equal to the mass associated with the kinetic energy of the object. As the object approaches the speed of light, the relativistic mass grows infinitely, because the kinetic energy grows infinitely and this energy is associated with mass.

The relativistic mass is always equal to the total energy (rest energy plus kinetic energy) divided by c^2. Because the relativistic mass is exactly proportional to the energy, relativistic mass and relativistic energy are nearly synonyms; the only difference between them is the units. If length and time are measured in natural units, the speed of light is equal to 1, and even this difference disappears. Then mass and energy have the same units and are always equal, so it is redundant to speak about relativistic mass, because it is just another name for the energy. This is why physicists usually reserve the useful short word "mass" to mean rest mass, or invariant mass, and not relativistic mass.

The relativistic mass of a moving object is larger than the relativistic mass of an object that is not moving, because a moving object has extra kinetic energy. The *rest mass* of an object is defined as the mass of an object when it is at rest, so that the rest mass is always the same, independent of the motion of the observer: it is the same in all inertial frames.

For things and systems made up of many parts, like an atomic nucleus, planet, or star, the relativistic mass is the sum of the relativistic masses (or energies) of the parts, because energies are additive in isolated systems. This is not true in open systems, however, if energy is subtracted. For example, if a system is *bound* by attractive forces, and the energy gained due to the forces of attraction in excess of the work done is removed from the system, then mass is lost with this removed energy. For example, the mass of an atomic nucleus is less than the total mass of the protons and neutrons that make it up, but this is only true after this energy from binding has been removed in the form of a gamma ray (which in this system, carries away the mass of the energy of binding). This mass decrease is also equivalent to the energy required to break up the nucleus into individual protons and neutrons (in this case, work and mass would need to be supplied). Similarly, the mass of the solar system is slightly less than the sum of the individual masses of the sun and planets.

For a system of particles going off in different directions, the invariant mass of the system is the analog of the rest mass, and is the same for all observers, even those in relative motion. It is defined as the total energy (divided by c^2) in the center of mass frame (where by definition, the system total momentum is zero). A simple example of an object with moving parts but zero total momentum is a container of gas. In this case, the mass of the container is given by its total energy (including the kinetic energy of the gas molecules), since the system total energy and invariant mass are the same in any reference frame where the momentum is zero, and such a reference frame is also the only frame in which the object can be weighed. In a similar way, the theory of special relativity posits that the thermal energy in all objects (including solids) contributes to their total masses and weights, even though this energy is present as the kinetic and potential energies of the atoms in the object, and it (in a similar way to the gas) is not seen in the rest masses of the atoms that make up the object.

In a similar manner, even photons (light quanta), if trapped in a container space (as a photon gas or thermal radiation), would contribute a mass associated with their energy to the container. Such an extra mass, in theory, could be weighed in the same way as any other type of rest mass. This is true in special relativity theory, even though individually photons have no rest mass. The property that trapped energy *in any form* adds weighable mass to systems that have no net momentum is one of the characteristic and notable consequences of relativity. It has no counterpart in classical Newtonian physics, in which radiation, light, heat, and kinetic energy never exhibit weighable mass under any circumstances.

Just as the relativistic mass of an isolated system is conserved through time, so also is its invariant mass. This property allows the conservation of all types of mass in systems, and also conservation of all types of mass in reactions where matter is destroyed (annihilated), leaving behind the energy that was associated with it (which is now in non-material form, rather than material form). Matter may appear and disappear in various reactions, but mass and energy are both unchanged in this process.

Applicability of the Strict Mass–energy Equivalence Formula, $E = mc^2$

As is noted above, two different definitions of mass have been used in special relativity, and also two different definitions of energy. The simple equation $E = mc^2$ is not generally applicable to all these types of mass and energy, except in the special case that the total additive momentum is zero for the system under consideration. In such a case, which is always guaranteed when observing the system from either its center of mass frame or its center of momentum frame, $E = mc^2$ is always true for any type of mass and energy that are chosen. Thus, for example, in the center of mass frame, the total energy of an object or system is equal to its rest mass times c^2, a useful equality. This is the relationship used for the container of gas in the previous example. It is *not* true in other reference frames where the center of mass is in motion. In these systems or for such an object, its total energy depends on both its rest (or invariant) mass, and its (total) momentum.

In inertial reference frames other than the rest frame or center of mass frame, the equation $E = mc^2$ remains true if the energy is the relativistic energy *and* the mass is the relativistic mass. It is also correct if the energy is the rest or invariant energy (also the minimum energy), *and* the mass is the rest mass, or the invariant mass. However, connection of the total or relativistic energy (E_r) with the rest or invariant mass (m_0) requires consideration of the system total momentum, in systems and reference frames where the total momentum has a non-zero value. The formula then required to connect the two different kinds of mass and energy, is the extended version of Einstein's equation, called the relativistic energy–momentum relation:

$$E_r^2 - | \vec{p} |^2 \ c^2 = m_0^2 c^4$$
$$E_r^2 - (pc)^2 = (m_0 c^2)^2$$

or

$$E_r = \sqrt{(m_0 c^2)^2 + (pc)^2}$$

Here the $(pc)^2$ term represents the square of the Euclidean norm (total vector length) of the various momentum vectors in the system, which reduces to the square of the simple momentum magnitude, if only a single particle is considered. This equation reduces to $E = mc^2$ when the momentum term is zero. For photons where $m_0 = 0$, the equation reduces to $E_r = pc$.

Meanings of the Strict Mass–energy Equivalence Formula, $E = mc^2$

The mass–energy equivalence formula was displayed on Taipei 101 during the event of the World Year of Physics 2005.

Mass–energy equivalence states that any object has a certain energy, even when it is stationary. In Newtonian mechanics, a motionless body has no kinetic energy, and it may or may not have other amounts of internal stored energy, like chemical energy or thermal energy, in addition to any potential energy it may have from its position in a field of force. In Newtonian mechanics, all of these energies are much smaller than the mass of the object times the speed of light squared.

In relativity, all the energy that moves with an object (that is, all the energy present in the object's rest frame) contributes to the total mass of the body, which measures how much it resists acceleration. Each bit of potential and kinetic energy makes a proportional contribution to the mass. As noted above, even if a box of ideal mirrors "contains" light, then the individually massless photons still contribute to the total mass of the box, by the amount of their energy divided by c^2.

In relativity, removing energy is removing mass, and for an observer in the center of mass frame, the formula $m = E/c^2$ indicates how much mass is lost when energy is removed. In a nuclear reaction, the mass of the atoms that come out is less than the mass of the atoms that go in, and the difference in mass shows up as heat and light with the same relativistic mass as the difference (and also the same invariant mass in the center of mass frame of the system). In this case, the E in the formula is the energy released and removed, and the mass m is how much the mass decreases. In the same way, when any sort of energy is added to an isolated system, the increase in the mass is equal to the added energy divided by c^2. For example, when water is heated it gains about 1.11×10^{-17} kg of mass for every joule of heat added to the water.

An object moves with different speed in different frames, depending on the motion of the observer, so the kinetic energy in both Newtonian mechanics and relativity is *frame dependent*. This means that the amount of relativistic energy, and therefore the amount of relativistic mass, that an object is measured to have depends on the observer. The *rest mass* is defined as the mass that an object has when it is not moving (or when an inertial frame is chosen such that it is not moving). The term also applies to the invariant mass of systems when the system as a whole is not "moving" (has no net momentum). The rest and invariant masses are the smallest possible value of the mass of the object or system. They also are conserved quantities, so long as the system is isolated. Because of the way they are calculated, the effects of moving observers are subtracted, so these quantities do not change with the motion of the observer.

The rest mass is almost never additive: the rest mass of an object is not the sum of the rest masses of its parts. The rest mass of an object is the total energy of all the parts, including kinetic energy, as measured by an observer that sees the center of the mass of the object to be standing still. The rest mass adds up only if the parts are standing still and do not attract or repel, so that they do not have any extra kinetic or potential energy. The other possibility is that they have a positive kinetic energy and a negative potential energy that exactly cancels.

Binding Energy and the "Mass Defect"

Whenever any type of energy is removed from a system, the mass associated with the energy is also removed, and the system therefore loses mass. This mass defect in the system may be simply calculated as $\Delta m = \Delta E/c^2$, and this was the form of the equation historically first presented by Einstein in 1905. However, use of this formula in such circumstances has led to the false idea that mass has been "converted" to energy. This may be particularly the case when the energy (and mass) removed from the system is associated with the *binding energy* of the system. In such cases, the binding energy is observed as a "mass defect" or deficit in the new system.

The fact that the released energy is not easily weighed in many such cases, may cause its mass to be neglected as though it no longer existed. This circumstance has encouraged the false idea of conversion of *mass* to energy, rather than the correct idea that the binding energy of such systems is relatively large, and exhibits a measurable mass, which is removed when the binding energy is removed..

The difference between the rest mass of a bound system and of the unbound parts is the binding energy of the system, if this energy has been removed after binding. For example, a water molecule weighs a little less than two free hydrogen atoms and an oxygen atom. The minuscule mass difference is the energy needed to split the molecule into three individual atoms (divided by c^2), which was given off as heat when the molecule formed (this heat had mass). Likewise, a stick of dynamite in theory weighs a little bit more than the fragments after the explosion, but this is true only so long as the fragments are cooled and the heat removed. In this case the mass difference is the energy/ heat that is released when the dynamite explodes, and when this heat escapes, the mass associated with it escapes, only to be deposited in the surroundings, which absorb the heat (so that total mass is conserved).

Such a change in mass may only happen when the system is open, and the energy and mass escapes. Thus, if a stick of dynamite is blown up in a hermetically sealed chamber, the mass of the chamber and fragments, the heat, sound, and light would still be equal to the original mass of the chamber and dynamite. If sitting on a scale, the weight and mass would not change. This would in theory also happen even with a nuclear bomb, if it could be kept in an ideal box of infinite strength, which did not rupture or pass radiation. Thus, a 21.5 kiloton (9×10^{13} joule) nuclear bomb produces about one gram of heat and electromagnetic radiation, but the mass of this energy would not be detectable in an exploded bomb in an ideal box sitting on a scale; instead, the contents of the box would be heated to millions of degrees without changing total mass and weight. If then, however, a transparent window (passing only electromagnetic radiation) were opened in such an ideal box after the explosion, and a beam of X-rays and other lower-energy light allowed to escape the box, it would eventually be found to weigh one gram less than it had before the explosion. This weight loss and mass loss would happen as the box was cooled by this process, to room temperature. However, any surrounding mass that absorbed the X-rays (and other "heat") would *gain* this gram of mass from the resulting heating, so the mass "loss" would represent merely its relocation. Thus, no mass (or, in the case of a nuclear bomb, no matter) would be "converted" to energy in such a process. Mass and energy, as always, would both be separately conserved.

Massless Particles

Massless particles have zero rest mass. Their relativistic mass is simply their relativistic energy, divided by c^2, or $m_{rel} = E/c^2$. The energy for photons is $E = hf$, where h is Planck's constant and f is the photon frequency. This frequency and thus the relativistic energy are frame-dependent.

If an observer runs away from a photon in the direction the photon travels from a source, and it catches up with the observer—when the photon catches up, the observer sees it as having less energy than it had at the source. The faster the observer is traveling with regard to the source when the photon catches up, the less energy the photon has. As an observer approaches the speed of light with regard to the source, the photon looks redder and redder, by relativistic Doppler effect (the Doppler shift is the relativistic formula), and the energy of a very long-wavelength photon approaches zero. This is why a photon is *massless*—this means that the rest mass of a photon is zero.

Massless Particles Contribute Rest Mass and Invariant Mass to Systems

Two photons moving in different directions cannot both be made to have arbitrarily small total energy by changing frames, or by moving toward or away from them. The reason is that in a two-photon system, the energy of one photon is decreased by chasing after it, but the energy of the other increases with the same shift in observer motion. Two photons not moving in the same direction exhibits an inertial frame where the combined energy is smallest, but not zero. This is called the center of mass frame or the center of momentum frame; these terms are almost synonyms (the center of mass frame is the special case of a center of momentum frame where the center of mass is put at the origin). The most that chasing a pair of photons can accomplish to decrease their energy is to put the observer in a frame where the photons have equal energy and are moving directly away from each other. In this frame, the observer is now moving in the same direction and speed as the center of mass of the two photons. The total momentum of the photons is now zero, since their momenta are equal and opposite. In this frame the two photons, as a system, have a mass equal to their total energy divided by c^2. This mass is called the invariant mass of the pair of photons together. It is the smallest mass and energy the system may be seen to have, by any observer. It is only the invariant mass of a two-photon system that can be used to make a single particle with the same rest mass.

If the photons are formed by the collision of a particle and an antiparticle, the invariant mass is the same as the total energy of the particle and antiparticle (their rest energy plus the kinetic energy), in the center of mass frame, where they automatically move in equal and opposite directions (since they have equal momentum in this frame). If the photons are formed by the disintegration of a *single* particle with a well-defined rest mass, like the neutral pion, the invariant mass of the photons is equal to rest mass of the pion. In this case, the center of mass frame for the pion is just the frame where the pion is at rest, and the center of mass does not change after it disintegrates into two photons. After the two photons are formed, their center of mass is still moving the same way the pion did, and their total energy in this frame adds up to the mass energy of the pion. Thus, by calculating the invariant mass of pairs of photons in a particle detector, pairs can be identified that were probably produced by pion disintegration.

A similar calculation illustrates that the invariant mass of systems is conserved, even when massive particles (particles with rest mass) within the system are converted to massless particles (such as photons). In such cases, the photons contribute invariant mass to the system, even though they individually have no invariant mass or rest mass. Thus, an electron and positron (each of which has rest mass) may undergo annihilation with each other to produce two photons, each of which is massless (has no rest mass). However, in such circumstances, no system mass is lost. Instead, the system of both photons moving away from each other has an invariant mass, which acts like a rest mass for any system in which the photons are trapped, or that can be weighed. Thus, not only the quantity of relativistic mass, but also the quantity of invariant mass does not change in transformations between "matter" (electrons and positrons) and energy (photons).

Relation to Gravity

In physics, there are two distinct concepts of mass: the gravitational mass and the inertial mass. The gravitational mass is the quantity that determines the strength of the gravitational field generated by an object, as well as the gravitational force acting on the object when it is immersed in a gravitational field produced by other bodies. The inertial mass, on the other hand, quantifies how much an object accelerates if a given force is applied to it. The mass–energy equivalence in special relativity refers to the inertial mass. However, already in the context of Newton gravity, the Weak Equivalence Principle is postulated: the gravitational and the inertial mass of every object are the same. Thus, the mass–energy equivalence, combined with the Weak Equivalence Principle, results in the prediction that all forms of energy contribute to the gravitational field generated by an object. This observation is one of the pillars of the general theory of relativity.

The above prediction, that all forms of energy interact gravitationally, has been subject to experimental tests. The first observation testing this prediction was made in 1919. During a solar eclipse, Arthur Eddington observed that the light from stars passing close to the Sun was bent. The effect is due to the gravitational attraction of light by the Sun. The observation confirmed that the energy carried by light indeed is equivalent to a gravitational mass. Another seminal experiment, the Pound–Rebka experiment, was performed in 1960. In this test a beam of light was emitted from the top of a tower and detected at the bottom. The frequency of the light detected was higher than the light emitted. This result confirms that the energy of photons increases when they fall in the gravitational field of the Earth. The energy, and therefore the gravitational mass, of photons is proportional to their frequency as stated by the Planck's relation.

Application to Nuclear Physics

Task Force One, the world's first nuclear-powered task force. *Enterprise*, *Long Beach* and *Bainbridge* in formation in the Mediterranean, 18 June 1964. *Enterprise* crew members are spelling out Einstein's mass–energy equivalence formula $E = mc^2$ on the flight deck.

Max Planck pointed out that the mass–energy equivalence formula implied that bound systems would have a mass less than the sum of their constituents, once the binding energy had been allowed to escape. However, Planck was thinking about chemical reactions, where the binding energy is too small to measure. Einstein suggested that radioactive materials such as radium would provide a test of the theory, but even though a large amount of energy is released per atom in radium, due to the half-life of the substance (1602 years), only a small fraction of radium atoms decay over an experimentally measurable period of time.

Once the nucleus was discovered, experimenters realized that the very high binding energies of the atomic nuclei should allow calculation of their binding energies, simply from mass differences. But it was not until the discovery of the neutron in 1932, and the measurement of the neutron mass, that this calculation could actually be performed. A little while later, the Cockcroft–Walton accelerator produced the first transmutation reaction ($_3^7\text{Li} + {}_1^1\text{p} \to 2\,{}_2^4\text{He}$), verifying Einstein's formula to an accuracy of ±0.5%. In 2005, Rainville et al. published a direct test of the energy-equivalence of mass lost in the binding energy of a neutron to atoms of particular isotopes of silicon and sulfur, by comparing the mass lost to the energy of the emitted gamma ray associated with the neutron capture. The binding mass-loss agreed with the gamma ray energy to a precision of ±0.00004%, the most accurate test of $E = mc^2$ to date.

The mass–energy equivalence formula was used in the understanding of nuclear fission reactions, and implies the great amount of energy that can be released by a nuclear fission chain reaction, used in both nuclear weapons and nuclear power. By measuring the mass of different atomic nuclei and subtracting from that number the total mass of the protons and neutrons as they would weigh separately, one gets the exact binding energy available in an atomic nucleus. This is used to calculate the energy released in any nuclear reaction, as the difference in the total mass of the nuclei that enter and exit the reaction.

Practical Examples

Einstein used the CGS system of units (centimeters, grams, seconds, dynes, and ergs), but the formula is independent of the system of units. In natural units, the numerical value of the speed of light is set to equal 1, and the formula expresses an equality of numerical values: $E = m$. In the SI system (expressing the ratio E/m in joules per kilogram using the value of c in meters per second):

$$E/m = c^2 = (299792458 \text{ m/s})^2 = 89875517873681764 \text{ J/kg } (\approx 9.0 \times 10^{16} \text{ joules per kilogram}).$$

So the energy equivalent of one kilogram of mass is

- 89.9 petajoules
- 25.0 million kilowatt-hours ($\approx$ 25 GW·h)
- 21.5 trillion kilocalories ($\approx$ 21 Pcal)
- 85.2 trillion BTUs
- 0.0852 quads

or the energy released by combustion of the following:

- 21 500 kilotons of TNT-equivalent energy ($\approx$ 21 Mt)
- 2630000000 litres or 695000000 US gallons of automotive gasoline

Any time energy is generated, the process can be evaluated from an $E = mc^2$ perspective. For instance, the "Gadget"-style bomb used in the Trinity test and the bombing of Nagasaki had an explosive yield equivalent to 21 kt of TNT. About 1 kg of the approximately 6.15 kg of plutonium in each of these bombs fissioned into lighter elements totaling almost exactly one gram less, after cooling. The electromagnetic radiation and kinetic energy (thermal and blast energy) released in this explosion carried the missing one gram of mass. This occurs because nuclear binding energy is released whenever elements with more than 62 nucleons fission.

Another example is hydroelectric generation. The electrical energy produced by Grand Coulee Dam's turbines every 3.7 hours represents one gram of mass. This mass passes to electrical devices (such as lights in cities) powered by the generators, where it appears as a gram of heat and light. Turbine designers look at their equations in terms of pressure, torque, and RPM. However, Einstein's equations show that all energy has mass, and thus the electrical energy produced by a dam's generators, and the resulting heat and light, all retain their mass—which is equivalent to the energy. The potential energy—and equivalent mass—represented by the waters of the Columbia River as it descends to the Pacific Ocean would be converted to heat due to viscous friction and the turbulence of white water rapids and waterfalls were it not for the dam and its generators. This heat would remain as mass on site at the water, were it not for the equipment that converted some of this potential and kinetic energy into electrical energy, which can move from place to place (taking mass with it).

Whenever energy is added to a system, the system gains mass:

- A spring's mass increases whenever it is put into compression or tension. Its added mass arises from the added potential energy stored within it, which is bound in the stretched chemical (electron) bonds linking the atoms within the spring.

- Raising the temperature of an object (increasing its heat energy) increases its mass. For example, consider the world's primary mass standard for the kilogram, made of platinum/iridium. If its temperature is allowed to change by 1 °C, its mass changes by 1.5 picograms (1 pg = 1×10^{-12} g).

- A spinning ball weighs more than a ball that is not spinning. Its increase of mass is exactly the equivalent of the mass of energy of rotation, which is itself the sum of the kinetic energies of all the moving parts of the ball. For example, the Earth itself is more massive due to its daily rotation, than it would be with no rotation. This rotational energy (2.14×10^{29} J) represents 2.38 billion metric tons of added mass.

Note that no net mass or energy is really created or lost in any of these examples and scenarios. Mass/energy simply moves from one place to another. These are some examples of the *transfer* of energy and mass in accordance with the *principle of mass–energy conservation*.

Efficiency

Although mass cannot be converted to energy, in some reactions matter particles (which contain a form of rest energy) can be destroyed and converted to other types of energy that are more usable and obvious as forms of energy—such as light and energy of motion (heat, etc.). However, the total amount of energy and mass does not change in such a transformation. Even when particles are not destroyed, a certain fraction of the ill-defined "matter" in ordinary objects can be destroyed, and its associated energy liberated and made available as the more dramatic energies of light and heat, even though no identifiable real particles are destroyed, and even though (again) the total energy is unchanged (as also the total mass). Such conversions between types of energy (resting to active energy) happen in nuclear weapons, in which the protons and neutrons in atomic nuclei lose a small fraction of their average mass, but this mass loss is not due to the destruction of any protons or neutrons (or even, in general, lighter particles like electrons). Also the mass is not destroyed, but simply removed from the system. in the form of heat and light from the reaction.

In nuclear reactions, typically only a small fraction of the total mass–energy of the bomb converts into the mass–energy of heat, light, radiation, and motion—which are "active" forms that can be used. When an atom fissions, it loses only about 0.1% of its mass (which escapes from the system and does not disappear), and additionally, in a bomb or reactor not all the atoms can fission. In a modern fission-based atomic bomb, the efficiency is only about 40%, so only 40% of the fission-able atoms actually fission, and only about 0.03% of the fissile core mass appears as energy in the end. In nuclear fusion, more of the mass is released as usable energy, roughly 0.3%. But in a fusion bomb, the bomb mass is partly casing and non-reacting components, so that in practicality, again (coincidentally) no more than about 0.03% of the total mass of the entire weapon is released as usable energy (which, again, retains the "missing" mass).

In theory, it should be possible to destroy matter and convert all of the rest-energy associated with matter into heat and light (which would of course have the same mass), but none of the theoret-ically known methods are practical. One way to convert all the energy within matter into usable energy is to annihilate matter with antimatter. But antimatter is rare in our universe, and must be made first. Due to inefficient mechanisms of production, making antimatter always requires far more usable energy than would be released when it was annihilated.

Since most of the mass of ordinary objects resides in protons and neutrons, converting all the ener-gy of ordinary matter into more useful energy requires that the protons and neutrons be converted to lighter particles, or particles with no rest-mass at all. In the Standard Model of particle physics, the number of protons plus neutrons is nearly exactly conserved. Still, Gerard 't Hooft showed that there is a process that converts protons and neutrons to antielectrons and neutrinos. This is the weak SU(2) instanton proposed by Belavin Polyakov Schwarz and Tyupkin. This process, can in principle destroy matter and convert all the energy of matter into neutrinos and usable energy, but it is normally extraordinarily slow. Later it became clear that this process happens at a fast rate at very high temperatures, since then, instanton-like configurations are copiously produced from thermal fluctuations. The temperature required is so high that it would only have been reached shortly after the big bang.

Many extensions of the standard model contain magnetic monopoles, and in some models of grand unification, these monopoles catalyze proton decay, a process known as the Callan–Rubakov ef-fect. This process would be an efficient mass–energy conversion at ordinary temperatures, but it requires making monopoles and anti-monopoles first. The energy required to produce monopoles is believed to be enormous, but magnetic charge is conserved, so that the lightest monopole is sta-ble. All these properties are deduced in theoretical models—magnetic monopoles have never been observed, nor have they been produced in any experiment so far.

A third known method of total matter–energy "conversion" (which again in practice only means conversion of one type of energy into a different type of energy), is using gravity, specifically black holes. Stephen Hawking theorized that black holes radiate thermally with no regard to how they are formed. So, it is theoretically possible to throw matter into a black hole and use the emitted heat to generate power. According to the theory of Hawking radiation, however, the black hole used radiates at a higher rate the smaller it is, producing usable powers at only small black hole masses, where usable may for example be something greater than the local background radiation. It is also worth noting that the ambient irradiated power would change with the mass of the black hole, increasing as the mass of the black hole decreases, or decreasing as the mass increases, at a rate where power is proportional to the inverse square of the mass. In a "practical" scenario, mass and energy could be dumped into the black hole to regulate this growth, or keep its size, and thus power output, near constant. This could result from the fact that mass and energy are lost from the hole with its thermal radiation.

Background

Mass–velocity Relationship

In developing special relativity, Einstein found that the kinetic energy of a moving body is

$$E_k = m_0(\gamma - 1)c^2 = \frac{m_0 c^2}{\sqrt{1 - \dfrac{v^2}{c^2}}} - m_0 c^2,$$

with v the velocity, m_0 the rest mass, and γ the Lorentz factor.

He included the second term on the right to make sure that for small velocities the energy would be the same as in classical mechanics, thus satisfying the correspondence principle:

$$E_k = \frac{1}{2} m_0 v^2 + \cdots$$

Without this second term, there would be an additional contribution in the energy when the particle is not moving.

Einstein found that the total momentum of a moving particle is:

$$P = \frac{m_0 v}{\sqrt{1 - \dfrac{v^2}{c^2}}}.$$

It is this quantity that is conserved in collisions. The ratio of the momentum to the velocity is the relativistic mass, m.

$$m = \frac{m_0}{\sqrt{1 - \dfrac{v^2}{c^2}}}$$

And the relativistic mass and the relativistic kinetic energy are related by the formula:

$$E_k = mc^2 - m_0 c^2.$$

Einstein wanted to omit the unnatural second term on the right-hand side, whose only purpose is to make the energy at rest zero, and to declare that the particle has a total energy, which obeys:

$$E = mc^2$$

which is a sum of the rest energy $m_0 c^2$ and the kinetic energy. This total energy is mathematically more elegant, and fits better with the momentum in relativity. But to come to this conclusion, Einstein needed to think carefully about collisions. This expression for the energy implied that matter at rest has a huge amount of energy, and it is not clear whether this energy is physically real, or just a mathematical artifact with no physical meaning.

In a collision process where all the rest-masses are the same at the beginning as at the end, either expression for the energy is conserved. The two expressions only differ by a constant that is the same at the beginning and at the end of the collision. Still, by analyzing the situation where particles are thrown off a heavy central particle, it is easy to see that the inertia of the central particle is reduced by the total energy emitted. This allowed Einstein to conclude that the inertia of a heavy particle is increased or diminished according to the energy it absorbs or emits.

Relativistic Mass

After Einstein first made his proposal, it became clear that the word mass can have two different meanings. Some denote the *relativistic mass* with an explicit index:

$$m_{rel} = \frac{m_0}{\sqrt{1 - \dfrac{v^2}{c^2}}}.$$

This mass is the ratio of momentum to velocity, and it is also the relativistic energy divided by c^2 (it is not Lorentz-invariant, in contrast to m_0). The equation $E = m_{rel} c^2$ holds for moving objects. When the velocity is small, the relativistic mass and the rest mass are almost exactly the same.

- $E = mc^2$ either means $E = m_0 c^2$ for an object at rest, or $E = m_{rel} c^2$ when the object is moving.

Also Einstein (following Hendrik Lorentz and Max Abraham) used velocity- and direction-dependent mass concepts (longitudinal and transverse mass) in his 1905 electrodynamics paper and in another paper in 1906. However, in his first paper on $E = mc^2$ (1905), he treated m as what would now be called the *rest mass*. Some claim that (in later years) he did not like the idea of "relativistic mass". When modern physicists say "mass", they are usually talking about rest mass, since if they meant "relativistic mass", they would just say "energy".

Considerable debate has ensued over the use of the concept "relativistic mass" and the connection of "mass" in relativity to "mass" in Newtonian dynamics. For example, one view is that only rest mass is a viable concept and is a property of the particle; while relativistic mass is a conglomeration of particle properties and properties of spacetime. A perspective that avoids this debate, due to Kjell Vøyenli, is that the Newtonian concept of mass as a particle property and the relativistic concept of mass have to be viewed as embedded in their own theories and as having no precise connection.

Low Speed Expansion

We can rewrite the expression $E = \gamma m_0 c^2$ as a Taylor series:

$$E = m_0 c^2 \left[1 + \frac{1}{2}\left(\frac{v}{c}\right)^2 + \frac{3}{8}\left(\frac{v}{c}\right)^4 + \frac{5}{16}\left(\frac{v}{c}\right)^6 + \ldots \right].$$

For speeds much smaller than the speed of light, higher-order terms in this expression get smaller and smaller because v/c is small. For low speeds we can ignore all but the first two terms:

$$E \approx m_0 c^2 + \frac{1}{2} m_0 v^2.$$

The total energy is a sum of the rest energy and the Newtonian kinetic energy.

The classical energy equation ignores both the $m_0 c^2$ part, and the high-speed corrections. This is appropriate, because all the high-order corrections are small. Since only *changes* in energy affect the behavior of objects, whether we include the $m_0 c^2$ part makes no difference, since it is constant. For the same reason, it is possible to subtract the rest energy from the total energy in relativity. By considering the emission of energy in different frames, Einstein could show that the rest energy has a real physical meaning.

The higher-order terms are extra correction to Newtonian mechanics, and become important at higher speeds. The Newtonian equation is only a low-speed approximation, but an extraordinarily good one. All of the calculations used in putting astronauts on the moon, for example, could have been done using Newton's equations without any of the higher-order corrections. The total mass energy equivalence should also include the rotational and vibrational kinetic energies as well as the linear kinetic energy at low speeds.

History

While Einstein was the first to have correctly deduced the mass–energy equivalence formula, he was not the first to have related energy with mass. But nearly all previous authors thought that the energy that contributes to mass comes only from electromagnetic fields.

Newton: Matter and Light

In 1717 Isaac Newton speculated that light particles and matter particles were interconvertible in "Query 30" of the *Opticks*, where he asks:

Are not the gross bodies and light convertible into one another, and may not bodies receive much of their activity from the particles of light which enter their composition?

Swedenborg: Matter Composed of "Pure and Total Motion"

In 1734 the Swedish scientist and theologian Emanuel Swedenborg in his *Principia* theorized that all matter is ultimately composed of dimensionless points of "pure and total motion." He described this motion as being without force, direction or speed, but having the potential for force, direction and speed everywhere within it.

Electromagnetic Mass

There were many attempts in the 19th and the beginning of the 20th century—like those of J. J. Thomson (1881), Oliver Heaviside (1888), and George Frederick Charles Searle (1897), Wilhelm Wien (1900), Max Abraham (1902), Hendrik Antoon Lorentz (1904) — to understand how the mass of a charged object depends on the electrostatic field. This concept was called electromagnetic mass, and was considered as being dependent on velocity and direction as well. Lorentz (1904) gave the following expressions for longitudinal and transverse electromagnetic mass:

$$m_L = \frac{m_0}{\left(\sqrt{1-\frac{v^2}{c^2}}\right)^3}, \quad m_T = \frac{m_0}{\left(\sqrt{1-\frac{v^2}{c^2}}\right)^3},$$

where

$$m_0 = \frac{4}{3}\frac{E_{em}}{c^2}$$

Radiation Pressure and Inertia

Another way of deriving some sort of electromagnetic mass was based on the concept of radiation pressure. In 1900, Henri Poincaré associated electromagnetic radiation energy with a "fictitious fluid" having momentum and mass

$$m_{em} = \frac{E_{em}}{c^2}.$$

By that, Poincaré tried to save the center of mass theorem in Lorentz's theory, though his treatment led to radiation paradoxes.

Friedrich Hasenöhrl showed in 1904, that electromagnetic cavity radiation contributes the "apparent mass"

$$m_0 = \frac{4}{3}\frac{E_{em}}{c^2}$$

to the cavity's mass. He argued that this implies mass dependence on temperature as well.

Einstein: Mass–energy Equivalence

Albert Einstein did not formulate exactly the formula $E = mc^2$ in his 1905 *Annus Mirabilis* paper "Does the Inertia of an object Depend Upon Its Energy Content?"; rather, the paper states that if a body gives off the energy L in the form of radiation, its mass diminishes by L/c^2. (Here, "radiation" means electromagnetic radiation, or light, and mass means the ordinary Newtonian mass of a slow-moving object.) This formulation relates only a change Δm in mass to a change L in energy without requiring the absolute relationship.

Objects with zero mass presumably have zero energy, so the extension that all mass is proportional to energy is obvious from this result. In 1905, even the hypothesis that changes in energy are accompanied by changes in mass was untested. Not until the discovery of the first type of antimatter (the positron in 1932) was it found that all of the mass of pairs of resting particles could be converted to radiation.

The First Derivation by Einstein (1905)

Already in his relativity paper "On the electrodynamics of moving bodies", Einstein derived the correct expression for the kinetic energy of particles:

$$E_k = mc^2 \left(\frac{1}{\sqrt{1 - \dfrac{v^2}{c^2}}} - 1 \right).$$

Now the question remained open as to which formulation applies to bodies at rest. This was tackled by Einstein in his paper "Does the inertia of a body depend upon its energy content?", where he used a body emitting two light pulses in opposite directions, having energies of E_0 before and E_1 after the emission as seen in its rest frame. As seen from a moving frame, this becomes H_0 and H_1. Einstein obtained:

$$\left(H_0 - E_0 \right) - \left(H_1 - E_1 \right) = E \left(\frac{1}{\sqrt{1 - \dfrac{v^2}{c^2}}} - 1 \right)$$

then he argued that $H - E$ can only differ from the kinetic energy K by an additive constant, which gives

$$K_0 - K_1 = E \left(\frac{1}{\sqrt{1 - \dfrac{v^2}{c^2}}} - 1 \right)$$

Neglecting effects higher than third order in v/c after a Taylor series expansion of the right side of this gives:

$$K_0 - K_1 = \frac{E}{c^2}\frac{v^2}{2}.$$

Einstein concluded that the emission reduces the body's mass by E/c^2, and that the mass of a body is a measure of its energy content.

The correctness of Einstein's 1905 derivation of $E = mc^2$ was criticized by Max Planck (1907), who argued that it is only valid to first approximation. Another criticism was formulated by Herbert Ives (1952) and Max Jammer (1961), asserting that Einstein's derivation is based on begging the question. On the other hand, John Stachel and Roberto Torretti (1982) argued that Ives' criticism was wrong, and that Einstein's derivation was correct. Hans Ohanian (2008) agreed with Stachel/Torretti's criticism of Ives, though he argued that Einstein's derivation was wrong for other reasons.

Alternative Version

An alternative version of Einstein's thought experiment was proposed by Fritz Rohrlich (1990), who based his reasoning on the Doppler effect. Like Einstein, he considered a body at rest with mass M. If the body is examined in a frame moving with nonrelativistic velocity v, it is no longer at rest and in the moving frame it has momentum $P = Mv$. Then he supposed the body emits two pulses of light to the left and to the right, each carrying an equal amount of energy $E/2$. In its rest frame, the object remains at rest after the emission since the two beams are equal in strength and carry opposite momentum.

However, if the same process is considered in a frame that moves with velocity v to the left, the pulse moving to the left is redshifted, while the pulse moving to the right is blue shifted. The blue light carries more momentum than the red light, so that the momentum of the light in the moving frame is not balanced: the light is carrying some net momentum to the right.

The object has not changed its velocity before or after the emission. Yet in this frame it has lost some right-momentum to the light. The only way it could have lost momentum is by losing mass. This also solves Poincaré's radiation paradox, discussed above.

The velocity is small, so the right-moving light is blueshifted by an amount equal to the nonrelativistic Doppler shift factor $1 - v/c$. The momentum of the light is its energy divided by c, and it is increased by a factor of v/c. So the right-moving light is carrying an extra momentum ΔP given by:

$$\Delta P = \frac{v}{c}\frac{E}{2c}.$$

The left-moving light carries a little less momentum, by the same amount ΔP. So the total right-momentum in the light is twice ΔP. This is the right-momentum that the object lost.

$$2\Delta P = v\frac{E}{c^2}.$$

The momentum of the object in the moving frame after the emission is reduced to this amount:

$$P' = Mv - 2\Delta P = \left(M - \frac{E}{c^2} \right)v.$$

So the change in the object's mass is equal to the total energy lost divided by c^2. Since any emission of energy can be carried out by a two step process, where first the energy is emitted as light and then the light is converted to some other form of energy, any emission of energy is accompanied by a loss of mass. Similarly, by considering absorption, a gain in energy is accompanied by a gain in mass.

Relativistic Center-of-mass Theorem (1906)

Like Poincaré, Einstein concluded in 1906 that the inertia of electromagnetic energy is a necessary condition for the center-of-mass theorem to hold. On this occasion, Einstein referred to Poincaré's 1900 paper and wrote:

Although the merely formal considerations, which we will need for the proof, are already mostly contained in a work by H. Poincaré[2], for the sake of clarity I will not rely on that work.

In Einstein's more physical, as opposed to formal or mathematical, point of view, there was no need for fictitious masses. He could avoid the *perpetuum mobile* problem because, on the basis of the mass–energy equivalence, he could show that the transport of inertia that accompanies the emission and absorption of radiation solves the problem. Poincaré's rejection of the principle of action–reaction can be avoided through Einstein's $E = mc^2$, because mass conservation appears as a special case of the energy conservation law.

Others

During the nineteenth century there were several speculative attempts to show that mass and energy were proportional in various ether theories. In 1873 Nikolay Umov pointed out a relation between mass and energy for ether in the form of $E = kmc^2$, where $0.5 \leq k \leq 1$. The writings of Samuel Tolver Preston, and a 1903 paper by Olinto De Pretto, presented a mass–energy relation. De Pretto's paper received recent press coverage when Umberto Bartocci discovered that there were only three degrees of separation linking De Pretto to Einstein, leading Bartocci to conclude that Einstein was probably aware of De Pretto's work.

Preston and De Pretto, following Le Sage, imagined that the universe was filled with an ether of tiny particles that always move at speed c. Each of these particles has a kinetic energy of mc^2 up to a small numerical factor. The nonrelativistic kinetic energy formula did not always include the traditional factor of $1/2$, since Leibniz introduced kinetic energy without it, and the $1/2$ is largely conventional in prerelativistic physics. By assuming that every particle has a mass that is the sum of the masses of the ether particles, the authors concluded that all matter contains an amount of kinetic energy either given by $E = mc^2$ or $2E = mc^2$ depending on the convention. A particle ether was usually considered unacceptably speculative science at the time, and since these authors did not formulate relativity, their reasoning is completely different from that of Einstein, who used relativity to change frames.

Independently, Gustave Le Bon in 1905 speculated that atoms could release large amounts of latent energy, reasoning from an all-encompassing qualitative philosophy of physics.

Radioactivity and Nuclear Energy

It was quickly noted after the discovery of radioactivity in 1897, that the total energy due to radioactive processes is about one *million times* greater than that involved in any known molecular change. However, it raised the question where this energy is coming from. After eliminating the idea of absorption and emission of some sort of Lesagian ether particles, the existence of a huge amount of latent energy, stored within matter, was proposed by Ernest Rutherford and Frederick Soddy in 1903. Rutherford also suggested that this internal energy is stored within normal matter as well. He went on to speculate in 1904:

If it were ever found possible to control at will the rate of disintegration of the radio-elements, an enormous amount of energy could be obtained from a small quantity of matter.

Einstein's equation is in no way an explanation of the large energies released in radioactive decay (this comes from the powerful nuclear forces involved; forces that were still unknown in 1905). In any case, the enormous energy released from radioactive decay (which had been measured by Rutherford) was much more easily measured than the (still small) change in the gross mass of materials, as a result. Einstein's equation, by theory, can give these energies by measuring mass differences before and after reactions, but in practice, these mass differences in 1905 were still too small to be measured in bulk. Prior to this, the ease of measuring radioactive decay energies with a calorimeter was thought possibly likely to allow measurement of changes in mass difference, as a check on Einstein's equation itself. Einstein mentions in his 1905 paper that mass–energy equivalence might perhaps be tested with radioactive decay, which releases enough energy (the quantitative amount known roughly by 1905) to possibly be "weighed," when missing from the system (having been given off as heat). However, radioactivity seemed to proceed at its own unalterable (and quite slow, for radioactives known then) pace, and even when simple nuclear reactions became possible using proton bombardment, the idea that these great amounts of usable energy could be liberated at will with any practicality, proved difficult to substantiate. Rutherford was reported in 1933 to have declared that this energy could not be exploited efficiently: "Anyone who expects a source of power from the transformation of the atom is talking moonshine."

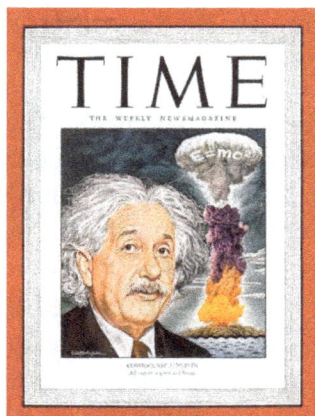

The popular connection between Einstein, $E = mc^2$, and the atomic bomb was prominently indicated on the cover of *Time* magazine in July 1946 by the writing of the equation on the mushroom cloud.

This situation changed dramatically in 1932 with the discovery of the neutron and its mass, allowing mass differences for single nuclides and their reactions to be calculated directly, and compared with the sum of masses for the particles that made up their composition. In 1933, the energy released from the reaction of lithium-7 plus protons giving rise to 2 alpha particles (as noted above by Rutherford), allowed Einstein's equation to be tested to an error of ±0.5%. However, scientists still did not see such reactions as a source of power.

After the very public demonstration of huge energies released from nuclear fission after the atomic bombings of Hiroshima and Nagasaki in 1945, the equation $E = mc^2$ became directly linked in the public eye with the power and peril of nuclear weapons. The equation was featured as early as page 2 of the Smyth Report, the official 1945 release by the US government on the development of the atomic bomb, and by 1946 the equation was linked closely enough with Einstein's work that the cover of *Time* magazine prominently featured a picture of Einstein next to an image of a mushroom cloud emblazoned with the equation. Einstein himself had only a minor role in the Manhattan Project: he had cosigned a letter to the U.S. President in 1939 urging funding for research into atomic energy, warning that an atomic bomb was theoretically possible. The letter persuaded Roosevelt to devote a significant portion of the wartime budget to atomic research. Without a security clearance, Einstein's only scientific contribution was an analysis of an isotope separation method in theoretical terms. It was inconsequential, on account of Einstein not being given sufficient information (for security reasons) to fully work on the problem.

While $E = mc^2$ is useful for understanding the amount of energy potentially released in a fission reaction, it was not strictly necessary to develop the weapon, once the fission process was known, and its energy measured at 200 MeV (which was directly possible, using a quantitative Geiger counter, at that time). As the physicist and Manhattan Project participant Robert Serber put it: "Somehow the popular notion took hold long ago that Einstein's theory of relativity, in particular his famous equation $E = mc^2$, plays some essential role in the theory of fission. Albert Einstein had a part in alerting the United States government to the possibility of building an atomic bomb, but his theory of relativity is not required in discussing fission. The theory of fission is what physicists call a non-relativistic theory, meaning that relativistic effects are too small to affect the dynamics of the fission process significantly." However the association between $E = mc^2$ and nuclear energy has since stuck, and because of this association, and its simple expression of the ideas of Albert Einstein himself, it has become "the world's most famous equation".

While Serber's view of the strict lack of need to use mass–energy equivalence in designing the atomic bomb is correct, it does not take into account the pivotal role this relationship played in making the fundamental leap to the initial hypothesis that large atoms were energetically *allowed* to split into approximately equal parts (before this energy was in fact measured). In late 1938, Lise Meitner and Otto Robert Frisch—while on a winter walk during which they solved the meaning of Hahn's experimental results and introduced the idea that would be called atomic fission—directly used Einstein's equation to help them understand the quantitative energetics of the reaction that overcame the "surface tension-like" forces that hold the nucleus together, and allowed the fission fragments to separate to a configuration from which their charges could force them into an energetic *fission*. To do this, they used *packing fraction*, or nuclear binding energy values for elements, which Meitner had memorized. These, together with use of $E = mc^2$ allowed them to realize on the spot that the basic fission process was energetically possible:

...We walked up and down in the snow, I on skis and she on foot. ...and gradually the idea took shape... explained by Bohr's idea that the nucleus is like a liquid drop; such a drop might elongate and divide itself... We knew there were strong forces that would resist, ..just as surface tension. But nuclei differed from ordinary drops. At this point we both sat down on a tree trunk and started to calculate on scraps of paper. ...the Uranium nucleus might indeed be a very wobbly, unstable drop, ready to divide itself... But, ...when the two drops separated they would be driven apart by electrical repulsion, about 200 MeV in all. Fortunately Lise Meitner remembered how to compute the masses of nuclei... and worked out that the two nuclei formed... would be lighter by about one-fifth the mass of a proton. Now whenever mass disappears energy is created, according to Einstein's formula $E = mc^2$, and... the mass was just equivalent to 200 MeV; it all fitted!

Minkowski Space

H. Minkowski

Hermann Minkowski (1864 – 1909) found that the theory of special relativity, introduced by his former student Albert Einstein, could best be understood in a four-dimensional space, since known as the Minkowski spacetime.

In mathematical physics, Minkowski space or Minkowski spacetime is a combination of Euclidean space and time into a four-dimensional manifold where the spacetime interval between any two events is independent of the inertial frame of reference in which they are recorded. Although initially developed by mathematician Hermann Minkowski for Maxwell's equations of electromagnetism, the mathematical structure of Minkowski spacetime was shown to be an immediate consequence of the postulates of special relativity.

Minkowski space is closely associated with Einstein's theory of special relativity, and is the most common mathematical structure on which special relativity is formulated. While the individual components in Euclidean space and time will often differ due to length contraction and time dilation, in Minkowski spacetime, all frames of reference will agree on the total distance in spacetime between events. Because it treats time differently than it treats the three spatial dimensions, Minkowski space differs from four-dimensional Euclidean space.

In Euclidean space, the isometry group (the maps preserving the regular inner product) is the Euclidean group. The analogous isometry group for Minkowski space, preserving intervals of spacetime equipped with the associated non-positive definite non-degenerate bilinear form (below variously called the *Minkowski metric*, the *Minkowski norm* or *Minkowski inner product* depending on the context) is the Poincaré group. The Minkowski inner product is defined as to yield the spacetime interval between two events when given their coordinate difference vector as argument.

History

Four-dimensional Euclidean Spacetime

In 1905–06 Henri Poincaré showed that by taking time to be an imaginary fourth spacetime coordinate ict, where c is the speed of light and i is the imaginary unit, a Lorentz transformation can formally be regarded as a rotation of coordinates in a four-dimensional space with three real coordinates representing space, and one imaginary coordinate representing time, as the fourth dimension. In physical spacetime special relativity stipulates that the quantity

$$-t^2 + x^2 + y^2 + z^2$$

is invariant under coordinate changes from one inertial frame to another, i. e. under Lorentz transformations. Here the speed of light c is, following Poincare, set to unity. In the space suggested by him (Poincare mentions this only in passing) where physical spacetime is *coordinatized* by $(t, x, y, z) \rightarrow (x, y, z, it)$, call it *coordinate space*, Lorentz transformations appear as ordinary rotations preserving the quadratic form

$$x^2 + y^2 + z^2 + t^2$$

on coordinate space. The naming and ordering of coordinates, with the same labels for space coordinates, but with the imaginary time coordinate as the *fourth coordinate*, is conventional. The above expression, while making the former expression more familiar, may potentially be confusing because it is *not* the same t that appears in the latter (*time coordinate*) as in the former (*time itself* in some inertial system as measured by clocks stationary in that system).

Rotations in planes spanned by two space unit vectors appear in coordinate space as well as in physical spacetime appear as Euclidean rotations and are interpreted in the ordinary sense. The "rotation" in a plane spanned by a space unit vector and a time unit vector, while formally still a rotation in coordinate space, is a Lorentz boost in physical spacetime with *real* inertial coordinates. The analogy with Euclidean rotations is thus only only partial.

This idea was elaborated by Hermann Minkowski, who used it to restate the Maxwell equations in four dimensions, showing directly their invariance under the Lorentz transformation. He further reformulated in four dimensions the then-recent theory of special relativity of Einstein. From this he concluded that time and space should be treated equally, and so arose his concept of events taking place in a unified four-dimensional spacetime continuum.

Minkowski Space

In a further development in his 1908 "Space and Time" lecture, he gave an alternative formulation of this idea that used a real time coordinate instead of an imaginary one, representing the four variables (x, y, z, t) of space and time in coordinate form in a four dimensional real vector space.

Points in this space correspond to events in spacetime. In this space, there is a defined light-cone associated with each point, and events not on the light-cone are classified by their relation to the apex as *spacelike* or *timelike*. It is principally this view of spacetime that is current nowadays, although the older view involving imaginary time has also influenced special relativity. An imaginary time coordinate is used also for more subtle reasons in quantum field theory than formal appearance of expressions. It this context, the transformation is called a Wick rotation.

In the English translation of Minkowski's paper, the Minkowski metric as defined below is referred to as the *line element*. The Minkowski inner product of below appears unnamed when referring to orthogonality (which he calls *normality*) of of certain vectors, and the Minkowski norm squared is referred to (somewhat cryptically, perhaps this is translation dependent) as "sum".

Minkowski's principal tool is the Minkowski diagram, and he uses it to define concepts and demonstrate properties of Lorentz transformations (e.g. proper time and length contraction) and to provide geometrical interpretation to the generalization of Newtonian mechanics to relativistic mechanics. For these special topics, as the presentation below will be principally confined to the mathematical structure (Minkowski metric and from it derived quantities and the Poincare group as symmetry group of spacetime) *following* from the invariance of the spacetime interval on the spacetime manifold as consequences of the postulates of special relativity, not to specific application or *derivation* of the invariance of the spacetime interval. This structure provides the background setting of all present relativistic theories, barring general relativity for which flat Minkowski spacetime still provides a springboard as curved spacetime is locally Lorentzian.

Minkowski, aware of the fundamental restatement of the theory which he had made, said

The views of space and time which I wish to lay before you have sprung from the soil of experimental physics, and therein lies their strength. They are radical. Henceforth space by itself, and time by itself, are doomed to fade away into mere shadows, and only a kind of union of the two will preserve an independent reality.

— *Hermann Minkowski, 1908, 1909*

Mathematical Structure

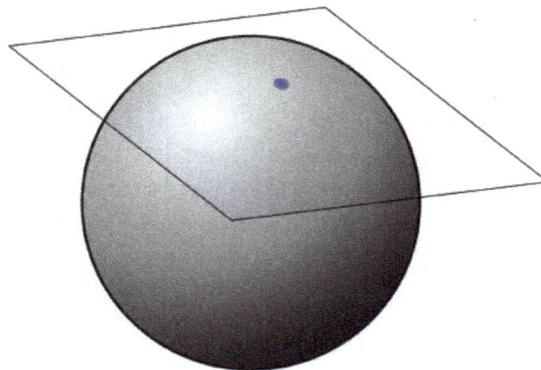

A pictorial representation of the tangent space at a point, *x*, on a sphere. This vector space can be thought of as a subspace of R^3 itself. Then vectors in it would be called *geometrical tangent vectors*. By the same principle, the tangent space at a point in flat spacetime can be thought of as a subspace of spacetime which happens to be *all* of spacetime.

It is assumed below that spacetime is endowed with a coordinate system corresponding to an inertial frame. This provides an *origin*, which is necessary in order to be able to refer to spacetime as being modeled as a vector space. This is not really *physically* motivated in that a canonical origin ("central" event in spacetime) should exist. One can get away with less structure, but this needlessly complicates the discussion and would not reflect how flat spacetime is normally treated mathematically in modern introductory literature.

For an overview, Minkowski space is a 4-dimensional real vector space equipped with a nondegenerate, symmetric bilinear form on the tangent space at each point in spacetime, here simply called the *Minkowski inner product*, with metric signature either $(+,-,-,-)$ or $(-,+,+,+)$. The tangent space at each event is a vector space of the same dimension as spacetime, 4.

Tangent Vectors

In practice, one need not be concerned with the tangent spaces. The vector space nature of Minkowski space allows for the canonical identification of vectors in tangent spaces at points (events) with vectors (points, events) in Minkowski space itself. These identifications are routinely done in mathematics. They can formally in Cartesian coordinates be expressed as

$$(x^0, x^1, x^2, x^3) \leftrightarrow x^0 \mathbf{e}_0 \big|_p + x^1 \mathbf{e}_1 \big|_p + x^2 \mathbf{e}_2 \big|_p + x^3 \mathbf{e}_3 \big|_p \leftrightarrow x^0 \mathbf{e}_0 \big|_q + x^1 \mathbf{e}_1 \big|_q + x^2 \mathbf{e}_2 \big|_c + x^3 \mathbf{e}_3 \big|_q,$$

with basis vectors in the tangent spaces defined by

$$\mathbf{e}_\mu \big|_p = \frac{\partial}{\partial x^\mu}\bigg|_p \quad \text{or} \quad \mathbf{e}_0 \big|_p = \begin{pmatrix} 1 \\ 0 \\ 0 \\ 0 \end{pmatrix}, \text{ etc.}$$

Here p and q are any two events and the last identification is referred to as parallel transport. The first identification is the canonical identification of vectors in the tangent space at any point with vectors in the space itself. The appearance of basis vectors in tangent spaces as first order differential operators is due to this identification. It is motivated by the observation that a geometrical tangent vector can be associated in a one-to-one manner with a directional derivative operator on the set of smooth functions. This is promoted to a *definition* of tangent vectors in manifolds *not* necessarily being embedded in R^n. This definition of tangent vectors is not the only possible one as ordinary n-tuples can be used as well.

[μ]For some purposes it is desirable to identify tangent vectors at a point p with *displacement vectors* at p, which is, of course, admissible by essentially the same canonical identification. The identifications of vectors referred to above in the mathematical setting can correspondingly be found in a more physical and explicitly geometrical setting in Misner, Thorne & Wheeler (1970). They offer various degree of sophistication (and rigor) depending on which part of the material one chooses to read.

Metric Signature

The metric signature refers to which sign the Minkowski inner product yields when given space (*spacelike* to be specific, defined further down) and time basis vectors (*timelike*) as arguments. Further discussion about this theoretically inconsequential, but practically necessary choice for purposes of internal consistency and convenience is deferred to the hide box below.

33Terminology

Mathematically associated to the bilinear form is a tensor of type (0,2) at each point in spacetime, called the *Minkowski metric*. The Minkowski metric, the bilinear form, and the Minkowski inner product are actually all the very same object; it is a bilinear function that accepts two (contravariant) vectors and returns a real number. In coordinates, this is the 4×4 matrix representing the bilinear form. Keeping this in mind may facilitate reading what follows.

For comparison, in general relativity, a Lorentzian manifold L is likewise equipped with a metric tensor g, which is a nondegenerate symmetric bilinear form on the tangent space T_pL at each point p of L. In coordinates, it may be represented by a 4×4 matrix *depending on spacetime position*. Minkowski space is thus a comparatively simple special case of a Lorentzian manifold. Its metric tensor is in coordinates the same symmetric matrix at every point of M, and its arguments can, per above, be taken as vectors in spacetime itself.

Introducing more terminology (but not more structure), Minkowski space is thus a pseudo-Euclidean space with total dimension $n = 4$ and signature (3, 1) or (1, 3). Elements of Minkowski space are called events. Minkowski space is often denoted $R^{3,1}$ or $R^{1,3}$ to emphasize the chosen signature, or just M. It is perhaps the simplest example of a pseudo-Riemannian manifold.

An interesting example of non-inertial coordinates for (part of) Minkowski spacetime are the Born coordinates.

Pseudo-Euclidean Metrics

The Minkowski metric η is the metric tensor of Minkowski space. It is a Pseudo-Euclidean metric, or more generally a *constant* Pseudo-Riemannian metric in Cartesian coordinates. As such it is a nondegenerate symmetric bilinear form, a type (0,2) tensor. It accepts two arguments u_p, v_p, vectors in T_pM, $p \in M$, the tangent space at p in M. Due to the above-mentioned canonical identification of T_pM with M itself, it accepts arguments u, v with both u and v in M.

As a notational convention, vectors v in M, called 4-vectors, are denoted in sans-serif italics, and not, as is common in the Euclidean setting, with boldface v. The latter is generally reserved for the 3-vector part (to be introduced below) of a 4-vector.

The definition

$$u \cdot v = \eta(u,v)$$

yields an inner product-like structure on M, previously and also henceforth, called the *Minkowski inner product*, similar to the Euclidean inner product, but it describes a different geometry. It is also called the *relativistic dot product*. If the two arguments are the same,

$$u \cdot u = \eta(u,u) \equiv \| u \|^2 \equiv u^2,$$

the resulting quantity will be called the *Minkowski norm squared*. The Minkowski inner product satisfies the following properties

- $\eta(au + v, w) = a\eta(u, w) + \eta(v, w), \quad \forall u, v \in M, \forall a \in \mathbb{R}$ (linearity in first slot)

- $\eta(u, v) = \eta(v, u)$ (symmetry)

- $\eta(u, v) = 0 \quad \forall v \in M \Rightarrow u = 0$ (non-degeneracy)

The first two conditions imply bilinearity. The defining *difference* between a pseudo-inner product and an inner product proper is that the former is *not* required to be positive definite, that is, $\eta(u, u) < 0$ is allowed.

The most important feature of the inner product and norm squared is that *these are quantities unaffected by Lorentz transformations*. In fact, it can be taken as the defining property of a Lorentz transformation that it preserves the inner product (i.e. the value of the corresponding bilinear form on two vectors). This approach is taken more generally for *all* classical groups definable this way in classical group. There, the matrix Φ is identical in the case O(3, 1) (the Lorentz group) to the matrix η to be displayed below.

Two vectors v and w are said to be orthogonal if $\eta(v, w) = 0$.

A vector e is called a unit vector if $\eta(e, e) = \pm 1$. A basis for M consisting of mutually orthogonal unit vectors is called an orthonormal basis.

For a given inertial frame, an orthonormal basis in space, combined by the unit time vector, forms an orthonormal basis in Minkowski space. The number of positive and negative unit vectors in any such basis is a fixed pair of numbers, equal to the signature of the bilinear form associated with the inner product. This is Sylvester's law of inertia.

More terminology (but not more structure): The Minkowski metric is a pseudo-Riemannian metric, more specifically, a Lorentzian metric, even more specifically, *the* Lorentz metric, reserved for 4-dimensional flat spacetime with the remaining ambiguity only being the signature convention.

Minkowski Metric

From the second postulate of special relativity, together with homogeneity of spacetime and isotropy of space, it follows that the spacetime interval between two arbitrary events called 1 and 2 is:

$$\pm \left[c^2 (t_1 - t_2)^2 - (x_1 - x_2)^2 - (y_1 - y_2)^2 - (z_1 - z_2)^2 \right].$$

The interval is independent of the inertial frame chosen, as is shown here. The factor ± 1 determines the choice of the metric signature as an arbitrary sign convention. The numerical values of η, viewed as a matrix representing the Minkowski inner product, follow from the theory of bilinear forms.

Just as the signature of the metric is differently defined in the literature, this quantity is not consistently named. The interval (as defined here) is sometimes referred to as the interval squared. Even the square root of the present interval occurs. When signature and interval are fixed, ambiguity still remains as which coordinate is the time coordinate. It may be the fourth, or it may be the zeroth. This is not an exhaustive list of notational inconsistencies. It is a fact of life that one has to check out the definitions first thing when one consults the relativity literature.

The invariance of the interval under coordinate transformations between inertial frames follows from the invariance of

$$\pm\left[c^2 t^2 - x^2 - y^2 - z^2 \right]$$

(with either sign $\pm$ preserved), provided the transformations are linear. This quadratic form can be used to define a bilinear form

$$u \cdot v = \pm\left[c^2 t_1 t_2 - x_1 x_2 - y_1 y_2 - z_1 z_2 \right].$$

via the polarization identity. This bilinear form can in turn be written as

$$u \cdot v = u^{\mathrm{T}} [\eta] v,$$

where $[\eta]$ is a 4×4 matrix associated with η. Possibly confusingly, denote $[\eta]$ with just η as is common practice. The matrix is read off from the explicit bilinear form as

$$\eta = \pm \begin{pmatrix} -1 & 0 & 0 & 0 \\ 0 & 1 & 0 & 0 \\ 0 & 0 & 1 & 0 \\ 0 & 0 & 0 & 1 \end{pmatrix},$$

and the bilinear form

$$u \cdot v = \eta(u, v),$$

with which this section started by assuming its existence, is now identified.

For definiteness and shorter presentation, the signature $(-,+,+,+)$ is adopted below. This choice (or the other possible choice) has no (known) physical implications. The symmetry group preserving the bilinear form with one choice of signature is isomorphic (under the map given here) with the symmetry group preserving the other choice of signature. This means that both choices are in accord with the two postulates of relativity. Switching between the two conventions is straightforward. If a the metric tensor η has been used in a derivation, go back to the earliest point where it was used, substitute η for $-\eta$, and retrace forward to the desired formula with the desired metric signature.

Standard Basis

A standard basis for Minkowski space is a set of four mutually orthogonal vectors $\{ e_0, e_1, e_2, e_3 \}$ such that

$$-\eta(e_0,e_0) = \eta(e_1,e_1) = \eta(e_2,e_2) = \eta(e_3,e_3) = 1.$$

These conditions can be written compactly in the form

$$\eta(e_\mu, e_\nu) = \eta_{\mu\nu}.$$

Relative to a standard basis, the components of a vector v are written (v^0, v^1, v^2, v^3) where the Einstein notation is used to write $v = v^\mu e_\mu$. The component v^0 is called the timelike component of v while the other three components are called the spatial components. The spatial components of a 4-vector v may be identified with a 3-vector $\mathrm{v} = (v_1, v_2, v_3)$.

In terms of components, the Minkowski inner product between two vectors v and w is given by

$$\eta(v, w) = \eta_{\mu\nu} v^\mu w^\nu = v^0 w_0 + v^1 w_1 + v^2 w_2 + v^3 w_3 = v^\mu w_\mu = v_\mu w^\mu,$$

and

$$\eta(v, v) = \eta_{\mu\nu} v^\mu v^\nu = v^0 v_0 + v^1 v_1 + v^2 v_2 + v^3 v_3 = v^\mu v_\mu.$$

Here lowering of an index with the metric was used.

Raising and Lowering of Indices

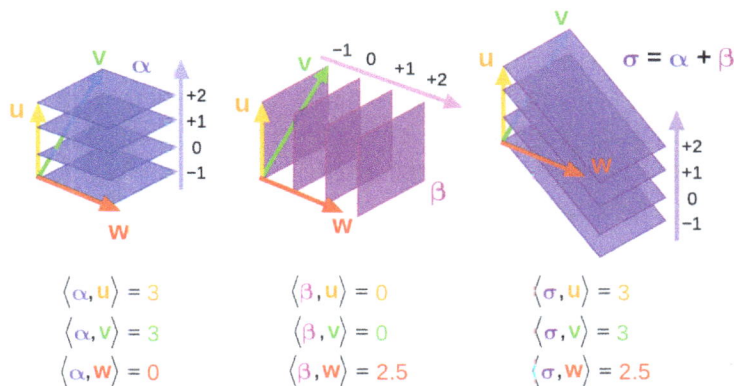

Linear functionals (1-forms) α, β and their sum σ and vectors **u**, **v**, **w**, in 3d Euclidean space. The number of (1-form) hyperplanes intersected by a vector equals the inner product.

Technically, a non-degenerate bilinear form provides a map between a vector space and its dual, in this context, the map is between the tangent spaces of M and the cotangent spaces of M. At a point in M, the tangent and cotangent spaces are dual vector spaces (so the dimension of the cotangent space at an event is also 4). Just as an authentic inner product on a vector space with one argument fixed, by Riesz representation theorem, may be expressed as the action of a linear functional on the vector space, the same holds for the Minkowski inner product of Minkowski space.

Thus if v^μ are the components of a vector in a tangent space, then $\eta_{\mu\nu}v^\mu = v_\nu$ are the components of a vector in the cotangent space (a linear functional). Due to the identification of vectors in tangent spaces with vectors in M itself, this is mostly ignored, and vectors with lower indices are referred to as covariant vectors. In this latter interpretation, the covariant vectors are (almost always implicitly) identified with vectors (linear functionals) in the dual of Minkowski space. The ones with upper indices are contravariant vectors. In the same fashion, the inverse of the map from tangent to cotangent spaces, explicitly given by the inverse of η in matrix representation, can be used to define raising of an index. The components of this inverse are denoted $\eta^{\mu\nu}$. It happens that $\eta^{\mu\nu} = \eta_{\mu\nu}$. These maps between a vector space and its dual can be denoted $\eta^\flat$ (eta-flat) and $\eta^\sharp$ (eta-sharp) by the musical analogy.

Contravariant and covariant vectors are geometrically very different objects. The first can and should be thought of as arrows. A linear functional can be characterized by two objects. It's kernel, which is a hyperplane passing through the origin, and its norm. Geometrically thus covariant vectors should viewed as a set of hyperplanes, with spacing depending on the norm (bigger = smaller spacing), with one of them (the kernel) passing through the origin. The mathematical term for a covariant vector is 1-covector or 1-form (though the latter is usually reserved for covector *fields*).

Misner, Thorne & Wheeler (1970) uses a vivid analogy with wave fronts of a de Broglie wave (scaled by a factor of Planck's reduced constant) quantum mechanically associated to a momentum four-vector to illustrate how one could imagine a covariant version of a contravariant vector. The inner product of two contravariant vectors could equally well be thought of as the action of the covariant version of one of them on the contravariant version of the other. The inner product is then how many time the arrow pierces the planes. The mathematical reference, Lee (2003), offers the same geometrical view of these objects (but mentions no piercing).

The electromagnetic field tensor is a differential 2-form, which geometrical description can as well be found in MTW.

One may, of course, ignore geometrical views all together (as is the style in e.g. Weinberg (2002) and Landau & Lifshitz 2002) and proceed algebraically in a purely formal fashion. The time-proven robustness of the formalism itself, sometimes referred to as index gymnastics, ensures that moving vectors around and changing from contravariant to covariant vectors and vice versa (as well as higher order tensors) is mathematically sound. Incorrect expressions tend to reveal themselves quickly.

The Formalism of the Minkowski Metric

The present purpose is to show semi-rigorously how *formally* one may apply the Minkowski metric to two vectors and obtain a real number, i.e. to display the rôle of the differentials, and how they disappear in a calculation. The setting is that of smooth manifold theory, and concepts such as convector fields and exterior derivatives are introduced.

Lorentz Transformations and Symmetry

The Poincaré group is the group of all transformations preserving the interval. The interval is quite easily seen to be preserved by the translation group in 4 dimensions. The other transformations are those that preserve the interval and leave the origin fixed. Given the bilinear form associated with the Minkowski metric, the appropriate group follows directly from the theory (in particular the definition) of classical groups. In the linked article, one should identify η (in its a matrix representation) with the matrix Φ.

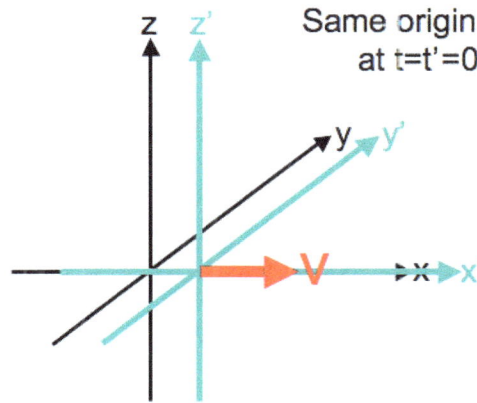

Standard configuration of coordinate systems for Lorentz transformations.

The appropriate group is O(3,1), in this context called the Lorentz group. Its elements are called (homogeneous) Lorentz transformations.

Among the simplest Lorentz transformations is a Lorentz boost. For reference, a boost in the x-direction is given by

$$
\begin{bmatrix} U'_0 \\ U'_1 \\ U'_2 \\ U'_3 \end{bmatrix} = \begin{bmatrix} \gamma & -\beta\gamma & 0 & 0 \\ -\beta\gamma & \gamma & 0 & 0 \\ 0 & 0 & 1 & 0 \\ 0 & 0 & 0 & 1 \end{bmatrix} \begin{bmatrix} U_0 \\ U_1 \\ U_2 \\ U_3 \end{bmatrix},
$$

where

$$
\gamma = \frac{1}{\sqrt{1 - \dfrac{v^2}{c^2}}}
$$

is the Lorentz factor, and

$$
\beta = \frac{v}{c}.
$$

Other Lorentz transformations are pure rotations, and hence elements of the SO(3) subgroup of O(3,1). A general homogeneous Lorentz transformation is a product of a pure boost and a pure rotation. An *inhomogeneous* Lorentz transformation is a homogeneous transformation followed by a translation in space and time. Special transformations are those that invert the space coordinates (P) and time coordinate (T) respectively, or both (PT).

All four-vectors in Minkowski space transform, by definition, according to the same formula under Lorentz transformations. Minkowski diagrams illustrate Lorentz transformations.

Causal Structure

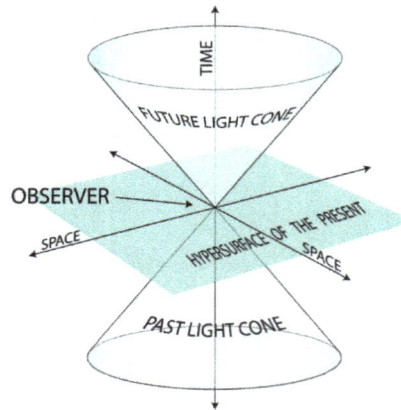

Subdivision of Minkowski spacetime with respect to an event in four disjoint sets. The light cone, the **absolute future**, the **absolute past**, and **elsewhere**. The terminology is from Sard (1970).

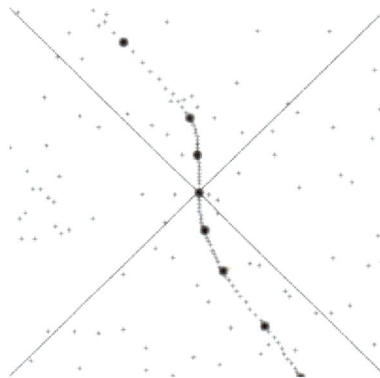

The momentarily co-moving inertial frames along the trajectory ("world line") of a rapidly accelerating observer (center). The vertical direction indicates time, while the horizontal indicates distance, the dashed line is the spacetime of the observer. The small dots are specific events in spacetime. Note how the momentarily co-moving inertial frame changes when the observer accelerates.

Vectors $v = (ct, x, y, z) = (ct, \mathbf{r})$ are classified according to the sign of $c^2t^2 - r^2$. A vector is timelike if $c^2t^2 > r^2$, spacelike if $c^2t^2 < r^2$, and null or lightlike if $c^2t^2 = r^2$. This can be expressed in terms of the sign of $\eta(v,v)$ as well, but depends on the signature. The classification of any vector will be the same in all frames of reference that are related by a Lorentz transformation (but not by a general Poincaré transformation because the origin may then be displaced) because of the invariance of the interval.

The set of all null vectors at an event of Minkowski space constitutes the light cone of that event. Given a timelike vector v, there is a worldline of constant velocity associated with it, represented by a straight line in a Minkowski diagram.

Once a direction of time is chosen, timelike and null vectors can be further decomposed into various classes. For timelike vectors one has

1. future-directed timelike vectors whose first component is positive, (tip of vector located in absolute future in figure) and

2. past-directed timelike vectors whose first component is negative (absolute past).

Null vectors fall into three classes:

1. the zero vector, whose components in any basis are (0,0,0,0) (origin),

2. future-directed null vectors whose first component is positive (upper light cone), and

3. past-directed null vectors whose first component is negative (lower light cone).

Spacelike vectors are in elsewhere. The terminology stems from the fact that spacelike separated events are connected by vectors requiring faster-than-light travel, and so cannot possibly influence each other. Together with spacelike and lightlike vectors there are 7 classes in all.

An orthonormal basis for Minkowski space necessarily consists of one timelike and three spacelike unit vectors. If one wishes to work with non-orthonormal bases it is possible to have other combinations of vectors. For example, one can easily construct a (non-orthonormal) basis consisting entirely of null vectors, called a null basis. Over the reals, if two null vectors are orthogonal (zero Minkowski tensor value), then they must be proportional. However, allowing complex numbers, one can obtain a null tetrad, which is a basis consisting of null vectors, some of which are orthogonal to each other.

Vector fields are called timelike, spacelike or null if the associated vectors are timelike, spacelike or null at each point where the field is defined.

Chronological and Causality Relations

Let $x, y \in M$. We say that

1. *x chronologically precedes* y if $y - x$ is future-directed timelike. This relation has the transitive property and so can be written x < y.

2. *x causally precedes* y if $y - x$ is future-directed null or future-directed timelike. It gives a partial ordering of space-time and so can be written x ≤ y.

Reversed Triangle Inequality

If v and w are both future-directed timelike four-vectors, then in the (+ - - -) sign convention for norm,

$$\|v + w\| \geq \|v\| + \|w\|.$$

Generalizations

A Lorentzian manifold is a generalization of Minkowski space in two ways. The total number of spacetime dimensions is not restricted to be 4 (2 or more) and a Lorentzian manifold need not be flat, i.e. it allows for curvature.

Generalized Minkowski Space

Minkowski space refers to a mathematical formulation in four dimensions. However, the mathematics can easily be extended or simplified to create an analogous generalized Minkowski space in any number of dimensions. If $n \geq 2$, n-dimensional Minkowski space is a vector space of real dimension n on which there is a constant Minkowski metric of signature $(n - 1, 1)$ or $(1, n - 1)$. These generalizations are used in theories where spacetime is assumed to have more or less than 4 dimensions. String theory and M-theory are two examples where $n > 4$. In string theory, there appears conformal field theories with $1 + 1$ spacetime dimensions.

de Sitter space can be formulated as a submanifold of generalized Minkowski space as can the model spaces of hyperbolic geometry.

Curvature

As a *flat spacetime*, the three spatial components of Minkowski spacetime always obey the Pythagorean Theorem. Minkowski space is a suitable basis for special relativity, a good description of physical systems over finite distances in systems without significant gravitation. However, in order to take gravity into account, physicists use the theory of general relativity, which is formulated in the mathematics of a non-Euclidean geometry. When this geometry is used as a model of physical space, it is known as curved space.

Even in curved space, Minkowski space is still a good description in an infinitesimal region surrounding any point (barring gravitational singularities). More abstractly, we say that in the presence of gravity spacetime is described by a curved 4-dimensional manifold for which the tangent space to any point is a 4-dimensional Minkowski space. Thus, the structure of Minkowski space is still essential in the description of general relativity.

Geometry

The meaning of the term *geometry* in the context of Minkowski space depends heavily on what is meant by geometry. Minkowski space is not endowed with a Euclidean geometry, not even with any of the generalized Riemannian geometries with intrinsic curvature, hyperbolic geometry and elliptic geometry. The reason is the indefiniteness of the Minkowski metric. It is, in particular, not a metric space.

It turns out that model spaces of hyperbolic geometry of low dimension, say 2 or 3, *cannot* be isometrically embedded in Euclidean space with one more dimension, i.e. R³ or R⁴ respectively, with the Euclidean metric g, disallowing easy visualization. By comparison, model spaces of elliptic geometry are just spheres in Euclidean space of one higher dimension. However, these spaces *can* be isometrically embedded in spaces of one more dimension when the embedding space is endowed with the Minkowski metric η.

Define $H_R^{1(n)} \subset M^{n+1}$ to be the upper sheet $(ct > 0)$ of the hyperboloid

$$\mathbf{H}_R^{1(n)} = \{(ct, x^1, \ldots, x^n) \in \mathbf{M}^n : c^2 t^2 - (x^1)^2 \cdots (x^n)^2 = R^2, ct > 0\}$$

in generalized Minkowski space M^{n+1} of spacetime dimension $n + 1$. This is one of the surfaces of transitivity of the generalized Lorentz group. The metric induced on this submanifold,

$$h_R^{1(n)} = \iota^*\eta,$$

the pullback of the Minkowski metric η under inclusion, is a Riemannian metric. With this metric $H_R^{1(n)}$ is a Riemannian manifold. It is one of the model spaces of Riemannian geometry, the hyperboloid model of hyperbolic space. It is a space of constant negative curvature $-1/R^2$. The 1 in the upper index refers to an enumeration of the different model spaces of hyperbolic geometry, and the n for its dimension. A 2(2) corresponds to the Poincaré disk model, while 3(n) corresponds to the Poincaré half-space model of dimension n.

Preliminaries

In the definition above $\iota\colon H_R^{1(n)} \to M^{n+1}$ is the inclusion map and the superscript star denotes the pullback. The present purpose is to describe this and similar operations as a preparation for the actual demonstration that H1(n)R actually is a hyperbolic space.

Hyperbolic Stereographic Projection

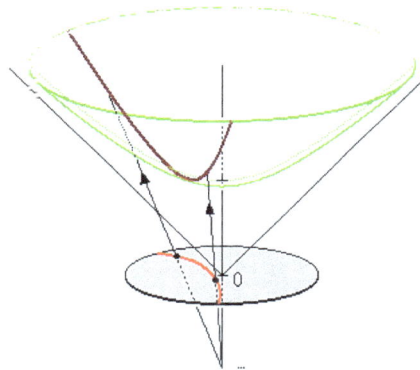

Red circular arc is geodesic in Poincaré disk model; it projects to the brown geodesic on the green hyperboloid.

In order to exhibit the metric it is necessary to pull it back via a suitable *parametrization*. A parametrization of a submanifold S of M is a map $U \subset R^m \to M$ whose range is an open subset of S. If S has the same dimension as M, a parametrization is just the inverse of a coordinate map $\varphi\colon M \to U \subset R^m$. The parametrization to be used is the inverse of *hyperbolic stereographic projection*. This is illustrated in the figure to the left for $n = 2$. It is instructive to compare to stereographic projection for spheres.

Stereographic projection $\sigma\colon$Hn
$R \to R^n$ and its inverse $\sigma^{-1}\colon R^n \to$ Hn
R are given by

$$\sigma(\tau, \mathbf{x}) = \mathbf{u} = \frac{R\mathbf{x}}{R+\tau},$$

$$\sigma^{-1}(\mathbf{u}) = (\tau, \mathbf{x}) = \left(R\frac{R^2 + |u|^2}{R^2 - |u|^2}, \frac{2R^2\mathbf{u}}{R^2 - |u|^2} \right),$$

where, for simplicity, $\tau \equiv ct$. The (τ, x) are coordinates on M^{n+1} and the u are coordinates on R^3.

Pulling Back the Metric

One has

$$h_R^{1(n)} = \eta \big|_{\mathbf{H}_R^{1(n)}} = (dx^1)^2 + \cdots + (dx^n)^2 - d\tau^2$$

and the map

$$\sigma^{-1} : \mathbb{R}^3 \to \mathbf{H}_R^{1(n)}; \sigma^{-1}(\mathbf{u}) = (\tau(\mathbf{u}), \mathbf{x}(\mathbf{u})) = \left(R\frac{R^2 + |u|^2}{R^2 - |u|^2}, \frac{2R^2\mathbf{u}}{R^2 - |u|^2} \right).$$

The pulled back metric can be obtained by straightforward methods of calculus;

$$(\sigma^{-1})^* \eta \big|_{\mathbf{H}_R^{1(n)}} = (dx^1(\mathbf{u}))^2 + \cdots + (dx^n(\mathbf{u}))^2 - (d\tau(\mathbf{u}))^2.$$

One computes according to the standard rules for computing differentials (though one is really computing the rigorously defined exterior derivatives),

$$dx^1(\mathbf{u}) = d\frac{2R^2 u^1}{R^2 - |u|^2} = \frac{\partial}{\partial u^1}\frac{2R^2 u^1}{R^2 - |u|^2}du^1 + \cdots + \frac{\partial}{\partial u^n}\frac{2R^2 u^1}{R^2 - |u|^2}du^n + \frac{\partial}{\partial \tau}\frac{2R^2 u^1}{R^2 - |u|^2}d\tau, \cdots$$

$$dx^n(\mathbf{u}) = d\frac{2R^2 u^1}{R^2 - |u|^2} = \cdots,$$

$$d\tau(\mathbf{u}) = d\left(R\frac{R^2 + |u|^2}{R^2 - |u|^2} \right) = \cdots,$$

and substitutes the results into the right hand side. This yields

$$(\sigma^{-1})^* h_R^{1(n)} = \frac{4R^2 \left[(du^1)^2 + \cdots + (du^n)^2 \right]}{(R^2 - |u|^2)^2} \equiv h_R^{2(n)}.$$

This last equation shows that the metric on the ball is identical to the Riemannian metric $h2(n)$ R in the Poincaré ball model, another standard model of hyperbolic geometry.

Tests of Special Relativity

Special relativity is a physical theory that plays a fundamental role in the description of all physical phenomena, as long as gravitation is not significant. Many experiments played (and still play) an important role in its development and justification. The strength of the theory lies in its unique ability to correctly predict to high precision the outcome of an extremely diverse range of experiments. Repeats of many of those experiments are still being conducted with steadily increased precision, with modern experiments focusing on effects such as at the Planck scale and in the neutrino sector. Their results are consistent with the predictions of special relativity. Collections of various tests were given by Jakob Laub, Zhang, Mattingly, Clifford Will, and Roberts/Schleif.

Special relativity is restricted to flat spacetime, *i.e.*, to all phenomena without significant influence of gravitation. The latter lies in the domain of general relativity and the corresponding tests of general relativity must be considered.

Experiments Paving the Way to Relativity

The predominant theory of light in the 19th century was that of the luminiferous aether, a *stationary* medium in which light propagates in a manner analogous to the way sound propagates through air. By analogy, it follows that the speed of light is constant in all directions in the aether and is independent of the velocity of the source. Thus an observer moving relative to the aether must measure some sort of "aether wind" even as an observer moving relative to air measures an apparent wind.

First-order Experiments

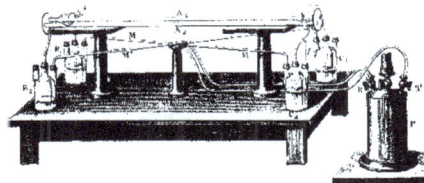

Fizeau experiment, 1851

Beginning with the work of François Arago (1810), a series of optical experiments had been conducted, which should have given a positive result for magnitudes to first order in v/c and which thus should have demonstrated the relative motion of the aether. Yet the results were negative. An explanation was provided by Augustin Fresnel (1818) with the introduction of an auxiliary hypothesis, the so-called "dragging coefficient", that is, matter is dragging the aether to a small extent. This coefficient was directly demonstrated by the Fizeau experiment (1851). It was later shown that all first-order optical experiments must give a negative result due to this coefficient. In addition, also some electrostatic first order experiments were conducted, again having a negative results. In general, Hendrik Lorentz (1892, 1895) introduced several new auxiliary variables for moving observers, demonstrating why all first-order optical and electrostatic experiments have produced null results. For example, Lorentz proposed a location-variable by which electrostatic fields contract in the line of motion and another variable ("local time") by which the time coordinates for moving observers depend on their current location.

Second-order Experiments

The stationary aether theory, however, would give positive results when the experiments are precise enough to measure magnitudes of second order in v/c. The first experiment of this kind was the Michelson's first experiment (1881) and the Michelson-Morley experiment (1887) where two rays of light, traveling for some time in different directions were brought to interfere, so that different orientations relative to the aether wind should lead to a displacement of the interference fringes. But the result was negative again. The way out of this dilemma was the proposal by George Francis FitzGerald (1889) and Lorentz (1892) that matter is contracted in the line of motion with respect to the aether (length contraction). That is, the older hypothesis of a contraction of electrostatic fields was extended to intermolecular forces. However, since there was no theoretical reason for that, the contraction hypothesis was considered ad hoc.

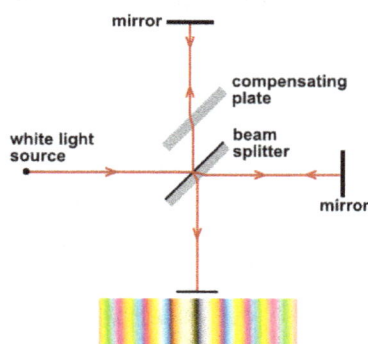

Michelson-Morley interferometer

Besides the optical Michelson–Morley experiment, its electrodynamic equivalent was also conducted, the Trouton–Noble experiment. By that it should be demonstrated that a moving condenser must be subjected to a torque. In addition, the Experiments of Rayleigh and Brace intended to measure some consequences of length contraction in the laboratory frame, for example the assumption that it would lead to birefringence. Though all of those experiments led to negative results. (The Trouton–Rankine experiment conducted in 1908 also gave a negative result when measuring the influence of length contraction on an electromagnetic coil.)

To explain all experiments conducted before 1904, Lorentz was forced to again expand his theory by introducing the complete Lorentz transformation. Henri Poincaré declared in 1905 that the impossibility of demonstrating absolute motion (principle of relativity) is apparently a law of nature.

Refutations of Complete Aether Drag

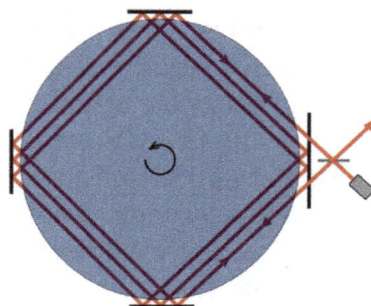

Lodge's ether machine. The steel disks were one yard in diameter. White light was split by a beam splitter and ran three times around the apparatus before reuniting to form fringes.

The idea that the aether might be completely dragged within or in the vicinity of Earth, by which the negative aether drift experiments could be explained, was refuted by a variety of experiments.

- Oliver Lodge (1893) found that rapidly whirling steel disks above and below a sensitive common path interferometric arrangement failed to produce a measurable fringe shift.

- Gustaf Hammar (1935) failed to find any evidence for aether dragging using a common path interferometer, one arm of which was enclosed by a thick-walled pipe plugged with lead, while the other arm was free.

- The Sagnac effect showed that the velocity of two light rays is unaffected by the rotation of the platform.

- The existence of the aberration of light was inconsistent with aether drag hypothesis.

- The assumption that aether drag is proportional to mass and thus only occurs with respect to Earth as a whole was refuted by the Michelson–Gale–Pearson experiment, which demonstrated the Sagnac effect through Earth's motion.

Lodge expressed the paradoxical situation in which physicists found themselves as follows: "… at no practicable speed does … matter [have] any appreciable viscous grip upon the ether. Atoms *must* be able to throw it into vibration, if they are oscillating or revolving at sufficient speed; otherwise they would not emit light or any kind of radiation; but in no case do they appear to drag it along, or to meet with resistance in any uniform motion through it."

Special Relativity

Overview

Eventually, Albert Einstein (1905) drew the conclusion that established theories and facts known at that time only form a logical coherent system when the concepts of space and time are subjected to a fundamental revision. For instance:

- Maxwell-Lorentz's electrodynamics (independence of the speed of light from the speed of the source),

- the negative aether drift experiments (no preferred reference frame),

- Moving magnet and conductor problem (only relative motion is relevant),

- the Fizeau experiment and the aberration of light (both implying modified velocity addition and no complete aether drag).

The result is special relativity theory, which is based on the constancy of the speed of light in all inertial frames of reference and the principle of relativity. Here, the Lorentz transformation is no longer a mere collection of auxiliary hypotheses but reflects a fundamental Lorentz symmetry and forms the basis of successful theories such as Quantum electrodynamics. Special relativity offers a large number of testable predictions, such as:

Principle of relativity	Constancy of the speed of light	Time dilation
Any uniformly moving observer in an inertial frame cannot determine his "absolute" state of motion by a co-moving experimental arrangement.	In all inertial frames the measured speed of light is equal in all directions (isotropy), independent of the speed of the source, and cannot be reached by massive bodies.	The rate of a clock C (= any periodic process) traveling between two synchronized clocks A and B at rest in an inertial frame is retarded with respect to the two clocks.

Also other relativistic effects such as length contraction, Doppler effect, aberration and the experimental predictions of relativistic theories such as the Standard Model can be measured.		

Fundamental Experiments

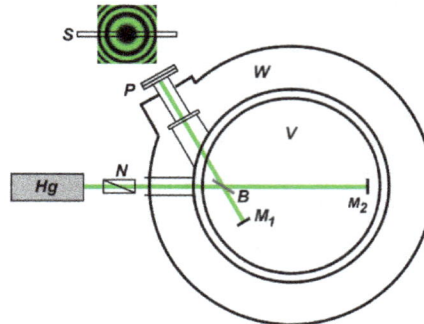

The Kennedy–Thorndike experiment

The effects of special relativity can phenomenologically be derived from the following three fundamental experiments:

- Michelson–Morley experiment, by which the dependence of the speed of light on the *direction* of the measuring device can be tested. It establishes the relation between longitudinal and transverse lengths of moving bodies.

- Kennedy–Thorndike experiment, by which the dependence of the speed of light on the *velocity* of the measuring device can be tested. It establishes the relation between longitudinal lengths and the duration of time of moving bodies.

- Ives–Stilwell experiment, by which time dilation can be directly tested.

From these three experiments and by using the Poincaré-Einstein synchronization, the complete Lorentz transformation follows, with $\gamma = 1/\sqrt{1 - v^2/c^2}$ being the Lorentz factor:

$$x' = \gamma(x - vt),\ y' = y,\ z' = z,\ t' = \gamma\left(t - \frac{vx}{c^2}\right)$$

Besides the derivation of the Lorentz transformation, the combination of these experiments is also important because they can be interpreted in different ways when viewed individually. For example, isotropy experiments such as Michelson-Morley can be seen as a simple consequence of the relativity principle, according to which any inertially moving observer can consider himself as at rest. Therefore, by itself, the MM experiment is compatible to Galilean-invariant theories like emission theory or the complete aether drag hypothesis, which also contain some sort of relativity principle. However, when other experiments that exclude the Galilean-invariant theories are considered (*i.e.* the Ives–Stilwell experiment, various refutations of emission theories and refutations of complete aether dragging), Lorentz-invariant theories and thus special relativity are the only theories that remain viable.

Constancy of the Speed of Light

Interferometers, Resonators

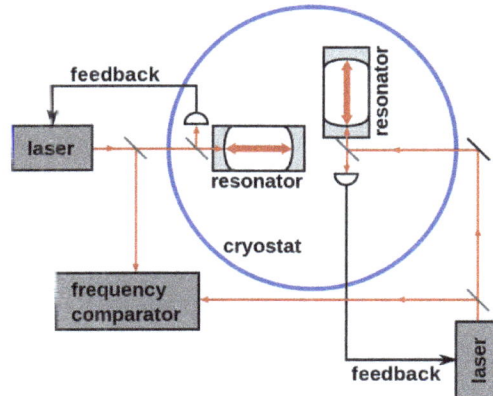

Michelson-Morley experiment with cryogenic optical resonators of a form such as was
used by Müller *et al.* (2003)

Modern variants of Michelson-Morley and Kennedy–Thorndike experiments have been conducted in order to test the isotropy of the speed of light. Contrary to Michelson-Morley, the Kenne-
dy-Thorndike experiments employ different arm lengths, and the evaluations last several months.
In that way, the influence of different velocities during Earth's orbit around the sun can be observed. Laser, maser and optical resonators are used, reducing the possibility of any anisotropy
of the speed of light to the 10^{-17} level. In addition to terrestrial tests, Lunar Laser Ranging Experiments have also been conducted as a variation of the Kennedy-Thorndike-experiment.

Another type of isotropy experiments are the Mössbauer rotor experiments in the 1960s, by which
the anisotropy of the Doppler effect on a rotating disc can be observed by using the Mössbauer
effect.

No Dependence on Source Velocity or Energy

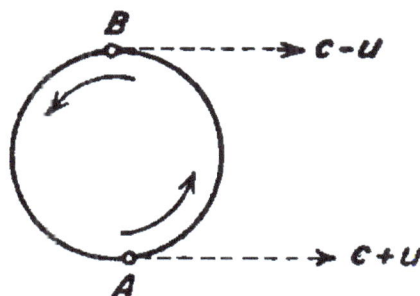

The de Sitter double star experiment, later repeated by Brecher under consideration of the extinction theorem.

Emission theories, according to which the speed of light depends on the velocity of the source,
can conceivably explain the negative outcome of aether drift experiments. It wasn't until the mid-
1960s that the constancy of the speed of light was definitively shown by experiment, since in 1965,
J. G. Fox showed that the effects of the extinction theorem rendered the results of all experiments
previous to that time inconclusive, and therefore compatible with both special relativity and emis-

sion theory. More recent experiments have definitely ruled out the emission model: the earliest were those of Filippas and Fox (1964), using moving sources of gamma rays, and Alväger et al. (1964), which demonstrated that photons didn't acquire the speed of the high speed decaying mesons which were their source. In addition, the de Sitter double star experiment (1913) was repeated by Brecher (1977) under consideration of the extinction theorem, ruling out a source dependence as well.

Observations of Gamma-ray bursts also demonstrated that the speed of light is independent of the frequency and energy of the light rays.

One-way Speed of Light

A series of one-way measurements were undertaken, all of them confirming the isotropy of the speed of light. However, it should be noted that only the two-way speed of light (from A to B back to A) can unambiguously be measured, since the one-way speed depends on the definition of simultaneity and therefore on the method of synchronization. The Poincaré-Einstein synchronization convention makes the one-way speed equal to the two-way speed. However, there are many models having isotropic two-way speed of light, in which the one-way speed is anisotropic by choosing different synchronization schemes. They are experimentally equivalent to special relativity because all of these models include effects like time dilation of moving clocks, that compensate any measurable anisotropy. However, of all models having isotropic two-way speed, only special relativity is acceptable for the overwhelming majority of physicists since all other synchronizations are much more complicated, and those other models (such as Lorentz ether theory) are based on extreme and implausible assumptions concerning some dynamical effects, which are aimed at hiding the "preferred frame" from observation.

Isotropy of Mass, Energy, and Space

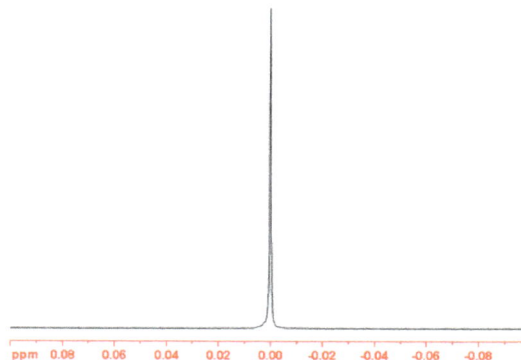

^{7}Li-NMR spectrum of LiCl (1M) in D_2O. The sharp, unsplit NMR line of this isotope of lithium is evidence for the isotropy of mass and space.

Clock-comparison experiments (periodic processes and frequencies can be considered as clocks) such as the Hughes–Drever experiments provide stringent tests of Lorentz invariance. They are not restricted to the photon sector as Michelson-Morley but directly determine any anisotropy of mass, energy, or space by measuring the ground state of nuclei. Upper limit of such anisotropies of 10^{-33} GeV have been provided. Thus these experiments are among the most precise verifications of Lorentz invariance ever conducted.

Time Dilation and Length Contraction

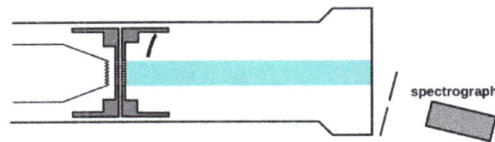

Ives–Stilwell experiment (1938).)

The transverse Doppler effect and consequently time dilation was directly observed for the first time in the Ives–Stilwell experiment (1938). In modern Ives-Stilwell experiments in heavy ion storage rings using saturated spectroscopy, the maximum measured deviation of time dilation from the relativistic prediction has been limited to $\leq 10^{-8}$. Other confirmations of time dilation include Mössbauer rotor experiments in which gamma rays were sent from the middle of a rotating disc to a receiver at the edge of the disc, so that the transverse Doppler effect can be evaluated by means of the Mössbauer effect. By measuring the lifetime of muons in the atmosphere and in particle accelerators, the time dilation of moving particles was also verified. On the other hand, the Hafele–Keating experiment confirmed the twin paradox, *i.e.* that a clock moving from A to B back to A is retarded with respect to the initial clock. However, in this experiment the effects of general relativity also play an essential role.

Direct confirmation of length contraction is hard to achieve in practice since the dimensions of the observed particles are vanishingly small. However, there are indirect confirmations; for example, the behavior of colliding heavy ions can only be explained if their increased density due to Lorentz contraction is considered. Contraction also leads to an increase of the intensity of the Coulomb field perpendicular to the direction of motion, whose effects already have been observed. Consequently, both time dilation and length contraction must be considered when conducting experiments in particle accelerators.

Relativistic Momentum and Energy

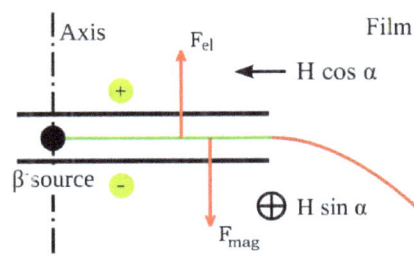

Bucherer's experimental setup for measuring the specific charge e/m of β^- electrons as a function of their speed v/c. (Cross-section through the axis of a circular capacitor with a beta-source at its center, at an angle α with respect to the magnetic field H)

Starting with 1901, a series of measurements was conducted aimed at demonstrating the velocity dependence of the mass of electrons. The results actually showed such a dependency but the precision necessary to distinguish between competing theories was disputed for a long time. Eventually, it was possible to definitely rule out all competing models except special relativity.

Today, special relativity's predictions are routinely confirmed in particle accelerators such as the

Relativistic Heavy Ion Collider. For example, the increase of relativistic momentum and energy is not only precisely measured but also necessary to understand the behavior of cyclotrons and synchrotrons etc., by which particles are accelerated near to the speed of light.

Sagnac and Fizeau

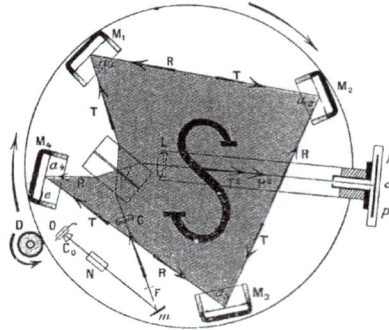

Original Sagnac interferometer

Special relativity also predicts that two light rays traveling in opposite directions around a spinning closed path (e.g. a loop) require different flight times to come back to the moving emitter/receiver (this is a consequence of the independence of the speed of light from the velocity of the source). This effect was actually observed and is called the Sagnac effect. Currently, the consideration of this effect is necessary for many experimental setups and for the correct functioning of GPS.

If such experiments are conducted in moving media (e.g. water, or glass optical fiber), it is also necessary to consider Fresnel's dragging coefficient as demonstrated by the Fizeau experiment. Although this effect was initially understood as giving evidence of a nearly stationary aether or a partial aether drag it can easily be explained with special relativity by using the velocity composition law.

Test Theories

Several test theories have been developed to assess a possible positive outcome in Lorentz violation experiments by adding certain parameters to the standard equations. These include the Robertson-Mansouri-Sexl framework (RMS) and the Standard-Model Extension (SME). RMS has three testable parameters with respect to length contraction and time dilation. From that, any anisotropy of the speed of light can be assessed. On the other hand, SME includes many Lorentz violation parameters, not only for special relativity, but for the Standard model and General relativity as well; thus it has a much larger number of testable parameters.

Other Modern Tests

Due to the developments concerning various models of Quantum gravity in recent years, deviations of Lorentz invariance (possibly following from those models) are again the target of experimentalists. Because "local Lorentz invariance" (LLI) also holds in freely falling frames, experiments concerning the weak Equivalence principle belong to this class of tests as well. The outcomes are analyzed by test theories (as mentioned above) like RMS or, more importantly, by SME.

- Besides the mentioned variations of Michelson–Morley and Kennedy–Thorndike experiments, Hughes–Drever experiments are continuing to be conducted for isotropy tests in the proton and neutron sector. To detect possible deviations in the electron sector, spin-polarized torsion balances are used.

- Time dilation is confirmed in heavy ion storage rings, such as the TSR at the MPIK, by observation of the Doppler effect of lithium, and those experiments are valid in the electron, proton, and photon sector.

- Other experiments use Penning traps to observe deviations of cyclotron motion and Larmor precession in electrostatic and magnetic fields.

- Possible deviations from CPT symmetry (whose violation represents a violation of Lorentz invariance as well) can be determined in experiments with neutral mesons, Penning traps and muons.

- Astronomical tests are conducted in connection with the flight time of photons, where Lorentz violating factors could cause anomalous dispersion and birefringence leading to a dependency of photons on energy, frequency or polarization.

- With respect to threshold energy of distant astronomical objects, but also of terrestrial sources, Lorentz violations could lead to alterations in the standard values for the processes following from that energy, such as Vacuum Cherenkov radiation, or modifications of synchrotron radiation.

- Neutrino oscillations and the speed of neutrinos are being investigated for possible Lorentz violations.

- Other candidates for astronomical observations are the Greisen–Zatsepin–Kuzmin limit and Airy disks. The latter is investigated to find possible deviations of Lorentz invariance that could drive the photons out of phase.

- Observations in the Higgs sector are under way.

Classical Electromagnetism and Special Relativity

The theory of special relativity plays an important role in the modern theory of classical electromagnetism. First of all, it gives formulas for how electromagnetic objects, in particular the electric and magnetic fields, are altered under a Lorentz transformation from one inertial frame of reference to another. Secondly, it sheds light on the relationship between electricity and magnetism, showing that frame of reference determines if an observation follows electrostatic or magnetic laws. Third, it motivates a compact and convenient notation for the laws of electromagnetism, namely the "manifestly covariant" tensor form.

Maxwell's equations, when they were first stated in their complete form in 1865, would turn out to be compatible with special relativity. Moreover, the apparent coincidences in which the same effect was observed due to different physical phenomena by two different observers would be shown to be not coincidental in the least by special relativity. In fact, half of Einstein's 1905 first paper on special relativity, "On the Electrodynamics of Moving Bodies," explains how to transform Maxwell's equations.

Transformation of the Fields between Inertial Frames

The E and B Fields

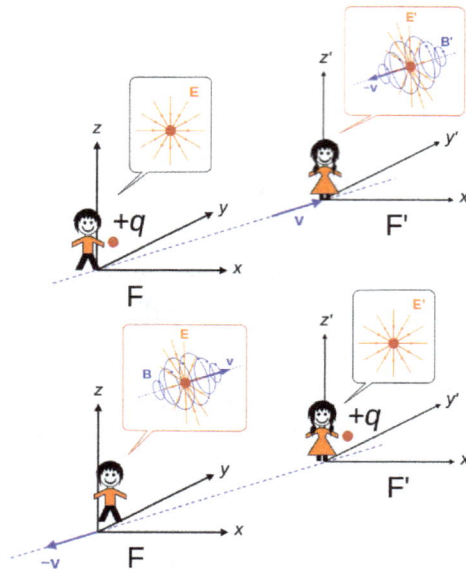

Lorentz boost of an electric charge. **Top:** The charge is at rest in frame F, so this observer notes a static electric field. An observer in another frame F′ moves with velocity **v** relative to F, and notices the charge to move with velocity −**v** with an altered electric field **E** due to length contraction and a magnetic field **B** due to the motion of the charge. **Bottom:** Similar setup, with the charge at rest in frame F′.

This equation, also called the Joules-Bernoulli equation, considers two inertial frames. As notation, the field variables in one frame are *unprimed*, and in a frame moving relative to the unprimed frame at velocity v, the fields are denoted with *primes*. In addition, the fields *parallel* to the velocity v are denoted by $\mathbf{E}_{\parallel}$ while the fields perpendicular to v are denoted as $\mathbf{E}_{\perp}$. In these two frames moving at relative velocity v, the E-fields and B-fields are related by:

$$\mathbf{E}_{\parallel}' = \mathbf{E}_{\parallel}$$

$$\mathbf{B}_{\parallel}' = \mathbf{B}_{\parallel}$$

$$\mathbf{E}_{\perp}' = \gamma \left(\mathbf{E}_{\perp} + \mathbf{v} \times \mathbf{B} \right)$$

$$\mathbf{B}_{\perp}' = \gamma \left(\mathbf{B}_{\perp} - \frac{1}{c^2} \mathbf{v} \times \mathbf{E} \right)$$

where

$$\gamma \overset{\text{def}}{=} \frac{1}{\sqrt{1 - v^2 / c^2}}$$

is called the Lorentz factor and c is the speed of light in free space. The inverse transformations are the same except $v \rightarrow -v$.

An equivalent, alternative expression is:

$$\mathbf{E}' = \gamma\left(\mathbf{E} + \mathbf{v} \times \mathbf{B}\right) - \left(\gamma - 1\right)\left(\mathbf{E} \cdot \hat{\mathbf{v}}\right)\hat{\mathbf{v}}$$

$$\mathbf{B}' = \gamma\left(\mathbf{B} - \frac{\mathbf{v} \times \mathbf{E}}{c^2}\right) - \left(\gamma - 1\right)\left(\mathbf{B} \cdot \hat{\mathbf{v}}\right)\hat{\mathbf{v}}$$

where $\hat{\mathbf{v}}$ is the velocity unit vector.

If one of the fields is zero in one frame of reference, that doesn't necessarily mean it is zero in all other frames of reference. This can be seen by, for instance, making the unprimed electric field zero in the transformation to the primed electric field. In this case, depending on the orientation of the magnetic field, the primed system could see an electric field, even though there is none in the unprimed system.

This does not mean two completely different sets of events are seen in the two frames, but that the same sequence of events is described in two different ways .

If a particle of charge q moves with velocity u with respect to frame S, then the Lorentz force in frame S is:

$$\mathbf{F} = q\mathbf{E} + q\mathbf{u} \times \mathbf{B}$$

In frame S', the Lorentz force is:

$$\mathbf{F}' = q\mathbf{E}' + q\mathbf{u}' \times \mathbf{B}'$$

If S and S' have aligned axes then:

$$u_{x'} = \frac{u_x + v}{1 + (v\,u_x)/c^2}$$

$$u_{y'} = \frac{u_y/\gamma}{1 + (v\,u_x)/c^2}$$

$$u_{z'} = \frac{u_z/\gamma}{1 + (v\,u_x)/c^2}$$

A derivation for the transformation of the Lorentz force for the particular case u = 0 is given here.

Component by component, for relative motion along the x-axis, this works out to be the following:

$$E'_x = E_x \qquad B'_x = B_x$$

$$E'_y = \gamma\left(E_y - vB_z\right) B'_y = \gamma\left(B_y + \frac{v}{c^2}E_z\right)$$

$$E'_z = \gamma\left(E_z + vB_y\right) B'_z = \gamma\left(B_z - \frac{v}{c^2}E_y\right).$$

The transformations in this form can be made more compact by introducing the electromagnetic tensor (defined below), which is a covariant tensor.

The D and H Fields

For the electric displacement D and magnetic intensity H, using the constitutive relations and the result for c^2:

$$\mathbf{D} = \epsilon_0\mathbf{E}, \quad \mathbf{B} = \mu_0\mathbf{H}, \quad c^2 = \frac{1}{\epsilon_0\mu_0},$$

gives

$$\mathbf{D}' = \gamma\left(\mathbf{D} + \frac{1}{c^2}\mathbf{v}\times\mathbf{H}\right) + (1-\gamma)(\mathbf{D}\cdot\hat{\mathbf{v}})\hat{\mathbf{v}}$$

$$\mathbf{H}' = \gamma\left(\mathbf{H} - \mathbf{v}\times\mathbf{D}\right) + (1-\gamma)(\mathbf{H}\cdot\hat{\mathbf{v}})\hat{\mathbf{v}}$$

Analogously for E and B, the D and H form the electromagnetic displacement tensor.

The φ and A Fields

An alternative simpler transformation of the EM field uses the electromagnetic potentials - the electric potential φ and magnetic potential A:

$$\varphi' = \gamma(\varphi - vA_\parallel)$$

$$A_{\parallel'} = \gamma(A_\parallel - v\varphi/c^2)$$

$$A_{\perp'} = A_\perp$$

where $A'_\parallel$ is the parallel component of A to the direction of relative velocity between frames v, and $A_\perp$ is the perpendicular component. These transparently resemble the characteristic form of other Lorentz transformations (like time-position and energy-momentum), while the transformations of E and B above are slightly more complicated. The components can be collected together as:

$$\mathbf{A}' = \mathbf{A} - \frac{\gamma\varphi}{c^2}\mathbf{v} + (\gamma - 1)(\mathbf{A}\cdot\hat{\mathbf{v}})\hat{\mathbf{v}}$$

$$\varphi' = \gamma(\varphi - \mathbf{A}\cdot\mathbf{v})$$

The ρ and J Fields

Analogously for the charge density ρ and current density J,

$$J_{\parallel'} = \gamma(J_{\parallel} - v\rho)$$

$$\rho' = \gamma(\rho - vJ_{\parallel}/c^2)$$

$$J_{\perp'} = J_{\perp}$$

Collecting components together:

$$\mathbf{J}' = \mathbf{J} - \gamma\rho\mathbf{v} + (\gamma - 1)(\mathbf{J}\cdot\hat{\mathbf{v}})\hat{\mathbf{v}}$$

$$\rho' = \gamma(\rho - \mathbf{J}\cdot\mathbf{v}/c^2)$$

Non-relativistic Approximations

For speeds $v \ll c$, the relativistic factor $\gamma \approx 1$, which yields:

$$\mathbf{E}' \approx \mathbf{E} + \mathbf{v}\times\mathbf{B}$$

$$\mathbf{B}' \approx \mathbf{B} - \frac{1}{c^2}\mathbf{v}\times\mathbf{E}$$

$$\mathbf{J}' \approx \mathbf{J} - \rho\mathbf{v}$$

$$\rho' \approx \left(\rho - \frac{1}{c^2}\mathbf{J}\cdot\mathbf{v}\right)$$

so that there is no need to distinguish between the spatial and temporal coordinates in Maxwell's equations.

Relationship between Electricity and Magnetism

Deriving Magnetism from Electrostatics

The chosen reference frame determines if an electromagnetic phenomenon is viewed as an effect of electrostatics or magnetism. Authors usually derive magnetism from electrostatics when special relativity and charge invariance are taken into account. The Feynman Lectures on Physics (vol. 2, ch. 13-6) uses this method to derive the "magnetic" force on a moving charge next to a current-carrying wire.

Fields Intermix in Different Frames

The above transformation rules show that the electric field in one frame contributes to the magnetic field in another frame, and vice versa. This is often described by saying that the electric field and magnetic field are two interrelated aspects of a single object, called the electromagnetic field. Indeed, the entire electromagnetic field can be encoded in a single rank-2 tensor called the electromagnetic tensor.

Moving Magnet and Conductor Problem

A famous example of the intermixing of electric and magnetic phenomena in different frames of reference is called the "moving magnet and conductor problem", cited by Einstein in his 1905 paper on Special Relativity.

If a conductor moves with a constant velocity through the field of a stationary magnet, eddy currents will be produced due to a *magnetic* force on the electrons in the conductor. In the rest frame of the conductor, on the other hand, the magnet will be moving and the conductor stationary. Classical electromagnetic theory predicts that precisely the same microscopic eddy currents will be produced, but they will be due to an *electric* force.

Covariant Formulation in Vacuum

The laws and mathematical objects in classical electromagnetism can be written in a form which is manifestly covariant. Here, this is only done so for vacuum (or for the microscopic Maxwell equations, not using macroscopic descriptions of materials such as electric permittivity), and uses SI units.

This section uses Einstein notation, including Einstein summation convention. The Minkowski metric tensor η here has metric signature $(+ - - -)$.

Field Tensor and 4-Current

The above relativistic transformations suggest the electric and magnetic fields are coupled together, in a mathematical object with 6 components: an antisymmetric second-rank tensor, or a bivector. This is called the electromagnetic field tensor, usually written as $F^{\mu\nu}$. In matrix form:

$$F^{\mu\nu} = \begin{pmatrix} 0 & -E_x/c & -E_y/c & -E_z/c \\ E_x/c & 0 & -B_z & B_y \\ E_y/c & B_z & 0 & -B_x \\ E_z/c & -B_y & B_x & 0 \end{pmatrix}$$

where c the speed of light - in natural units $c = 1$.

There is another way of merging the electric and magnetic fields into an antisymmetric tensor, by replacing $E/c \rightarrow B$ and $B \rightarrow - E/c$, to get the dual tensor $G^{\mu\nu}$.

$$G^{\mu\nu} = \begin{pmatrix} 0 & -B_x & -B_y & -B_z \\ B_x & 0 & E_z/c & -E_y/c \\ B_y & -E_z/c & 0 & E_x/c \\ B_z & E_y/c & -E_x/c & 0 \end{pmatrix}$$

In the context of special relativity, both of these transform according to the Lorentz transformation according to

$$F'^{\alpha\beta} = \Lambda^{\alpha}{}_{i}\Lambda^{\beta}{}_{i}F^{\mu\nu},$$

where $\Lambda^{\alpha}{}_{\nu}$ is the Lorentz transformation tensor for a change from one reference frame to another. The same tensor is used twice in the summation.

The charge and current density, the sources of the fields, also combine into the four-vector

$$J^{\alpha} = \begin{pmatrix} c\rho & J_x & J_y & J_z \end{pmatrix}$$

called the four-current.

Maxwell's Equations in Tensor Form

Using these tensors, Maxwell's equations reduce to:

> **Maxwell's equations** *(Covariant formulation)*
>
> $$\frac{\partial F^{\alpha\beta}}{\partial x^{\alpha}} = \mu_0 J^{\beta}$$
>
> $$\frac{\partial G^{\alpha\beta}}{\partial x^{\alpha}} = 0$$

where the partial derivatives may be written in various ways. The first equation list-ed above corresponds to both Gauss's Law (for $\beta = 0$) and the Ampère-Maxwell Law (for $\beta = 1, 2, 3$). The second equation corresponds to the two remaining equations, Gauss's law for magnetism (for $\beta = 0$) and Faraday's Law (for $\beta = 1, 2, 3$).

These tensor equations are manifestly-covariant, meaning the equations can be seen to be co-variant by the index positions. This short form of writing Maxwell's equations illustrates an idea shared amongst some physicists, namely that the laws of physics take on a simpler form when written using tensors.

By lowering the indices on $F^{\alpha\beta}$ to obtain $F_{\alpha\beta}$:

$$F_{\alpha\beta} = \eta_{\alpha\lambda}\eta_{\beta\mu}F^{\lambda\mu}$$

the second equation can be written in terms of $F_{\alpha\beta}$ as:

$$\epsilon^{\delta\alpha\beta\gamma}\frac{\partial F_{\beta\gamma}}{\partial x^{\alpha}} = \frac{\partial F_{\alpha\beta}}{\partial x^{\gamma}} + \frac{\partial F_{\gamma\alpha}}{\partial x^{\beta}} + \frac{\partial F_{\beta\gamma}}{\partial x^{\alpha}} = 0$$

where $\epsilon^{\alpha\beta\gamma\delta}$ is the contravariant Levi-Civita symbol. Notice the cyclic permutation of indices in this

equation: $\begin{array}{c} \alpha \to \beta \\ \nwarrow \searrow \swarrow \\ \gamma \end{array}$.

Another covariant electromagnetic object is the electromagnetic stress-energy tensor, a covariant rank-2 tensor which includes the Poynting vector, Maxwell stress tensor, and electromagnetic energy density.

4-Potential

The EM field tensor can also be written

$$F^{\alpha\beta} = \frac{\partial A^{\beta}}{\partial x_{\alpha}} - \frac{\partial A^{\alpha}}{\partial x_{\beta}},$$

where

$$A^{\alpha} = (\varphi / c, A_x, A_y, A_z),$$

is the four-potential and

$$x_{\alpha} = (ct, -x, -y, -z)$$

is the four-position.

Using the 4-potential in the Lorenz gauge, an alternative manifestly-covariant formulation can be found in a single equation (a generalization of an equation due to Bernhard Riemann by Arnold Sommerfeld, known as the Riemann–Sommerfeld equation, or the covariant form of the Maxwell equations):

Maxwell's equations *(Covariant Lorenz gauge formulation)*
$$\Box A^{\mu} = \mu_0 J^{\mu}$$

---

where $\Box$ is the d'Alembertian operator, or four-Laplacian.

References

- Albert Einstein (2001). Relativity: The Special and the General Theory (Reprint of 1920 translation by Robert W. Lawson ed.). Routledge. p. 48. ISBN 0-415-25384-5.

- Richard Phillips Feynman (1998). Six Not-so-easy Pieces: Einstein's relativity, symmetry, and space–time (Reprint of 1995 ed.). Basic Books. p. 68. ISBN 0-201-32842-9.

- Steane, Andrew M. (2012). Relativity Made Relatively Easy (illustrated ed.). OUP Oxford. p. 226. ISBN 978-0-19-966286-9. Extract of page 226.

- Edwin F. Taylor & John Archibald Wheeler (1992). Spacetime Physics: Introduction to Special Relativity. W. H. Freeman. ISBN 0-7167-2327-1.

- Rindler, Wolfgang (1977). Essential Relativity: Special, General, and Cosmological (illustrated ed.). Springer Science & Business Media. p. §1,11 p. 7. ISBN 978-3-540-07970-5.

- Schutz, J. (1997) Independent Axioms for Minkowski Spacetime, Addison Wesley Longman Limited, ISBN 0-582-31760-6.

- Yaakov Friedman (2004). Physical Applications of Homogeneous Balls. Progress in Mathematical Physics. 40.

pp. 1–21. ISBN 0-8176-3339-1.

- David Morin (2007) Introduction to Classical Mechanics, Cambridge University Press, Cambridge, chapter 11, Appendix I, ISBN 1-139-46837-5.

- Max Jammer (1997). Concepts of Mass in Classical and Modern Physics. Courier Dover Publications. pp. 177–178. ISBN 0-486-29998-8.

- Wesley C. Salmon (2006). Four Decades of Scientific Explanation. University of Pittsburgh. p. 107. ISBN 0-8229-5926-7., Section 3.7 page 107.

Permissions

Index